Lecture Notes in Earth Sciences

41

Editors:
S. Bhattacharji, Brooklyn
G. M. Friedman, Brooklyn and Troy
H. J. Neugebauer, Bonn
A. Seilacher, Tuebingen

Lecture Notes in Earth Sciences

41

Editors:
S. Bhattacharji, Brooklyn
G. M. Friedman, Brooklyn and Troy
H. J. Neugebauer, Bonn
A. Seilacher, Tübingen

Reinhard Pflug   John W. Harbaugh (Eds.)

Computer Graphics in Geology

Three-Dimensional Computer Graphics
in Modeling Geologic Structures and
Simulating Geologic Processes

Springer-Verlag

Berlin Heidelberg New York
London Paris Tokyo
Hong Kong Barcelona
Budapest

Editors

Reinhard Pflug
Geologisches Institut, Albert-Ludwigs-Universität Freiburg
Albertstraße 23 B, W-7800 Freiburg, FRG

John W. Harbaugh
Department of Applied Earth Sciences, Stanford University
Stanford, CA 94305-2225, USA

"For all Lecture Notes in Earth Sciences published till now please see final page of the book"

ISBN 3-540-55190-5 Springer-Verlag Berlin Heidelberg New York
ISBN 0-387-55190-5 Springer-Verlag New York Berlin Heidelberg

© Springer-Verlag Berlin Heidelberg 1992
Printed in Germany

Typesetting: Camera ready by author
Printing and binding: Druckhaus Beltz, Hemsbach/Bergstr.
32/3140-543210 - Printed on acid-free paper

To Dan Merriam and Renier Vinken,
indefatigable promoters of
computer applications in geology

Preface

This volume contains the proceedings of a symposium held at Freiburg im Breisgau, October 7-11, 1990. The symposium was sponsored mainly by the Deutsche Forschungsgemeinschaft (DFG), by the Geological Institute of the University of Freiburg, and by the International Association of Mathematical Geology. We thank these and all other sponsors of the meeting.

The symposium whose participants came from more then twenty countries was the first international meeting dedicated entirely to geological applications of three-dimensional computer graphics, a rapidly growing field of scientific visualization in geology. The selection of papers in this volume covers a wide range of methods developed in the last decade.

We thank Springer Verlag for supporting the production of this volume and the authors for their contributions. We are grateful for donations by IBM and Silicon Graphics to promote the publication of these proceedings.

December 1991

Reinhard Pflug

John W. Harbaugh

CONTENTS

LIST OF CONTRIBUTORS

Massimiliano Barchi   Dipartimento di Scienze della Terra, Università di Perugia, Piazza dell'Università, I-06100 Perugia, Italy

Jens-Uwe Berthold   Universität Greifswald, Sektion Geologische Wissenschaften, Friedrich-Ludwig-Jahn-Str. 17A, O-2200 Greifswald, Germany

Raymonde Blanchin   Bureau de Recherches Géologiques et Minières, B.P. 6009, F-45060 Orléans Cédex 2, France

Kerry L. Burns   Los Alamos National Laboratory, ESS-4, MS-D44, P.O. Box 1663, Los Alamos, NM 87545, U.S.A.

Jean-Paul Chilès   Bureau de Recherches Géologiques et Minières, B.P. 6009, F-45060 Orléans Cédex 2, France

Isobel Clark   Geostokos Ltd., 36 Baker Street, London W1M 1DG, United Kingdom

Costanzo Federico   Dipartimento di Scienze della Terra, Università di Perugia, Piazza dell'Università, I-06100 Perugia, Italy

Thomas R. Fisher   Radian Corporation, P.O. Box 201088, Austin, Texas 78720-1088, U.S.A.

Martin Genter   Geologisches Institut der ETH, ETH Zentrum, CH-8092 Zürich, Switzerland

Giovanni Guglielmo Jr.   Earth Sciences Board, University of California, Santa Cruz, CA 95064, U.S.A.

Fausto Guzzetti   C.N.R., Istituto di Ricerca per la Protezione Idrogeologica nell'Italia centrale, V. Madonna Alta 126, I-06100 Perugia, Italy

John W. Harbaugh   Department of Applied Earth Sciences, Stanford University, Stanford, CA 94305, U.S.A.

William W. Hay   GEOMAR, Wischhofstr. 1-3, D-2300 Kiel 14, Germany

John M. Herzog   Department of Geology and CIRES, Campus Box 216, University of Colorado, Boulder, CO 80309, U.S.A.

Simon W. Houlding   Lynx Geosystems INC, 1199 West Pender St., Vancouver, British Columbia V6E 2R1, Canada

Yungao Huang   LIAD, Ecole Nationale Supérieure de Géologie de Nancy, B.P. 40, F-54501 Vandoeuvre-les-Nancy Cédex, France

Herbert Klein   terra tec, Am Sulzbach 26, D-7843 Heitersheim, Germany

Kenneth E. Kolm   Department of Geology and Geological Engineering, Colorado School of Mines, 1500 Illinois Street, Golden, CO 80401, U.S.A.

Thomas J. Lasseter   Tech·Logic Inc., 15325 189th Avenue NE, Woodinville, WA 98072, U.S.A.

Young-Hoon Lee   Department of Applied Earth Sciences, Stanford University, Stanford, CA 94305, U.S.A.

Stefan M. Luthi — Schlumberger-Doll Research, Old Quarry Road, Ridgefield, CT 06877, U.S.A.

Jean-Laurent Mallet — LIAD, Ecole Nationale Supérieure de Géologie de Nancy, B.P. 40, F-54501 Vandoeuvre-les-Nancy Cédex, France

Paul A. Martinez — Department of Applied Earth Sciences, Stanford University, Stanford, CA 94305, U.S.A.

Giorgio Minelli — Dipartimento di Scienze della Terra, Università di Perugia, Piazza dell'Università, I-06100 Perugia, Italy

Yoshikazu Ohashi — Arco Oil and Gas Company, Plano Research Center, 2300 West Plano Parkway, Plano, Texas 75075, U.S.A.

Maria Teresa Pareschi — IBM Scientific Center, Via Santa Maria 67, I-56100 Pisa, Italy

Gerald J. Peschel — Universität Greifswald, Sektion Geologische Wissenschaften, Friedrich-Ludwig-Jahn-Str. 17A, O-2200 Greifswald, Germany

Reinhard Pflug — Geologisches Institut, Universität Freiburg, Albertstr. 23-B, D-7800 Freiburg, Germany

René Prissang — Freie Universität Berlin, Institut für Geologie, Geophysik und Geoinformatik, FR Geoinformatik, Malteserstr. 74-100, D-1000 Berlin 46, Germany

Christoph Ramshorn — Department of Applied Earth Sciences, Stanford University, Stanford, CA 94305, U.S.A.

Jonathan F. Raper — Department of Geography, Birkbeck College, 7-15 Gresse St., London W1P 1PA, United Kingdom

Pauli Saksa — Saanio & Riekkola Consulting Engineers Ltd., Laulukuja 4, SF-00420 Helsinki, Finland

Edmund J. Sides — SOMINCOR, Apartado 12, P-7780 Castro Verde, Portugal

Achim Stärk — Geologisches Institut, Universität Freiburg, Albertstr. 23-B, D-7800 Freiburg, Germany

Mark A. Stoakes — Lynx Geosystems INC, 1199 West Pender St., Vancouver, British Columbia V6E 2R1, Canada

John C. Tipper — Department of Geology, Australian National University, GPO Box 4, Canberra, ACT 2601, Australia

A. Keith Turner — Department of Geology and Geological Engineering, Colorado School of Mines, 1500 Illinois Street, Golden, CO 80401, U.S.A.

J. W. Dominik Ulmer — Geologisches Institut, Universität Freiburg, Albertstr. 23-B, D-7800 Freiburg, Germany

Mark Verschuren — Renard Centre of Marine Geology, Gent University, Krijgslaan 281-S8, B-9000 Gent, Belgium

Robert K. Wales — Intergraph Corporation, Suite 200, 675 Bering Drive, Houston, Texas 77057, U.S.A.

Johannes Wendebourg — Department of Applied Earth Sciences, Stanford University, Stanford, CA 94305, U.S.A.

Christopher N. Wold — GEOMAR, Wischhofstr. 1-3, D-2300 Kiel 14, Germany

THREE-DIMENSIONAL COMPUTER GRAPHICS IN GEOLOGY

REINHARD PFLUG

Geologisches Institut der Universität Freiburg
D-7800 Freiburg, Germany

The use of three-dimensional computer graphics in geology has many facets as this volume demonstrates. Presently, geologists are participating in a technical revolution that has dramatically enlarged the possibilities of scientific visualization and interpretation by use of sophisticated three-dimensional rendering techniques. Hardware and software are available at the PC-level for many applications. Other applications need UNIX workstations, and at the high end, graphic supercomputers. The fascinating possibilities are best shown in the contribution by Lasseter (p. 189-198, this volume), which was also a highlight of the Freiburg symposium. Lasseter´s "single consistent model" approach (Figures on p. 191 and 196) uses 3-D seismics and well data that are interactively interpreted in a multiple windows environment, and forms a landmark contribution. Lasseter´s "seismic-display-on-ribbon-sections" between deviated wells is an exciting new technique for interpreting complex subsurface geological situations. The intersection of a curved surface formed by a ribbon with a 3-D seismic block, can be effectively achieved only with advanced three-dimensional computer graphic methods. It is difficult to transmit adequately to the reader the properties of this exceptional modeling and interpretation system. The interaction of the interpreting geologist with the model can only be appreciated fully by looking over his shoulders. The audience at Freiburg, however, could follow Lasseter´s technique by viewing a large screen video projector linked directly to a Silicon Graphics superworkstation.

Lasseter´s approach is not yet applicable in all fields of geology for many reasons. One of the most important is inherent to geological data, which are generally sparse and unevenly distributed. A principal problem involves transforming sparse data so as to form a "block" of data comparable to the contents of a 3-D seismic block. Unfortunately, simple methods to create consistent and realistic spatial models over a range of scales and for any geological applications do not yet exist. Promising methods have been proposed by Fisher and Wales (p. 17-28), Houlding, Stokes and Clark (p. 199-212), Huang and Mallet (p. 3-16), Prissang (p. 213-228), and Sides (p.213-228), to cite a few, but they require large amounts of spatial data that in many geological field and laboratory applications are not available, or are to expensive to collect.

We will have to wait for methods that permit three-dimensional sampling of data at any scale and level of detail, at reasonable costs. Serial sectioning is one possibility, but the detail visible in sections from microscopic to macroscopic scales very often cannot be correlated satisfactorily and with a consistent level of detail from section to section. Our group at Freiburg tried to construct three-dimensional geometric models of complicated sequences of carbonate cements formed during diagenesis of reef limestones, but the attempts failed because it was impossible to make reasonable connections between the the two-dimensional detail observable on a section with similar detail on adjacent sections. Furthermore, there remained the impossibility of rendering the details three-dimensionally with computer graphics methods available to us. We need tomographic methods such as those involving 3-D seismic surveys, that can be applied at microscopic scales to permit measurements of spatial positions and infererences about physical properties at points on surfaces within an object at any degree of detail without destroying the object.

From the same project at Freiburg, and from many other 3-D modeling experiments, we have also learned that it is desirable to develop geometric models that can contain as many surfaces, horizons, and faults as possible. But, we have learned additionally that we should avoid trying to render such models completely because the human visual system is not able to comprehend more than two or three surfaces at a time when they are represented simultaneously on a monitor as perspective views. The user of geological rendering programs should be able to concentrate on selected horizons of a complex geometric model by interactively "fading out" the other surfaces. Also, multiple windows should be available for zooming and simultaneous viewing from different directions. An option to view objects stereoscopically would also be very convenient and may be expected as a future development.

Prospective users of three-dimensional computer graphics for the visualization of geologic structures should be aware of the following:

What is seen on a screen is only as good as the modeling effort that has been invested in developing the 3-D geometric model, particularly if only sparse observations are available. If the modeling is performed solely by a fully automatic program, the results generally will not be satisfactory because automatic modeling programs do not yet follow geologic reasoning. Also, the user should be aware that development of a consistent 3-D geometric model that fulfills expectations will require considerable time and effort.

The resolution of computer monitors (which is 1.3×10^6 pixels for standard graphic workstation monitors which have pixel diameters of about 0.3 mm) is still far below the resolution abilities of the eye. Therefore the user should avoid overloading 3-D pictures with too much detail, because the detail often cannot be distinguished due to restrictions on the resolution of monitors. For the same reasons the user should avoid trying to show too many objects at a time.

Unfortunately, the data format for 3-D geometric models is not standardized. Trying to render the same model with different programs generally requires more or less complicated transformations of data. The development of spatial databases for geological objects is being undertaken by many working groups, although standardization is not yet in sight.

While three-dimensional computer graphics first appeared in geologic publications shortly after automatic plotters became available, with rare exceptions for more then two decades only simple block diagrams with fishnet representation of an upper surface were used. Astonishing changes have occurred in the last few years, as documented by papers of this volume or by extended abstracts of the Freiburg meeting (Freiburger Geowissenschaftliche Beiträge, vol. 2, viii and 131 p., 1990, ISSN 0936-6571).

Major Issues in Simulating Geologic Processes

JOHN W. HARBAUGH
Department of Applied Earth Sciences, Stanford University
Stanford, CA 94305 - 2225

In overviewing the papers at the Freiburg conference that combine geologic process simulation and graphic display, I have not confined my remarks to the papers themselves, and instead, discourse on the general status of geologic process simulation and attempt to forecast where it is going in the next decade.

In science, as in business, forecasting is an essential but difficult game. My remarks here are conditioned in part by my involvement in geologic process simulation in the past quarter century. Beginning in the late 1960's, my attempts to project the status of process simulation were flawed (as seen in hindsight) because they were excessively optimistic and failed to appreciate the difficulties that were later encountered. Time will tell if the forecasts outlined here will prove closed to reality.

While my remarks on major issues pertain to process simulation overall in geology, it is convenient to focus on applications to sedimentary basins. Most present large-scale efforts in geologic process simulation are focused on sedimentary basins. Oil companies provide the impetus because they explore for and produce oil and gas from sedimentary basins. If simulation promises improved understanding and predictive capabilities for sedimentary basins (which it does), there is ample justification for focusing on sedimentary basins.

Number of dimensions: The number of dimensions to be represented in simulations is a major issue. Few, if any, geologic process simulation models are truly three-dimensional. Some are quasi 3-D in that they provide 3-D responses when simulation experiments are carried out, but representation of geologic processes is not fully 3-D. For example, several papers in this volume (Wendebourg and Ulmer; Martinez; Lee and Harbaugh) pertain to SEDSIM (Stanford University's SEDimentary Basin SIMulation project), which while three-dimensional in many respects, adopts shortcuts such as representing flow in open bodies of water with constant velocity with respect to depth. In other words, flow velocity varies from place to place geographically in SEDSIM, but not in the vertical dimension. This is a serious deficiency in many respects, although justified in view of large problems in simulating flow in full 3-D; furthermore, SEDSIM has worked well with respect to those geologic problems to which it has been applied.

Most sedimentary basin models being developed by oil companies are 2-D and represent geologic processes and responses in a vertical plane only, such as that developed by Shell. They perform realistically and are highly useful in spite of confinement to a vertical plane. Their limitations, however, are large because they cannot represent variations outside the plane and therefore are precluded for use for making predictions about facies relationships or structural features in a spatial context within a basin.

Scale and expanse: The range of scale and geographic expanse over which process models can be applied is presumably large, but the limits have not been tested. Stanford's SEDSIM has been operated over expanses ranging from a few hundred meters to several hundred kilometers, and the Shell model has generally been applied over ranges measured in hundreds of kilometers. The real issue, however, is the degree of resolution, rather than scale or expanse. For example, the simulation experiments performed with the WAVE component of SEDSIM (Martinez, this volume) involve expanses ranging from a few hundred meters to tens of kilometers, but the number of geographic cells has remained the same. In other words, the resolution has been in inverse proportion to the scale or expanse; as expanse increased, resolution with respect to scale decreased proportionally. There seems to be no actual limit to the expanse over which process models can be operated, except as posed by the processes themselves. For example, wave processes are not effective at the centimeter scale.

Graphic display: Graphic display is vitally important for simulation models. The responses of models must generally be examined visually, although it is possible that statistical or parametric measures will be developed for comparison purposes. Graphic capabilities for 2-D basin simulation models are mature, involving geologic sections generally shown in color, and sometimes in dynamic form to represent evolution of a basin through geologic time.

The situation with respect to display of 3-D models differs in that full 3-D display is required to properly show relationships. The use of older forms of display involving maps and sections (which are essentially two-dimensional in form) is inadequate. Fortunately, the computer industry is currently engaged in developing 3-D graphics workstations that are capable of generating 3-D displays in a fraction of a second. These workstations are also capable of showing textural qualities of materials (such as sand) by combining large palettes of colors with high-resolution screens. Much developmental work remains to be done, however, because procedures for displaying components of 3-D solid bodies (such as facies relationships in a sedimentary sequence) are still inadequate. However, these workstations are capable of showing processes in action, such as motions of pore fluids during compaction of sediment. Displays linking processes and responses in video--like form should have major impact upon geology.

Calibration: Calibration of process models is a key issue, and one that is largely unresolved. Problems of calibration are not confined to process simulation for they pertain to geology as a whole. For example, in a simulation experiment involving a river entering a basin, discharge rates of water and sediment supplied to the basin are critical. Is the fluid discharge steady or cyclic, and does sediment discharge vary as climatic changes affect the drainage basin? These are relevant questions for virtually any sedimentary basin. They are difficult for modern basins such as the Gulf of Mexico, and become increasingly difficult as we deal with sedimentary basins of progressively increasing geologic age.

Calibration is necessarily based on modern processes. With rare exceptions, there is no direct way of extracting information on process rates from ancient sedimentary features. For example, radiometric dates applied to sedimentary sequences provide estimates of gross sedimentation rates, but have little direct relevance for process rates. Calibration requires rates derived from observations of modern processes and then applying them in series of simulation experiments with the hope of finding ranges over which the assumed process rates provide responses that compare reasonably well with observed features in sedimentary basins.

First principles versus empirical relationships: An ideal would be to represent geological relationships in terms of "first principles" consisting of fundamental physical and chemical principles, such as thermodynamic laws and Newtonian mechanics. This is a difficult quest. While first principles are broadly applicable in geology, as they are everywhere in nature, it is difficult or impossible to represent any geologic process solely in terms of first principles. For example, transport of clastic sediment by running water is a fundamental component of virtually all geologic processes that affect the earth's crust. Sediment transport, however, cannot be represented solely by first principles because the physical and mathematical relationships that link flow velocities, transport capacities, and sediment characteristics are mostly empirical. It is true that first principles are represented in sediment transport through incorporation of the conservation laws for materials, energy, and momentum. Unfortunately, however, we cannot formulate effective relationships for sediment transport solely in terms of fundamental laws, and we must blend empirical relationships with first principles. Furthermore, this situation is likely to exist for decades to come in geology, and we probably never will be able to escape from it.

If we focus on the spectrum of empirical relationships, we find a continuum ranging from "well-established relationships" based on extensive observations and of broad applicability, to relationships that are vaguely defined and for which observational data are rare or non existent, for which rates cannot be objectively estimated, and that are of generally limited applicability. For example, there is abundant observational data that bears on rates and controls for transport of clastic sediment, but there is much less observational data on rates and controls for production of carbonate sediment. In part these contrasts reflect the difficulties in making the observations, but they also probably reflect fundamental differences in our basic ability to make the observations. Some observations are virtually impossible.

These differences also affect our simulation models. For example, consider 2-D basin models such as Shell's model, which are based mostly on empirical relationships. Clastic sediment transport is represented simply by utilizing the diffusion equation, and rules for carbonate production are totally empirical. These models perform well, however, because the rules embodied in them are geologically reasonable, albeit greatly simplified. But, these models are of limited applicability geologically because of their highly empirical nature. Probably models will evolve from them or replace them that will be more detailed and fundamental and less empirical, progressively reflecting advances in geology as a whole.

Deterministic or stochastic: Some simulation models are deterministic in that the responses of the model are completely predetermined, there being no random or "stochastic" components. By contrast, a stochastic model has components that are not predetermined and that behave randomly. Both forms of models are feasible for incorporation in process simulation models, with accompanying advantages and disadvantages. Stochastic models are usually simpler to formulate because the simulated behavior of a geologic process component can be represented by one or more probability distributions that are sampled at random with psuedorandom numbers. By contrast, a deterministic model is generally represented with equations that maybe simple or involved, and which are generally solved numerically. Relationships in stochastic models are generally highly empirical, whereas those in deterministic models may range from highly empirical to those based largely on first principles. Papers in this volume provide examples of both empirical and deterministic models. Guglielmo's pluton emplacement model, for example, is deterministic, whereas Tipper's model is probabilistic.

The philosphical differences between stochastic and deterministic models might seem large, but in practice the differences are less pronounced. Some purely deterministic models behave as if they were stochastic. For example, in SEDSIM (Lee and Harbaugh, this volume), flow in an open body of water such as a river is represented by the motion of large numbers of individual fluid elements whose positions, velocities, and directions are continuously tracked. While the process of flow represented in this manner is purely deterministic because the locations and motions of all fluid elements are completely predetermined, their behavior is essentially random when paths of individual elements are considered.

Computing power: The final issue concerns computing power and the insatiable demand for ever-increasing computing speed. One of the axioms of process simulation is that computers never provide enough computing power for our needs. Will this situation never change? If we examine our progress in the past quarter century, we find that affordable computing facilities have grown about four or five orders of magnitude in capability (i.e., computing speeds have increased between 10,000 and 100,000-fold in this period), but the increased level of detail and degree of resolution of our present models require about as much computational "effort" as they did a quarter century ago. In other words, relevant to our needs, we are still as short on computer power as were then. Will the next quarter century hold fundamental changes for us in terms of our needs versus improvements in programming efficiency and the ability of computers to reach ever-higher speeds?

First, affordable computing power will probably increase at an accelerated rate. It seems likely that a four to five order-of- magnitude increase will be attained well before another quarter century elapses, and probably within a decade. In spite of this trend, process simulators will still be frustrated because their demands are basically insatiable - there is no limit to the level of resolution that may be ultimately desirable. On the other hand, process simulation in the next quarter century, if not the next decade, should advance so that will be increasingly useful as an extension of the scientific method overall.

Chapter 1

Modeling and Rendering of Structures

CONVERSION OF 3D GRID INTO T-SURFACES

Y.G. Huang and J.L. Mallet*

Abstract

We propose a method to represent the isovalue surfaces, which are defined on a 3D regular grid, by a set of triangular facets (T-surfaces). Generally an isovalue surface is split into many pieces by the grid, and we can approximate each of these pieces by the triangular facets. Using a small memory, this method can handle all kinds of isovalue surfaces and works efficiently. In the case where the T-surface produced by this method has local roughness, we use DSI (see [1]) method to smooth it. This algorithm has been fully implemented and tested on various examples. Some resultats will be presented.

Keywords : isovalue surface, T-surface, grid, cube, face.

Introduction

Representation of a complex surface by triangular facets (abbreviated "T-surface") has been used for quite a number of years. For instance, in Medecine and Computer Vision, a surface is always known by a set of points; in order to obtain a model of the surface which has to be both simple and precise, one chooses usually to approximate the surface by a set of triangles. This is a good compromise between the precision of approximation and the simplicity of modelling.

Several methods have been proposed in the literature to approximate complex surfaces by T-surfaces when the data are scattered points (see [6]) or a sequence of cross-sections (see [2]). These methods are widely applied in many applications; however, they are no longer adequate when a complex surface is defined at the nodes of a regular grid (2D or 3D). In practice, many applications supply this kind of data. For example, in mining application, the ore body is always split into regular little blocks, and the mean ore grade of each block is attached to the center of the block in order to get a regular 3D grid of data. The grade $t = \varphi(x, y, z)$ at each node of the grid is estimated by a *kriging* method and the ore body G is defined as the set of points having the grade greater than a given "cut-off" t_0. In many cases, G is composed of several pieces in the shape of "lens", so that the boundary of G is a very complex surface. A similar problem occurs in the oil industry when one wants to study the variation in 3D space of physical parameter such as velocities, porosities described by a function $\varphi(x, y, z)$.

In this paper, we propose a new method which converts a surface defined by a regular 3D grid and a given "cut-off" t_0 into a T-surface. The input data is a list of values corresponding to the values at each node of the grid. At first, our method computes the points of intersection

---

*Centre de Recherche en Informatique de Nancy & Ecole Nationale Superieure de Geologie

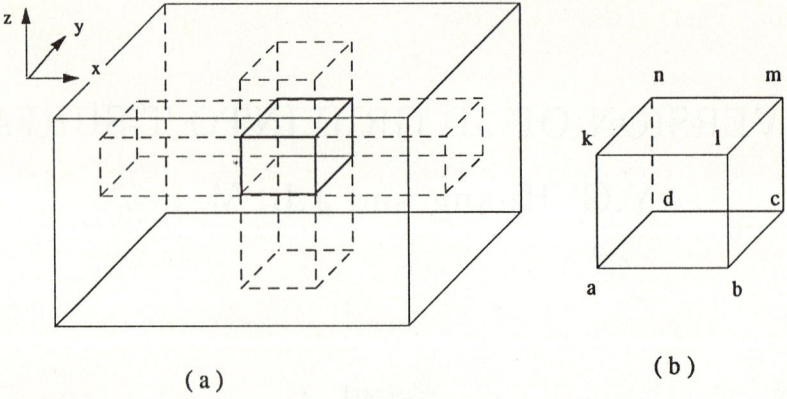

(a)

(b)

Figure 1 *The research domain is split into little elementary cubes.*

of the surface with the grid; secondly, we make a triangulation of these points of intersections. This triangulation can be done in such a way that at any time we use only 8 neighbour nodes of the grid, so that our method uses little memory and works efficiently.

In some cases, the T-surface obtained by our algorithm is not smooth; for this reason, we propose to use a method called DSI (see [1]), in order to smooth the T-surface.

1 Notion of isovalue surface defined by a regular 3D grid

Let B be a parallelepipedic domain imbedded in a 3-dimensional Euclidean space and let $t = \varphi(x, y, z)$ be a function defined on B and assumed to be non constant in B. For a given value t_0 called "cut-off", we can define a surface $S(t_0)$ in the following way :

$$(x, y, z) \in S(t_0) \Longleftrightarrow \varphi(x, y, z) = t_0$$

In this case, $S(t_0)$ is an isovalue surface (see [3]).

In the following, we will assume that $t = \varphi(x, y, z)$ is defined by its values known at each node of a regular grid R filling B (see Figure 2); if needed, in geological applications, these values may be interpolated from scattered data by using for example a *krigging* or DSI method.

The purpose of the following work is to propose a method which approximates $S(t_0)$ by a triangulated surface using only the values at the nodes of R.

2 Automatic triangulation of $S(t_0)$

As shown in Figure 1, in order to simplify the discussion, we suppose that the edges of the research domain B are parallel with the coordinate's axis and the grid R splits B into little elementary "cubes". Moreover, we assume that the following hypotheses are always satisfied for any elementary cube $\square$:

- The grid R is small enough in order that $S(t_0)$ intersects any edge of $\square$ at most one time.

- If two vertices (x', y', z') and (x'', y'', z'') of $\square$ satisfy :

$$\varphi(x', y', z') > t_0 \text{ and } \varphi(x'', y'', z'') < t_0$$

then $\square$ is intersected by $S(t_0)$.

- If none of the edge of $\square$ is intersected by $S(t_0)$, then $S(t_0)$ is certainly outside of $\square$.

- The surface $S(t_0)$ does not strictly go through any node of R corresponding to the vertices of the elementary cubes.

As shown in Figure 1, let $\{a, b, c, d, k, l, m, n\}$ be the eight vertices of an elementary cube $\square$ of the grid R and let P be the part of $S(t_0)$ located inside of $\square$:

$$P = \square \cap S(t_0)$$

This part P may consist of several disjoint pieces but, in any cases, we will note C this boundary constitued by the intersection of the six faces $\{F_i(\square) : i = 1, 6\}$ of $\square$ with $S(t_0)$:

$$C = \bigcup_{i=1}^{6} \{F_i(\square) \cap S(t_0)\}$$

Therefore, approximating $S(t_0)$ by a T-surface S reduces to approach P by a piece of this T-surface S for each elementary cube of R. Fore that purpose, we assume first that the boundary C of P is a polygonal line which, in some cases, may be composed of several disjoints pieces.

As shown in Figure 2, taking into account the previous hypotheses, we deduce that the intersection of a face $F_i(\square)$ of $\square$ with $S(t_0)$ consists of, at most, two pieces of curves. These pieces of curves can be approached by two straight segments, so that C can be represented by a set of polygonal contours. Actually, as shown in Figure 3, we have only to consider 13 different topological cases corresponding to the different types of possible borders C; among these 13 cases, we will distinguish simple cases from complex ones :

- The simple cases correspond to the ones where each face $F_i(\square)$ of the cube $\square$ contains at most one segment (edge) of C. In this case, the contour C may be either void or composed of one (several) closed polygonal line.

- The complex cases correspond to the ones where at least one face $F_i(\square)$ of the cube $\square$ contains two segments (edges) of C. In this case, the contour C may be composed of several closed polygonal lines which are disjoint from each other.

In the simple cases, C can be obtained simply by joining any pair of segments having a common vertex but, in complex cases, there are several possibilities for building C. In fact, for complex cases, the main difficulty consists of choosing the right way to link the edges of C each time that a same face of $\square$ contains two segments of C (see Figure 2); we will propose a solution for solving this problem in the next section.

Once the polygonal contour C is determined, the only remaining problem consists of building a T-surface $S(\square)$ satisfying the following conditions :

- C is the boundary of $S(\square)$[1].

---

[1] the boundary of a T-surface is composed of the edges each of which belongs to only one triangle of the T-surface

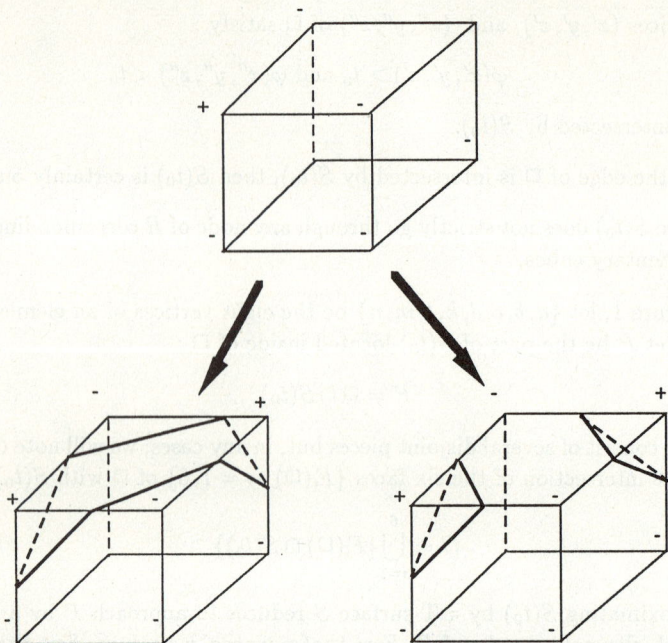

Figure 2 *There are two possibilities to build C because the top face of the cube □ contains two edges of C.*

- the vertices of $S(\square)$ are also the ones of C.

- $S(\square)$ is contained in $\square$.

In any case, it should be noted that the number of vertices of C is at most 12 and C may only be a contour as shown in Figure 4 according to the fact that P is a piece of $S(t_0)$; we conclude that there is always a T-surface $S(\square)$ satisfying the above conditions. For example considering the following procedure (in pseudo-C):

```
LIST_OF_TRIANGLES Triangulating_Contour(C)
{



    let T be the list of triangles to be built;
    let p_{i-1}, p_i and p_{i+1} be the 3 current consecutive vertices of C;

    T = φ ;
    while( C is not a triangle)
    {
        for( all vertex p_i of C )
        {
            if( p_{i-1} and p_{i+1} are not located in the same face of F_i(□) )
```

number of vertices +	topological configurations	possible contour (c)	number of cases
0 (8)	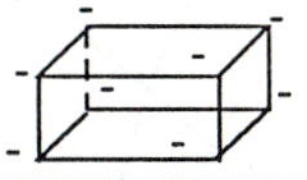	C = 0	1 (1)
1 (7)	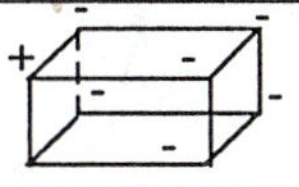	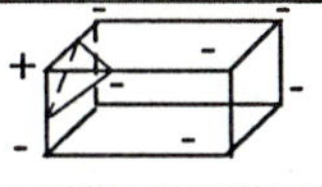	8 (8)
2 (6)	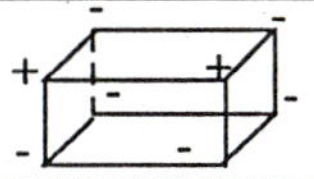	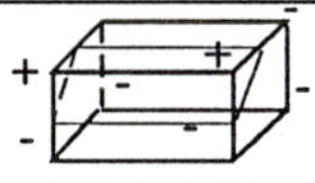	12 (12)
	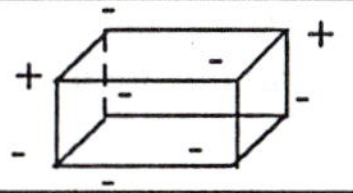	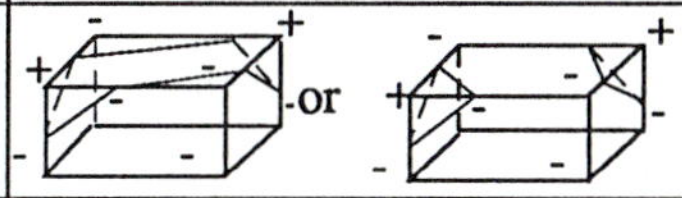 -or	12 (12)
	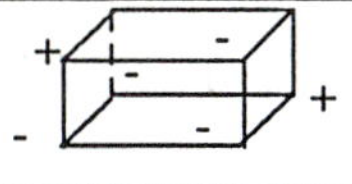	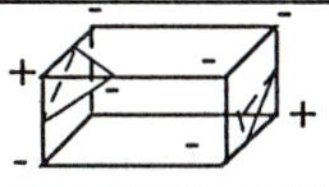	4 (4)
3 (5)	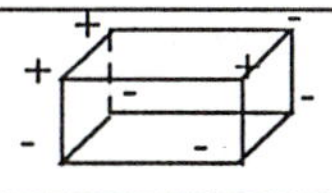	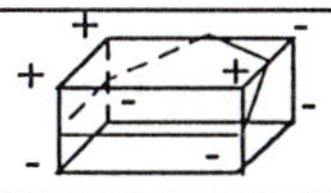	24 (24)
	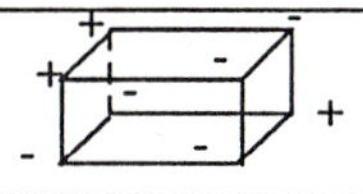		24 (24)
	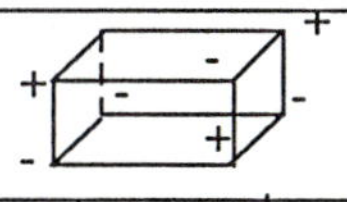		8 (8)
4	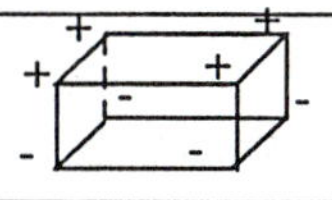	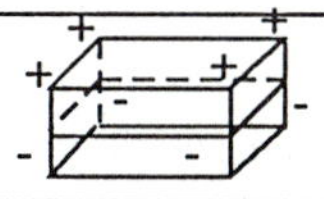	6 (6)
	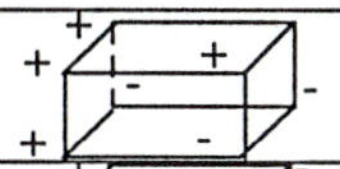	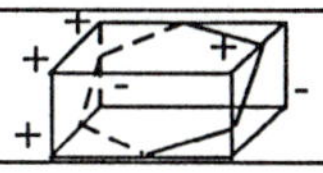	8 (8)
	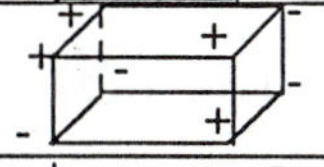		16 (16)
	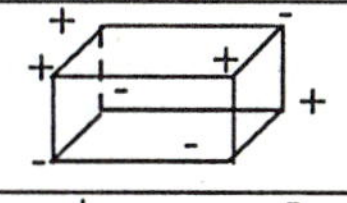		24 (24)
	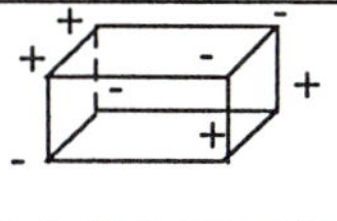		6 (6)
	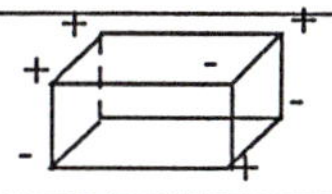		2 (2)

Figure 3 *Generation of a contour C in terms of values at the vertices of an elementary cube □. These values are noted "+" when they are greater than the "cut-off" t_0 and "−" when they are lower.*

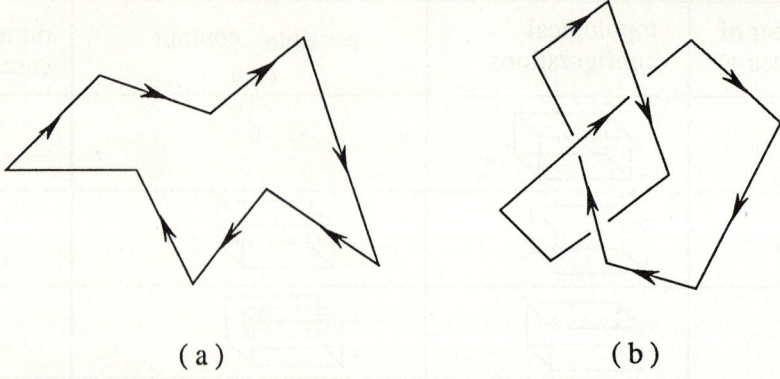

(a)   (b)

Figure 4 *(a) C is always a simple polygonal contour and C cannot be a contour as the one shown in (b).*

```
        {
            build the triangle T(p_{i-1}, p_i, p_{i+1});
            add T(p_{i-1}, p_i, p_{i+1}) to T: T += T(p_{i-1}, p_i, p_{i+1});
            remove the vertex p_i from C : C -= p_i;
        }
    }
}

add C to T ;

return( T ) ;
}
```

The output T-surface S is the one which satisfies the conditions (see Figure 5). generally, for a given C several T-surfaces can be produced from it; in this case we choose the one which minimizes the area of the generated T-surfaces $S(\square)$ (see Figure 6).

3   Determination of $F_i(\square) \cap S(t_0)$

Let $F_i(\square)$ be a face of a cube $\square$ defined by its 4 vertices (k, l, m, n), let $H_i(\square)$ be the plane containing this face and let C_i be the intersection of $H_i(\square)$ by C. As shown in Figure 7, we have to consider 3 topological cases :

- $S(t_0)$ crosses two opposite edges of $F_i(\square)$.

- $S(t_0)$ crosses two adjacent edges of $F_i(\square)$.

- $S(t_0)$ crosses the 4 edges of $F_i(\square)$.

In the first cases, the problem of building C_i has a trivial solution; as shown in Figure 7, C_i can be obtained by just joining the two crossing points. In the case where the 4 edges are crossed

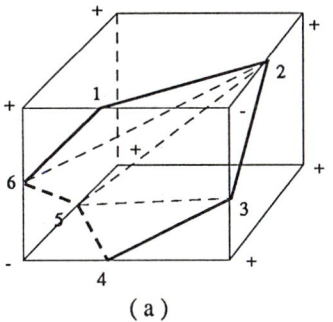 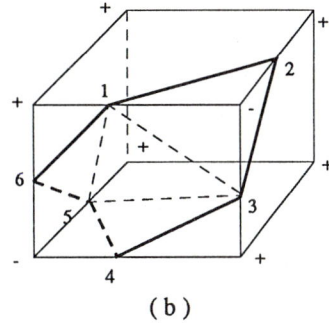

(a)　　　　　　　　　　　(b)

Figure 5 *(a) a correct triangulation of C; (b) an incorrect triangulation of C because the vertices 1 and 3 are located in the same face of $\square$.*

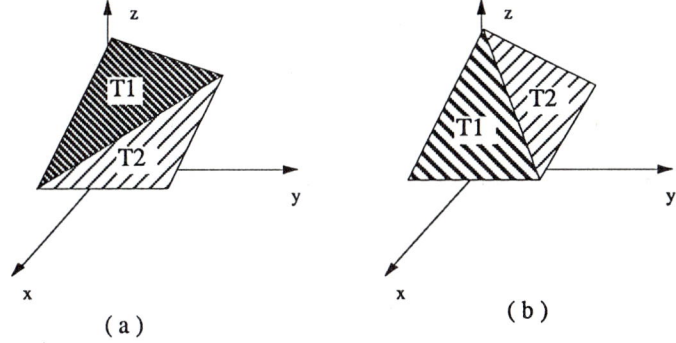

(a)　　　　　　　　　　　(b)

Figure 6 *$S(t_0)$ obtained in (b) is the one whose area is smaller than the one shown in (a), therefore $S(t_0)$ in (b) is considered to be better.*

by $S(t_0)$, however, the trivial solution is no longer acceptable since there are two possibilities to building C_i. In order to decide which one has to be chosen, it is necessary to check the variation of $\varphi(x, y, z)$ in the neighbourhood of $F_i(\square)$.

As shown in Figure 8, let us assume that the 4 edges of $F_i(\square)$ are crossed by $S(t_0)$ and let p_0 be the point of intersection on edge (kl) located in a plane parallel to the (ox, oy) plane. Given a step $h > 0$ small regarding to the size of $F_i(\square)$, we can build the series of points $\{\vec{p_0}, \vec{p_1}, \ldots, \vec{p_n}\}$ such as :

$$
\begin{cases}
p_i\vec{p}_{i+1} = h \cdot \dfrac{\vec{\theta_i}}{\|\vec{\theta_i}\|} \\[3ex]
with \quad \vec{\theta_i} = \varepsilon \cdot \begin{bmatrix} -\dfrac{\partial\varphi(\vec{p_i})}{\partial y} \\[2ex] \dfrac{\partial\varphi(\vec{p_i})}{\partial x} \end{bmatrix} \\[4ex]
\varepsilon = \pm 1 \text{ such as} : \vec{\theta_i} \cdot \vec{oy} > 0
\end{cases}
$$

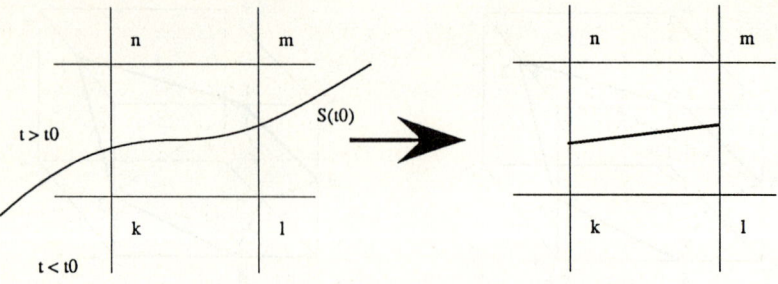

S(t0) crosses two opposite edges

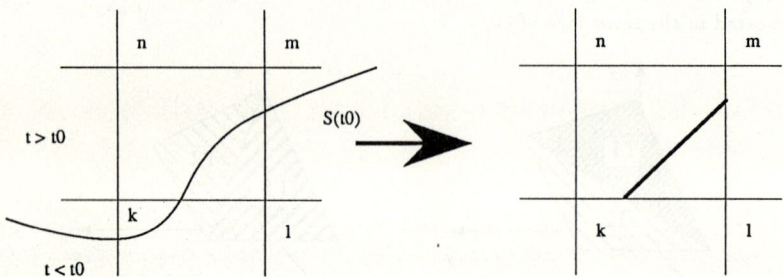

S(t0) crosses two adjacent edges

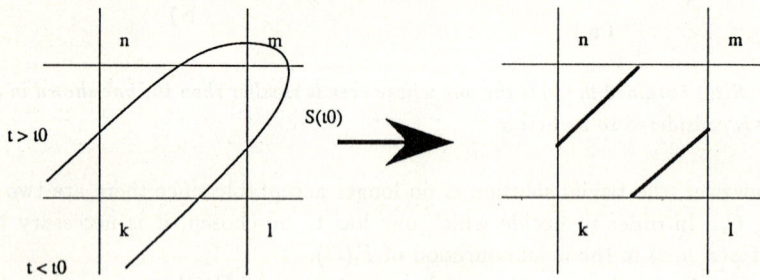

S(t0) crosses the 4 edges

Figure 7 *Intersection of a face $F_i(\Box)$ with the isovalue surface $S(t_0)$ and generation of straight segments belonging to C.*

In this expression, $\frac{\partial(\varphi(\vec{p_i}))}{\partial x}$ and $\frac{\partial(\varphi(\vec{p_i}))}{\partial y}$ denoting the partial derivative of $\varphi()$ at the point $\vec{p_n}$ which is computed with the help of a local interpolation method. This interpolation is based on the values of $\varphi()$ at the node of the grid R in the neighbourhood surrounding $F_i(\Box)$.   8

Let $\vec{p_n}$ be the first point of the series which goes through the straight line (kn) or (lm). If we notice that the series $\{\vec{p_0}, \vec{p_1}, \ldots, \vec{p_n}\}$ is approximately located on the surface $S(t_0)$, we can now decide if C_i belongs to the case 1 or case 2 represented in the Figure 8. Therefore, we have only to check if $\vec{p_n}$ is located on the left of the straight (kn) or on the right of the straight line (lm) :

$$\left\{ \begin{array}{l} \vec{p_n} \text{ is on the left of } (kn) \implies case\ 1 \\ \vec{p_n} \text{ is on the right of } (lm) \implies case\ 2 \end{array} \right.$$

One can notice that the decision taken this way is better if the local interpolation method uses a bigger neighbourhood of $\Box$ in the regular grid R.

4 Smoothing the T-surface $S(t_0)$ with DSI

In some cases where the T-surfaces $S(t_0)$ obtained may have some local roughness (sometimes these local roughness may be intense). In practice, this roughness may become a big disadvantage for the applications. In fact, these irregularities occur in the following cases :

- the grid R is not small enough.

- the function $t = \varphi(x, y, z)$ is not regular.

In order to avoid these cases and prevent the generation of local roughness, we need to either make the grid R smaller or regularize the function $t = \varphi(x, y, z)$. If this is not possible, we propose to use a method called "Discrete Smooth Interpolation" and abbreviated DSI (see [1]) to reduce the local roughness and make the T-surface $S(t_0)$ smooth. In the following, we will present shortly the DSI method and show some T-surfaces treated by DSI.

4.1 Introducing the DSI method

Let us assume that vertices of the T-surface $S(t_0)$ have been numbered from 1 to N et let Ω be the set of these N vertices; in the following we will identify the "vertex number" with "k" so that :

$$\Omega = \{1, 2, \ldots, N\}$$

For any vertex "k" belonging to Ω, we define the neighbourhood $N(k)$ as the subset of Ω such that :

$$\alpha \in N(k) \iff (\alpha, k) \text{ is the edge of a triangle}$$

the vector joining the origin of the 3D space to a given vertex k is noted $\vec{\varphi_k}$ and φ is the collection of all these vertices :

$$\varphi = \{\vec{\varphi_1}, \vec{\varphi_2}, \ldots, \vec{\varphi_N}\}$$

the DSI method is based on a local roughness criterion $R(\varphi|k)$ defined as follows at each node $k \in \Omega$:

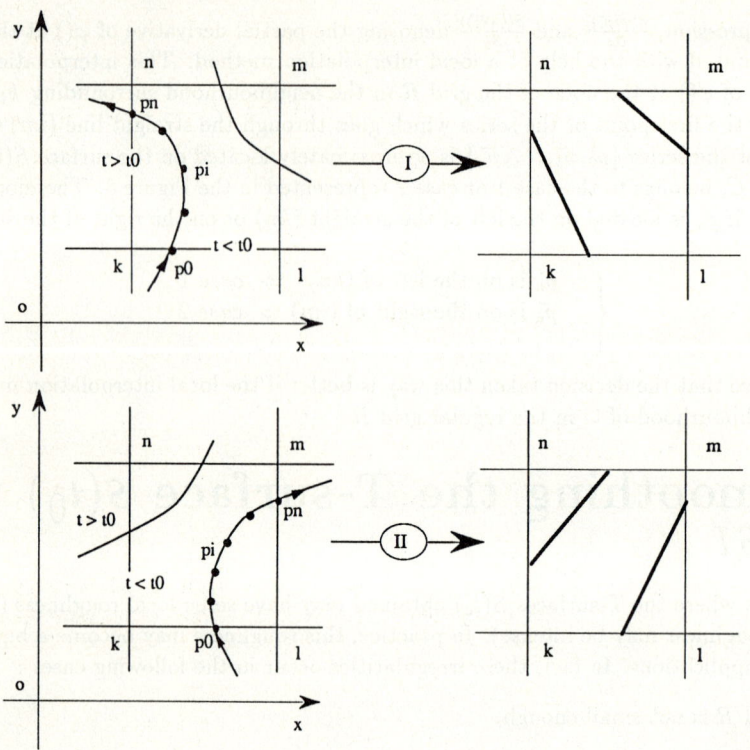

Figure 8

$$R(\varphi|k) = || \sum_{\alpha \in N(k)} v^{\alpha}(k) \cdot \vec{\varphi_{\alpha}} ||^2$$

the coefficients $\{v^{\alpha}(k)\}$ are assumed to be given weighting coefficients. For instance, in the examples represented in this paper, we have chosen these coefficients as follows :

$$\left[\begin{array}{lll} v^{\alpha}(k) & = & \left\{ \begin{array}{lll} -|\Lambda(k)| & \text{if} & \alpha = k \\ 1 & \text{if} & \alpha \in \Lambda(k) \end{array} \right. \\ \\ with & : & \left\{ \begin{array}{lll} \Lambda(k) & = & N(k) - \{k\} \\ |\Lambda(k)| & = & \text{number of elements of } \Lambda(k) \end{array} \right. \end{array} \right.$$

The local roughness criterion $R(\varphi|k)$ are used to build a global roughness criterion $R(\varphi)$ such that :

$$R(\varphi) = \sum_{k \in \Omega} R(\varphi|k)$$

In practice, the surface $S(t_0)$ has to respect a given set of constraints and each of these contraints induces a constraint on the set φ; the DSI method assume that each of these constraints can be expressed as :

$$C_i(\varphi) = 0$$

The goal of the DSI method is to look for the set φ of vertices minimizing the criterion $R^*(\varphi)$ such that :

$$R^*(\varphi) = \sum_{k \in \Omega} R(\varphi|k) + \sum_i \varpi_i^2 \cdot |C_i(\varphi)|^2$$

the coefficients ϖ_i^2 are called "certainty factor" and are given weighting coefficients that must be chosen in order to modulate the relative importance of each constraint.

4.2   Smoothing a T-surface

Let $S^0(t_0)$ be the initial rough surface yielded by our algorithm and let φ^0 be the collection of its associated vertices $\vec{\varphi_k^0}$:

$$\varphi^0 = \{\vec{\varphi_1^0}, \vec{\varphi_2^0}, \ldots, \vec{\varphi_N^0}\}$$

we would like to transform this surface into a smooth T-surface $S(t_0)$ close to $S^0(t_0)$; in other word, each node φ_i of $S(t_0)$ must satisfy the following constraints :

$$\vec{\varphi_i} \simeq \vec{\varphi_i^0}$$

this suggests us to choose $R^*(\varphi)$ such that :

$$R^*(\varphi) = \sum_{k \in \Omega} R(\varphi|k) + \sum_{i \in \Omega} \varpi_i^2 \cdot ||\vec{\varphi_i} - \vec{\varphi_i^0}||^2$$

Minimizing $R^*(\varphi|k)$ allows to compute the position of the vertices of the smooth surface $S(t_0)$.

5   Conclusion

In this paper, we presented an algorithm to create T-surfaces from 3D regular grids. Using automatic mapping method (see [4]), we transformed complex 3D topological problems into 2D ones, which makes this algorithm very efficient. In Figure 9, some T-surfaces obtained by our algorithm are shown. As one can see, the T-surface in Figure 9.a is very rough; after the treatment by the *DSI* method (Fig 9.b) the T-surface becomes very smooth and its triangles are much more equilateral.

Acknowledgements

This research was developped in the frame of the *GOCAD* project and we would like to thank the sponsors who are currently supporting this project :

ELF Aquitaine	CFP TOTAL
Compagnie Générale de Géophysique	Phillips
Standford University	IPG (Paris)
BRGM (France)	Chevron
Geologistes Institut da Universitat (Freiburg)	Amoco

6   BIBLIOGRAPHY

1. J.L.MALLET. *Discrete Smooth Interpolation.* In : ACM Transactions on Graphics, Vol.8, num.2,April 1889, pages 121-144.

2. J.D.BOISSONNAT. *Shape reconstruction from planar cross-section.* In : Rapport de recherche. Num.546. Juillet 1986. Institut Nationale en Informatique et en Automatique.

3. H.KLEIN, R.PFLUG and C.RAMSHORN. *Shaded Perspective View by Computer: A new tool for geologists.* In : Geobyte, August 1989, pages 16-24.

4. J.L.MALLET. *Présentation d'un ensemble de méthodes de la cartographie automatique.* In : Sciences de la terre, num.4, Octobre 1974.

5. P.SHAMOS. *Voronoi diagram.* In : Computational Geometry. pages 198-205.

6. Y.CHIPOT. *Smoothing a surface with DSI.* In : Revue Internationale de CFAO et d'Infographie. Vol.4, pages 10-20.

7. R.M.SRIVASTAVA. *G2T : A program for the triangulation of faulted grids.* In : Report of research. LIAD. ENSG. Nancy, France.

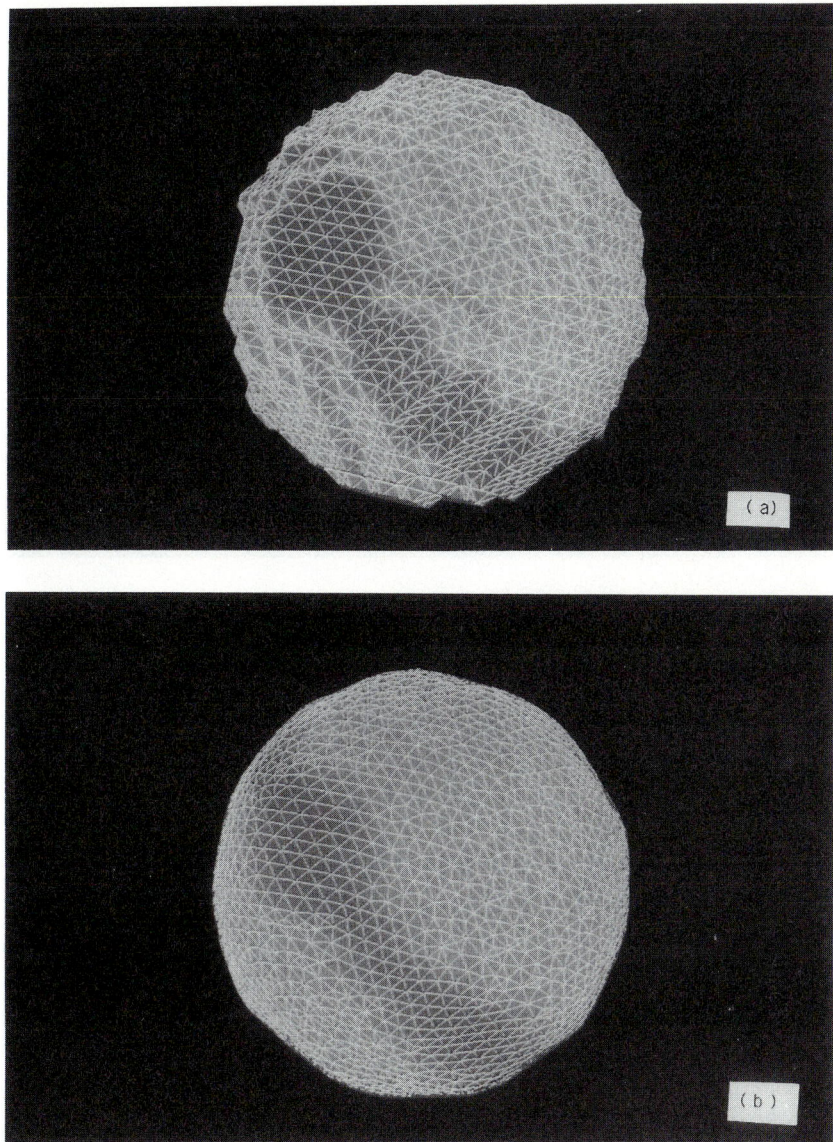

Figure 9 *(a) Exemple of isovalue surface obtained by our method; (b) the same surface after the treatment by DSI.*

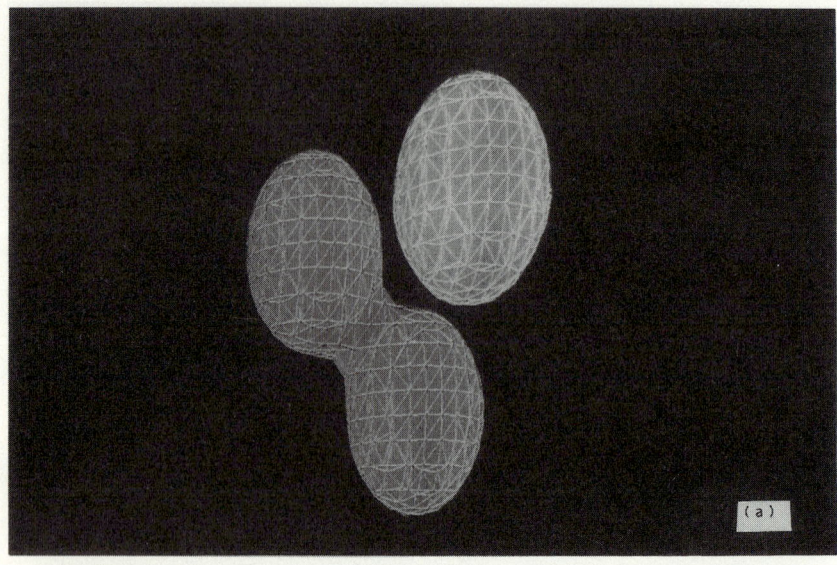

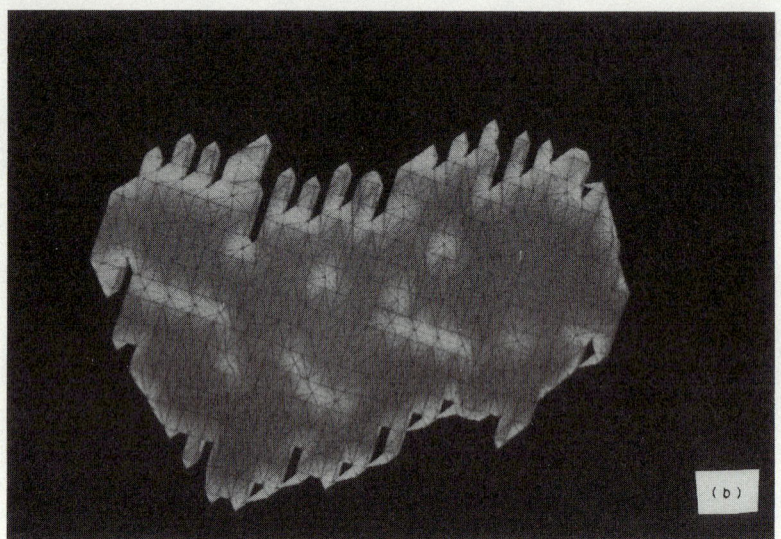

Figure 10 *(a) Isovalue surface corresponding to a mathmatical function; (b) Isovalue surface corresponding to the border of a deposit.*

RATIONAL SPLINES AND MULTIDIMENSIONAL GEOLOGIC MODELING

T. R. FISHER and R. Q. WALES

Intergraph Corporation, Mail Stop IW-17a5, Huntsville, Alabama, 35894, U.S.A
Intergraph Corporation, Suite 200, 675 Bering Drive, Houston, Texas, 77057, U.S.A.

ABSTRACT

Traditional 2D and 2.5D methods of geologic interpretation are no longer adequate to visualize and interpret the wealth of multidimensional data available to the geologist. This has created a need for computer-based methods of 3D analysis (3D GIS) with capabilities which allow the geologist to integrate the data, interpret geologic features, and visualize attributes in their true 3D spatial relationships.

These requirements are met by solid volume modeling techniques. Most use a semi-transparent depiction of both surface and internal features, and have no manual equivalents. One system which has been used to create stratigraphic and structural geologic models is based on non-uniform rational B-splines (NURBS). This technology is an alternative to more well known solids modeling approaches (e.g., polygon meshes, 3D grid and isosurfaces, voxels), and holds promise of providing a unified basis for geologic modeling.

NURBS are piecewise, parametric polynomials that can describe all large complex surfaces and solids, and they provide a single, uniform and precise mathematical form capable of representing the free-form curves, surfaces, and solids necessary for geologic modeling. Splines are n-dimensional, making them extensible to representation of distributed attributes within a defined solid. Functional integration between graphic elements (e.g., points, lines, wireframes, surfaces, solids) is also possible because the elements share a common mathematical basis.

KEYWORDS

3D GIS, geologic modeling systems, solid volume modeling, NURBS, splines, rational curve and surface descriptions, distributed attributes

INTRODUCTION

The advent of sophisticated instrumentation and computerized analyses has supplemented the usual geologic field observations with more data than can now be effectively interpreted by the ordinary geologist. Most of this geologic data is multidimensional; having spatial and time-dependent qualities. Traditional methods (e.g., contour maps, cross sections, fence diagrams, isometric surfaces), however limit visualization and analysis of this data to two or maybe quasi-three (2.5D) dimensions. In most cases these traditional methods make it impossible to visualize combinations of more than a few different attributes at a time, and the introduction of new information into previous interpretations becomes a laborious, time consuming task. If testing of alternate models or hypotheses is required, the effort increases several fold. Computer-based, true 3D analysis or 3D GIS (Geographic Information Systems) with continuous volumetric data structures and appropriate analytical functions would give geologists the tools to integrate this mass of data (Van Driel, 1989), and allow efficient visualization, modeling and interpretation of multiple geologic or other attributes in their true 3D spatial relationships.

Fortunately, some possible solutions to the problem of visualization and modeling of spatial relationships between large quantities of data already exist. Geographic information systems (GIS) currently aid in the collection, management, and analysis of locationally or spatially defined 2D data. Their counterpart in geology and related fields, is known as 3D GIS, or alternatively "GSIS" for *geoscientific* information systems (Raper, 1989 and Turner, 1989); these add the third dimension to accommodate subsurface data. These newly emerging computer-based technologies range from the simple to the sophisticated, come in many varieties (Fried and Leonard, 1990), and incorporate methods of solid volume modeling or rendering. Most use a semi-transparent depiction of both surface and internal features and have no manual equivalents (Van Driel, 1989).

Many of these systems had their origins elsewhere in CAD/CAM (computer-aided design and manufacture) applications such as mechanical design or architectural modeling. The present paper focuses on a technology known as NURBS (non-uniform rational B-splines), which has roots in systems for design of complex machine and industrial parts. The purpose of the paper is to 1) review the capabilities of this method for modeling of geo-objects; 2) introduce some of the underlying mathematical concepts; and 3) present the results of two case studies using NURBS for modeling stratigraphy and structure.

NURBS TECHNOLOGY

Non-uniform rational B-splines present an alternative to more well known approaches to solid volume modeling, such as those summarized by Fried and Leonard (1990). They hold promise of yielding a unified basis for geologic modeling; for both static and time-dependent (dynamic) modeling applications. Originally used to design complex machine and industrial parts, NURBS technology grew from the need for a modeling system that had a common internal method of representing and storing the many different geometric entities required for design applications (Piegl, 1990). Splines, and rational curve and surface descriptions have their roots in a very old branch of mathematics known as approximation theory. However, they have only recently found application outside of mechanical design; representing data obtained from medical, geological, physical and other natural phenomena (Rogers and Adams, 1990).

NURBS can describe all large complex surfaces and solids, and provide a single uniform and precise mathematical form capable of representing common analytical shapes, primitive quadrics, and free-form curves and surfaces necessary to geologic modeling. This single mathematical representation is extensible to representation of distributed attributes within a defined solid (Herring, 1990a,b). Functional integration is also possible because all graphical elements share a common mathematical basis.

Mathematical Elements of NURBS

A thorough and detailed mathematical discussion of splines is beyond the scope of this paper, but some of the basic concepts are necessary to understanding a NURBS-based modeling system. A detailed overview of the foundations of NURBS can be found in Tiller (1983). Comprehensive expositions on the mathematics of general splines can be found in Bartels *et al.* (1987) and de Boor (1978). Rogers and Adams (1990) provide excellent discussions on the use of splines in computer-based graphics and modeling. Fisher and Wales (1990b) cover, in detail, the foundations of NURBS and their application to geologic modeling.

Interpolation and Approximation. Geologic investigations like most scientific studies result in the output of some finite amount of data. If we view this data as a series of points through which we wish to fit a curve (or surface), we arrive at a *data fitting problem*. The curve or surface can be arbitrarily approximated by a series of short, straight line segments (or planar polygons, in the case of a surface) to yield a visual representation. However, this method works only for the simplest geometric forms. Use of "primitives" (i.e., line segments, planar polygons, etc.) to approximate complex free-form objects such as those found in nature, require large amounts of data to obtain the necessary smoothness of fit, and they become awkward to manipulate with the computer (Bartels *et al.*, 1987).

It is more efficient if we instead, represent the curve or surface analytically. This reduces storage requirements, increases precision, and ease of calculation of intermediate points. Analytically fitting a curve through all known data points then becomes a problem in *classical interpolation*, but interpolation can be shown to be just a special case of the more general approximation problem (i.e., creating a curve or surface which comes near the data points, but does not necessarily pass through them). Commonly, it will be found that most geologic data is in form a that lends itself to approximation, because such data are subject to error or noise.

If a curve or surface is relatively simple, a single polynomial function will suffice to describe the shape. Complex shapes, however require either single, high-order polynomials (which may introduce unwanted oscillations) or approximation in *piecewise* fashion, by a series of short, simple *curve segments*, each defined by a low order polynomial. Likewise, a surface can be represented by a collection of smaller *surface patches*. Unfortunately, piecewise polynomials are not necessarily continuous functions, and there is no guarantee of smoothness of the curve at the points where the segments join. Additionally, there is the problem of representing multiple z values at the same x, y coordinates.

The Multiple "Z" Problem. Parametric polynomials are a functional form that solve the multiple z problem. These functions allow us to represent each point on a curve or surface as a function of a single parameter. A planar (2D) curve can be described by a set of two single-valued functions $x=x(u)$ and $y=y(u)$ of a parameter u. The functions $x(u)$ and $y(u)$ yield the x- and y- coordinates, respectively, of a point on the curve for any value of u (Bartels *et al.*, 1987).

The parameter simply acts as a coordinate label for points on the curve. The position of the point is fixed by the value of the parameter, and accordingly the position of the point is given by a vector, $P(u)=[x(u),y(u)]$. Similarly, the position of point on a space (3D) curve is given by $C(u)=[x(u),y(u),z(u)]$. A surface is represented in a similar fashion, by a set of two parameters u and v, such that we have $S(u)=[x(u,v),y(u,v),z(u,v)]$.

Basic Spline Functions. Combining the previously outlined concepts, we introduce a special form of piecewise, parametric polynomial; the *spline*. In its classical definition, this functional form is a piecewise polynomial of degree k with *continuous* derivatives of order k-1 at the common joints between segments (Rogers and Adams, 1990). This form permits us to overcome the problem of undesirable oscillations found in higher order polynomials, and the discontinuities found in normal piecewise polynomials.

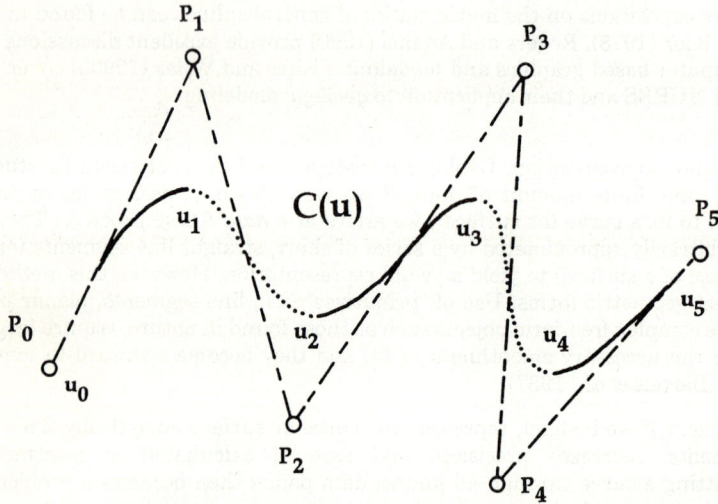

Fig. 1.   Example spline curve. P_is are poles or control vertices
            on the *control polygon* (dashed lines). Segments of the splines
            are shown by the alternately solid and dotted curve. u_is
            are locations of knots, or joints between segments.

Because a cubic polynomial is the lowest degree function with which we can adequately describe a space curve, the *uniform cubic B-spline* is introduced. The general equation of this family of splines is given by

$$C(u) = \sum_{i=1}^{n} w_i (u) P_i \qquad\qquad (1)$$

where

P_i are 3D points called *poles* on a control polygon. u is the parameter, where $u_0 \leq u \leq u_l$ and $w_i(u)$ are scalar-valued polynomials of the variable u, of order k (degree k-1). These being the so-called *basis* (weight) functions, defined by the order k and a *knot vector*.

If the knots (certain integer values u_i, where the polynomials of the spline tie together) are a constant distance apart then the knot vector is said to be uniform (hence the term "uniform" cubic B-spline), otherwise it is non-uniform. The knot vector is given by the sequence of values

$$u_0, \ldots, u_i, \ldots, u_j, \ldots, u_l, \ldots, u_{last} \qquad (2)$$

In uniform B-splines (and other more advanced spline functions) the basis function is generally *non-global*. That is, each pole P_i, is associated with a unique basis function which controls the influence of any one pole on the geometry of the curve or surface. One may visualize the influence of the poles on the curve as being somewhat analogous to the manner in which a point mass would affect the path of a free-falling object. This property of splines is known as *local control* and permits modification of a portion of a curve or surface without affecting the entire model (Fisher and Wales, 1990b).

A uniform B-spline surface is described in a manner similar to the parametric approach describe previously. Beginning with (1), we substitute a second set of spline functions for P_i, such that

$$P_i = \sum_{j=1}^{m} w_j(v) Q_{ij} \qquad (3)$$

and then derive a *tensor product* surface B-spline

$$S(u,v) = \sum_{i=1}^{n} \sum_{j=1}^{m} w_i(u) w_j(v) Q_{ij} \qquad (4)$$

where

Q_{ij} forms an n times m array of control points

We now have two knot vectors and directions (u and v), so that our B-spline curve in Fig. 1 is displaced along the this new vector, producing a surface similar to what is shown in Fig. 2. Once the parametric surface is created it is mapped to the Euclidian target space. The technique for this step is given in Rogers and Adams (1990).

Definition of NURBS. Because our goal has been to create a single unified foundation which will permit the *precise* mathematical representation of both free-form shapes as well as circles, conic and quadric primitives, we need a functional form of spline whose domain contains all of these. Non-uniform rational B-splines have this characteristic (Tiller, 1983).

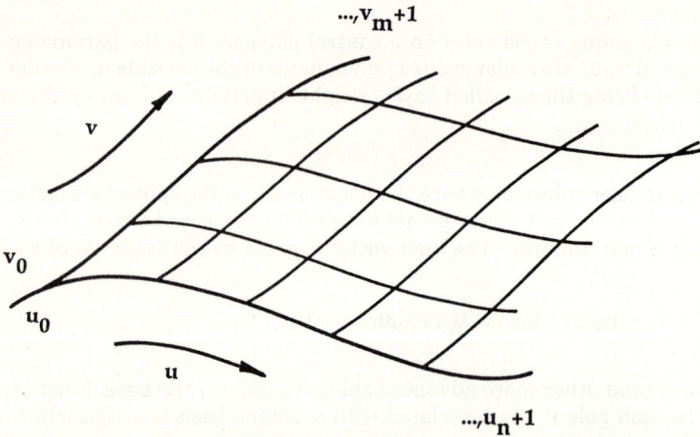

Fig. 2.   Three-dimensional parametric surface produced by
tensor product B-spline.

A rational B-spline is defined by Rogers and Adams (1990), as the projection of a non-rational (polynomial) B-spline curve described in 4D *homogeneous coordinate space* (if $P = (x,y,z)$ is a point in 3D space, the corresponding point in 4D homogeneous space is given by $P^h = [hx, hy, hz, h]$ where $h > 0$), back into 3D physical space. We have then by analogy to (1)

$$C^h(u) = \sum_{i=1}^{n} w_i(u) P_i^h \qquad (5)$$

where

w_i(u) are the kth-order B-spline basis functions and P_i^h the poles or control vertices in homogeneous space. Implicit in this equation is the previously defined knot vector.

The set of points created by projecting $C^h(u)$ into Euclidian space, and obtained by dividing the first three coordinates of each point by its homogeneous coordinate (Tiller, 1983), we call C(u). C(u) then becomes the non-uniform rational B-spline defined by

$$C^h(u) = \frac{\sum_{i=1}^{n} w_i(u) h_i P_i}{\sum_{i=1}^{n} w_i(u) h_i} \qquad (6)$$

Similarly, the rational generalization of the *tensor product* B- spline is given by

$$S^h(u,v) = \frac{\sum_{i=1}^{n} \sum_{j=1}^{m} w_i(u)\, w_j(v)\, h_{ij}\, P_{ij}^h}{\sum_{i=1}^{n} \sum_{j=1}^{m} w_i(u)\, w_{ij}(v)\, h_{ij}} \qquad (7)$$

Use of Splines for Distribution of Attributes. Examination of the mathematical characteristics of splines reveals that they are n-dimensional functions. This permits us to extend the use of NURBS from simply rendering or modeling the "shell" of a geo-object, to distributing attributes (e.g., rock properties, fluid flow trajectories) within the defined solid. Recent investigations by Herring (1990a,b) notes this is possible because, in splines a *tensor sum* model is created that makes no implementation distinction between geometry, and other numerically measurable attributes. Herring (1990b) proposes two approaches to representing distributed attributes using splines:

> 1) definition of the attributes with the geometry in a single spline
> 2) use of multiple splines over a single parameter space

In the first case, a spline is generated in which the first three range coordinates represent the position of a point and the trailing range coordinates are for attributes along the spline. For the second approach, attributes are generated by separate spline functions sharing a common parameter domain. The first functions map the geometry in parameter space, each additional spline describes an attribute or set of attributes within the geometry.

Herring's methods (Herring, 1990b) are far reaching and suggest that in using splines, we can set attributes to vary in time and space, and include complex mathematical structures such as vector fields or sets of trajectories for differential equations (e.g., fluid flow). Such attributes can also be represented to any degree of accuracy required via use of standard spline functions.

Advantages of NURBS for Geologic Modeling. From our studies we have determined that there are many advantages to using NURBS to model natural phenomenon, such as geologic structures and processes. Amongst these are:

> 1)   a unified math basis capable of representing and describing all known geometries; including free-form shapes and solids
> 2)   a unified data base for storage of analytic and free- form objects (stored as spline functions)
> 3)   local control of the model; data points may be changed without affecting entire model
> 4)   fast computation times; coordinates of points are contained within the spline function and carried through the computations
> 5)   reduced data storage via parameterization of elements
> 6)   functional integration of graphic elements
> 7)   invariance under scaling, rotation, translation, shear, parallel and perspective projection
> 8)   distribution of attributes within a defined solid

CASE STUDIES: GEOLOGIC MODELING WITH NURBS

Two case studies, one of a producing oil field in Colorado, and a second of a proposed nuclear waste disposal site in Nevada follow. In the first, NURBS were used to model stratigraphic traps in aggradational stream deposits. The nuclear waste site study focuses on structural relationships in a series of faulted, ash flow tuffs. Both cases were experimental and utilized "real world" data sets to test the capability of an existing NURBS-based system for geologic modeling. The studies helped point out both the strengths and weaknesses of the system, and give direction for future development.

Noonen Ranch Oil Field

Noonen Ranch Field Geology. Noonen Ranch is a producing oil field in the Denver basin of Colorado, and produces from stratigraphic traps in the Cretaceous D and J Sandstones. Our study highlights the D Sandstone, which at Noonen Ranch is a series of east-west trending aggradational stream deposits in a valley-fill complex (Fisher and Wales, 1990a).

The sandstones modeled occur in four distinct and mappable units deposited in a fault-controlled stream valley eroded into the underlying Huntsman Shale. Individual sandbodies measure approximately 15 to 20 feet (4 to 6m) thick, 5400 feet (1645m) in length and 2600 feet (790m) in width. Our interest in modeling the sandbodies was to determine if more would be revealed about their geometry from solid models than by conventional contour maps. At this scale and level of detail, we also expected to make some initial statements about sandbody continuity, and interconnectedness.

Modeling Process. A data set consisting of 21 wireline well logs, plus core descriptions and analyses, provided the main body of information for Noonen Ranch. Base maps digitized from U.S. Geological Survey 7.5 minute topographic quadrangles were posted with well locations. This data was pre-processed using a workstation-based, interactive geologic interpretation system. Several steps, as illustrated in Fig. 3, were necessary to create the solid models.

Well logs were entered into the system by digitizing from paper hard copy. We utilized the interactive interpretation system to correlate the well logs and "pick" geologic horizons of interest. The tops were posted on the digital base map and "manually" contoured. The resulting contour maps were appropriately combined to densify the data. This resulted in 125 to 250 new data points in addition tops from the 21 wells.

Once the data was densified, contoured surfaces were created for the upper and lower halves of each sandbody; an additional surface was created for the Huntsman Shale. In each instance the lower surface of the sandbody was assumed to mirror the upper surface of the unit immediately beneath. The next step was to "regularize" the data for handling by the NURBS-based system. This was done by creating a series of profiles or sections, through the previously described contoured surfaces. These were sent to the solids modeling system where B-spline surfaces were fitted. The final step was to combine the B-spline surfaces to form true 3D solid models of the morphology of each sandbody.

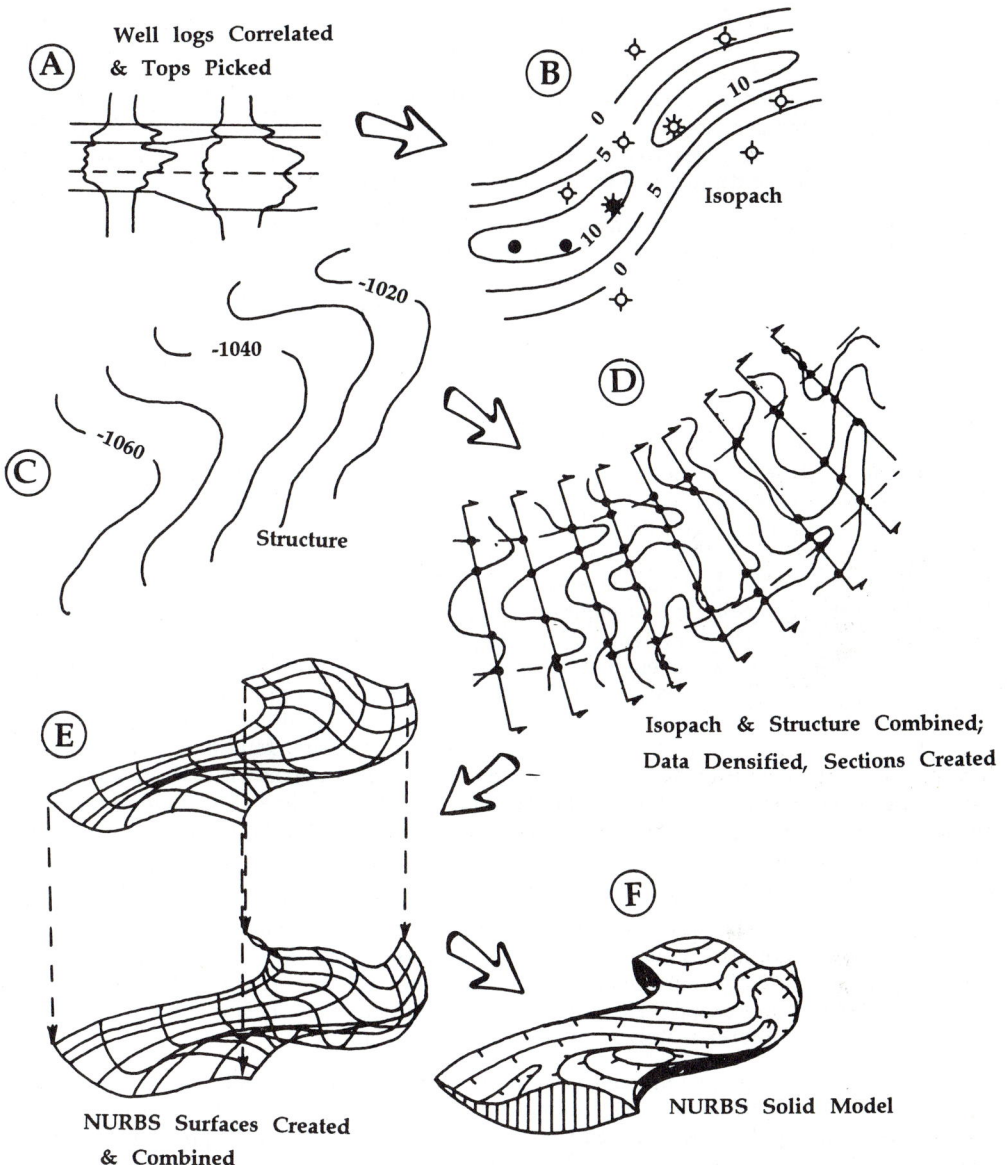

A Well logs Correlated & Tops Picked

B Isopach

C Structure -1020 -1040 -1060

D Isopach & Structure Combined; Data Densified, Sections Created

E NURBS Surfaces Created & Combined

F NURBS Solid Model

Fig. 3. Graphic representation of the steps used to create solid models of the Noonen Ranch reservoirs.

Interpretation of the Models. Our display began with the Huntsman surface. Effects of local and regional structure was not removed. The Huntsman surface model was then overlain with the solid models of the sandbodies, giving us a clear picture of the spatial relationships between the Huntsman and the sandstones, and the sandbodies to one another. In this mode on-lapping relationships between sandstones and the role the Huntsman plays in trapping hydrocarbons at this particular location are clearly seen. The coarseness of the models made it difficult to determine much about the continuity and interconnectedness of the sandstones.

Only inferences based on the "revealed" geometry of individual sandbodies, such as the serpentine trough visible over the length of some models can be made. The trough and mounds which develop laterally to it are believed to reflect the position of a clay-filled channel facies; the mound like build ups are interpreted to be point bar complexes.

Yucca Mountain Project

Yucca Mountain is a mesa of faulted, Tertiary ash-flow tuff, about 70 miles (112 km) northwest of Las Vegas, Nevada, and has been proposed by the U. S. Department of Energy as the nation's first repository for commercial high-level nuclear waste (Borns, et al., 1990). The site is in the Basin and Range Structural Province of the western U.S., and is part of the Southwest Nevada Volcanic Field (SWNVF). As a nuclear waste repository the site and the surrounding region is the subject of intense geologic and engineering studies. Of primary interest are the relationships between ground water systems, fluid flow, and tectonic and structural features of the region. Distribution of rock and engineering properties (e.g., mineralogy, porosity, permeability, water saturation) of the rock slated to contain the waste is also of great importance; the presence of certain clays (e.g., zeolite) and minerals may inhibit migration of radionucleids, while presence of water may threaten integrity of the containment site. Multidimensional geologic models such as the one describe below, are expected to become important tools in determining the suitability of the site, providing a framework for dynamic modeling of fluids movement, and prediction of future geologic events which may compromise the integrity of the site.

Yucca Mountain Modeling Sequence. The generation of the models for the YMP followed a different sequence than the one used for the Noonen Ranch. Data was received as x-y-z locations for surface locations of test boreholes and downhole sample points. A map of the area with traces of 31 of the most significant faults, shown at sea level (0m) and the 6000 foot (1828m) elevation, was also provided. The surface elevations were derived from a grid of 124 columns by 196 rows of 1:250,000, 3 arc-second digital elevation model (DEM) data. This grid is shown in Plate 2, and covers an area approximately 11 by 6 miles (18 x 10 km).

The fault traces were digitized into a 3D file in their true Cartesian (x-y) locations as a series of line strings. Each pair of fault traces, one representing the fault at 6000 feet (1828m) and the other showing the fault location at sea level, was then converted to a B-spline, and a surface was generated using the two B-splines. To provide versatility of display, each fault was placed on a different level in the system, thus permitting examination of individual aspects of a single fault. Relationships between faults can be analyzed by displaying several layers at once.

The grid was input and converted to a spline surface correctly located with the fault surfaces. The capability of the system to find intersections between two surfaces was used to generate the traces of the faults at ground level. When the model is shaded, and just the surface and the fault traces are displayed, there is a good visual correlation for several of the faults; especially some of the long north-south trending major fractures bounding ridge lines of the mesa and other topographic highs. Next in the process, was "clipping" the fault surfaces at the elevation of the ground. This was accomplished by regenerating the fault surfaces using the ground trace of the fault as the upper limit, instead of the projected 6000 foot (1828m) elevation used earlier.

Downhole control points are yet to be entered. These will provide true locations of the locations of each sample point provided for the 30 or more horizons penetrated by each borehole. This information will provide the basis for creating surfaces on each horizon.

Once the downhole data is entered as both discrete points representing geologic horizon depths, and traces representing actual boreholes, the locations where the boreholes intersect the fault surfaces can be determined with great accuracy. The final solid model of zones of having similar geologic or engineering properties (e.g., zeolite bearing rock), will be generated in a manner similar to the Noonen Ranch project. Depth maps for the bounding horizons will be generated using either manual or machine contouring depending on the data sampling and fault complexity of the block being processed. The upper and lower contours will be converted into surfaces; the two resultant surfaces will then be processed to generate the solid. The resulting solid model outlining the limits of the attribute of interest can then be merged with the solid model of the faulted "layer cake" that is the geology of Yucca Mountain.

SUMMARY AND CONCLUSIONS

NURBS and other forms of rational splines are an emergent technology which has proven useful for true multidimensional geologic modeling. The inherent characteristics of this technology already meet many of the criteria set out by Raper (1989), Turner (1989) and Van Driel (1989). A NURBS-based system when fully developed, will give the geologist considerable freedom in interpretation and model creation.

Amongst the many advantages of NURBS are the common and unified mathematical basis which allows representation of any known geometry and spatial relationships, including complex geologic models and processes, functional integration between graphical elements, and distribution of attributes within defined solid models. NURBS additionally provide a unified data base for storage (via parametric polynomial splines) of both analytic and free-form shapes. Their parameterized form also reduces data storage requirements and speeds computation time.

REFERENCES

Bartels, R. H., J. C. Beatty and B. A. Barsky (1987). *An Introduction to Splines for Use in Computer Graphics and Geometric Modeling*. Morgan Kaufmann, Los Altos.

Borns, D. J., J. H. Sass and R. A. Schweickert (1990). Proposed study of the Basin and Range from Death Valley to Yucca Flat. *EOS Transactions-AGU*, 71, 1012-1013.

de Boor, Carl (1978). *A Practical Guide to Splines*. Springer-Verlag, New York.

Fisher, T. R. and R. Q. Wales (1990a). 3D solid modeling of sandstone reservoirs using NURBS: a case study of Noonen Ranch field, Denver basin, Colorado. *Geobyte*, 5:1, 39-41.

Fisher, T. R. and R. Q. Wales (1990b). Three dimensional solid modeling of geo-objects using non-uniform rational B-splines (NURBS). In: *Three-Dimensional Modeling with Geoscientific Information Systems* (A. K. Turner, ed.). Kluwer Academic Publishers, Boston.

Fried, C. C. and J. E. Leonard (1990). Petroleum 3-D models come in many flavors. *Geobyte*, 5:1, 27-30.

Herring, J. R. (1990a). The mathematical modeling of spatial and non-spatial information ia a geographic information system. In: *Cognitive and Linguistic Aspects of Geographic Space* (D. Mark et al, eds.), (to be published).

Herring, J. R. (1990b). Using spline functions to represent distributed attributes. (to appear).

Piegl, Les (1990). NURBS - a survey. *IEEE Computer Graphics and Applications,* 10, (to appear).

Raper, J. F. (1989). The 3-dimensional geoscientific mapping and modelling system: a conceptual design. In: *Three Dimensional Applications in Geographical Information Systems* (J. F. Raper, ed.), Taylor and Francis, London.

Rogers, D. F. and J. A. Adams (1990). *Mathematical Elements for Computer Graphics,* (2d ed.). McGraw-Hill, New York.

Tiller, Wayne (1983). Rational B-splines for curve and surface representation. *IEEE Computer Graphics and Applications,* 3:6, 61-69.

Turner, A. K. (1989). The role of three-dimensional geographic information systems in subsurface characterization for hydrogeological applications. In: *Three Dimensional Applications in Geographic Information Systems* (J. F. Raper, ed.), pp. 115-129. Taylor and Francis, London.

Van Driel, J. N. (1989). Three dimensional display of geologic data. In: *Three Dimensional Applications in Geographic Information Systems* (J. F. Raper, ed.), pp.1-9. Taylor and Francis, London.

3-D VISUALIZATION OF GEOLOGIC STRUCTURES AND PROCESSES

R.Pflug, H.Klein, Ch.Ramshorn, M.Genter, and A.Stärk

Geologisches Institut, Albertstr.23-B, D-7800 Freiburg i.Br., West Germany

ABSTRACT

Interactive 3-D computer graphics techniques are used to visualize geologic structures and simulated geologic processes. Geometric models that serve as input to 3-D viewing programs are generated from contour maps, from serial sections, or directly from simulation program output. Choice of viewing parameters strongly affects the perception of irregular surfaces. An interactive 3-D rendering program and its graphical user interface provide visualization tools for structural geology, seismic interpretation, and visual post-processing of simulations. Dynamic display of transient ground-water simulations and sedimentary process simulations can visualize processes developing through time.

KEYWORDS

3-D computer graphics; geometric modeling; scientific visualization.

INTRODUCTION

Presentation of geologic information in graphic form plays a central role in geology. It is certainly desirable to make use of present day graphic workstations for producing 3-D views that cannot be easily provided otherwise. At the same time, programs simulating geologic processes require adequate graphics for visualizing computed results.

Prior to rendering any geologic object a geometric model of the object must be created, either using data from geologic surveys, geophysics, or simulation program output. This process is the actual link between geology and computer graphics. The present paper gives examples of using 3-D computer graphics to visualize geologic structures and processes. Visual post-processing of a ground-water simulation is discussed in detail. Viewing programs and tools for generating geometric models have been written by the authors unless stated otherwise.

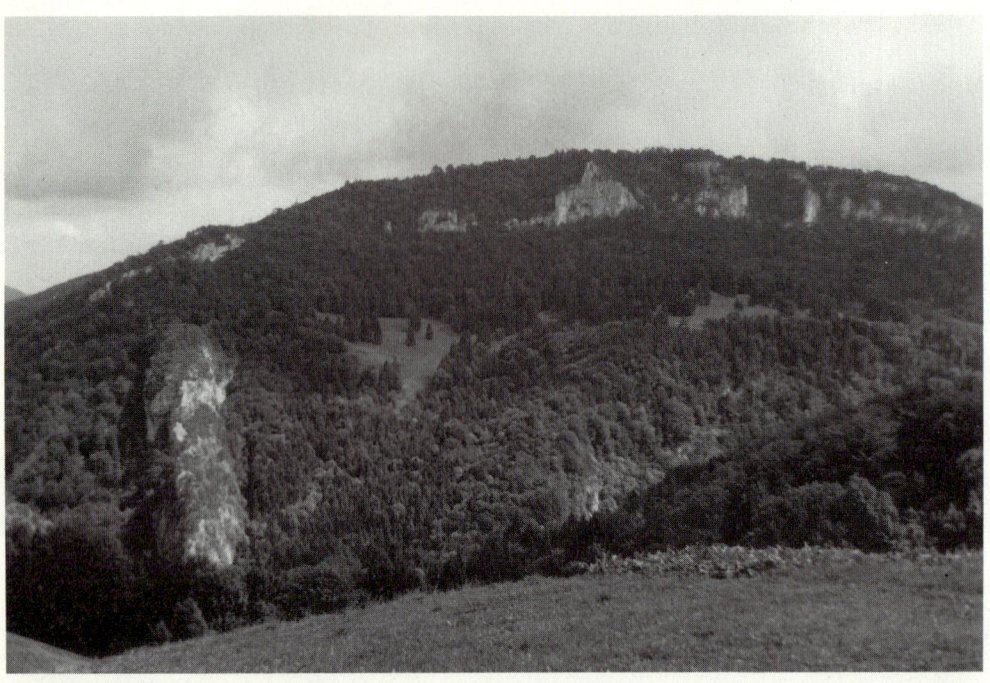

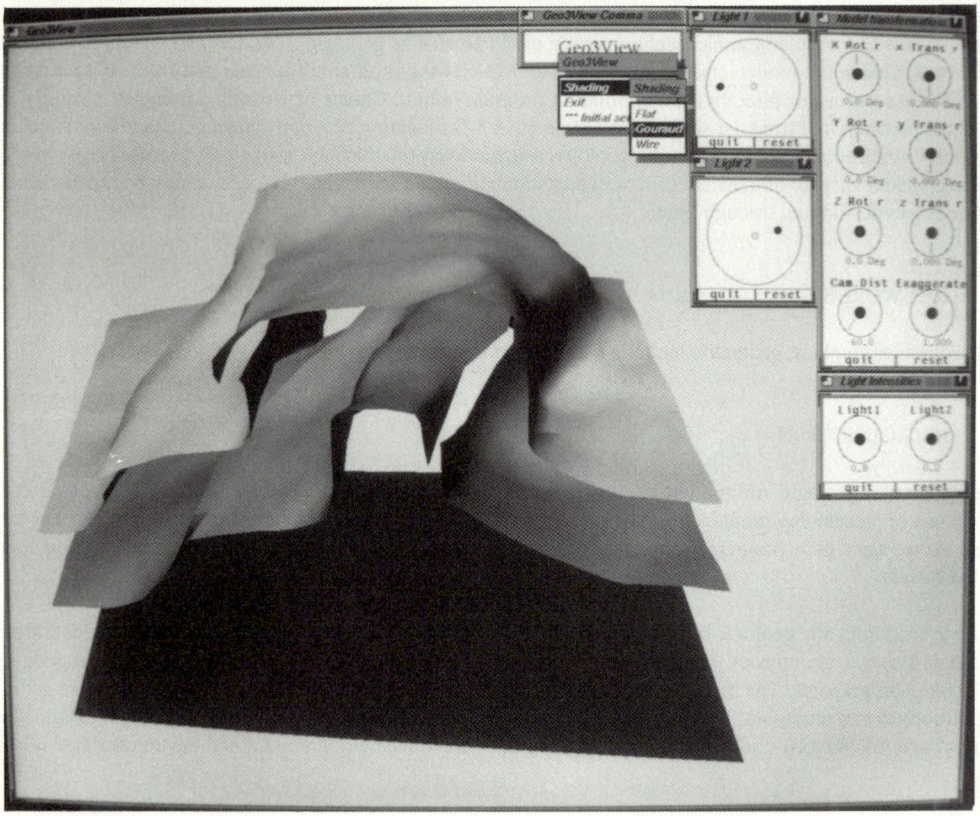
Geo3View
Shading
Exit
Shading
Flat
Gouraud
Wire
Light 1
quit
reset
Light 2
quit
reset
X Rot r
x Trans r
Y Rot r
y Trans r
Z Rot r
z Trans r
Cam.Dist
Exaggerate
quit
reset
Light Intensities
Light1
Light2
quit
reset

VISUALIZING GEOLOGIC STRUCTURES

Geologic structures can be depicted by cross-sections, and a series of cross-sections can be drawn to present a structure three-dimensionally. For example, the Vellerat anticline (Jura Mountains, Switzerland; Fig. 1) is presented in a number of parallel sections (Heckendorn, 1974). Section data can then be digitized and triangulated to reconstruct structural surfaces three-dimensionally (Tipper, 1977; Klein *et al.*, 1989). Figure 3 schematically shows the triangulation process. Using an interactive graphics program, the geologist may specify which features must be connected between adjacent sections and have the program perform the triangulation accordingly. Regarding the reconstruction of geologic surfaces, experience shows that fully automatic triangulation algorithms (Christiansen, 1978) can produce satisfying results with some structures while they completely fail with others.

Once triangle meshes are generated that form the geometric model of a structure (Fig. 3), the structure can be displayed in shaded views (Fig. 2) and observed from any angle. This provides an overview and helps to reveal structural features that otherwise may be difficult to detect.

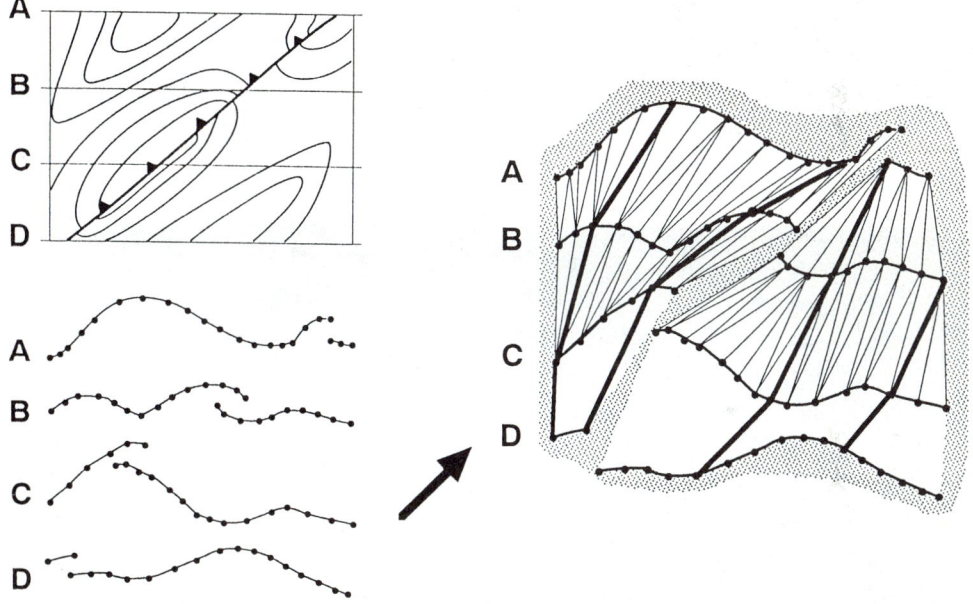

Fig.3. Creating a triangle mesh from serial sections. Key connections (thick lines) govern the triangulation pattern

Some seismic packages let the geologist digitize lines into seismic sections for interpretation. For each interpreted horizon, a grid of sections is generated (Fig. 4). The sections perpendicular to structural strike are triangulated and the generated surfaces are displayed (Fig. 5), providing visual control for the interpretation.

Fig.1. Vellerat anticline, Jura Mountains, Switzerland
Fig.2. Shaded view of Vellerat anticline; reconstructed from four sections

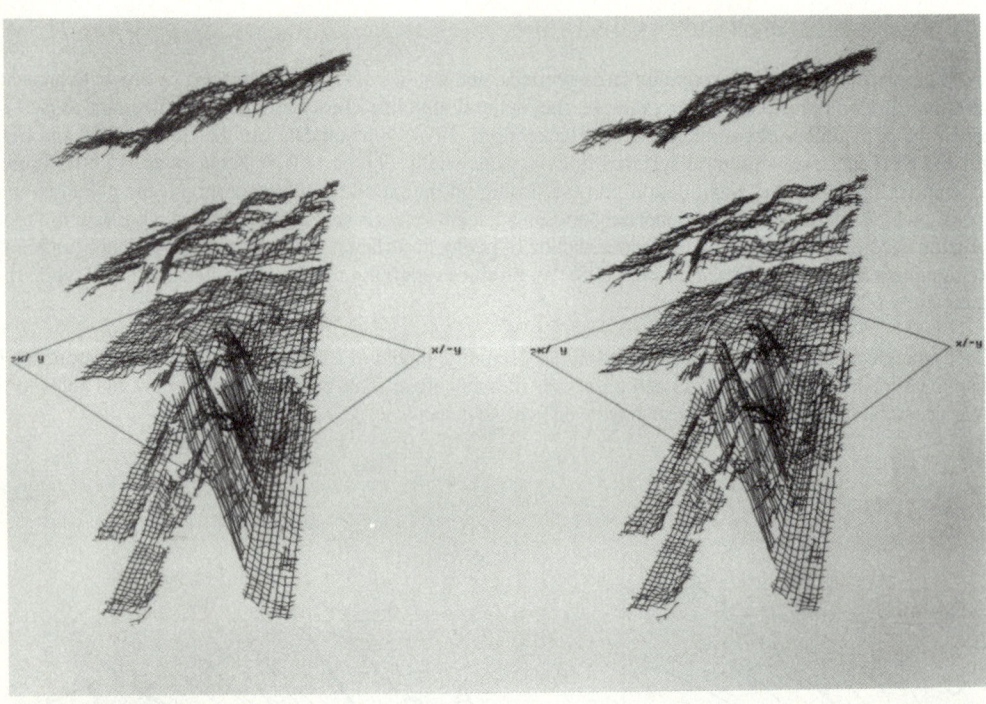

INTERACTIVE VIEWING PROGRAMS

Perceiving geologic structures from shaded images may be difficult because geologic structures usually have highly irregular shapes, and because subsurface structures do not have much in common with objects which our visual system is familiar with. Whether or not a view is understandable depends heavily on the viewing parameters choosen. Viewing angle and light source positions are examples of sensitive parameters, because geometric features may be enhanced or almost completely obscured due to lighting and viewing angle. Complex geometries are more easily perceived in stereo displays (Fig. 4) or when the observer can interactively rotate them in real-time. Switching visibility of portions of a picture (Fig. 7, 10) improves perception of details while preserving spatial relationships.

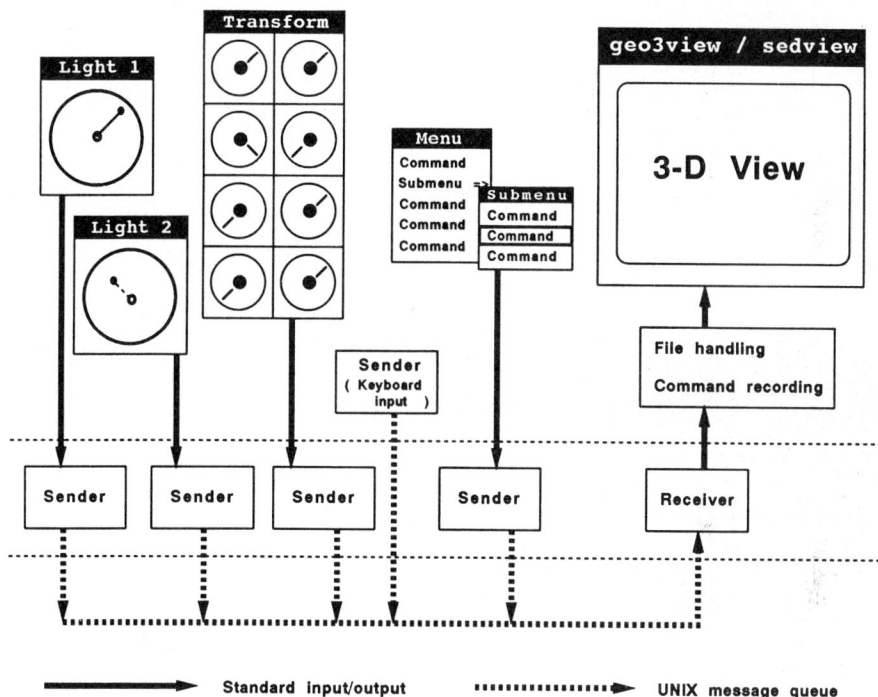

Fig.6.   Geo3View/SedView modules

Geo3View is an interactive 3-D viewing program for visualizing geologic structures which are represented by triangle meshes (Ramshorn, Klein & Pflug, in press). The program consists of a number of modules that co-operatively provide data rendering tools that are controlled through a graphical user interface. This design is depicted schematically in Fig. 6. Shaded views shown in this paper were rendered using Geo3View and Sed-View (see below) on a Silicon Graphics IRIS 4D-series workstation.

Fig.4.   Seismic interpretation; horizons are represented by lines traced on vertical sections

Fig.5.   Shaded view of structure depicted in Fig. 4

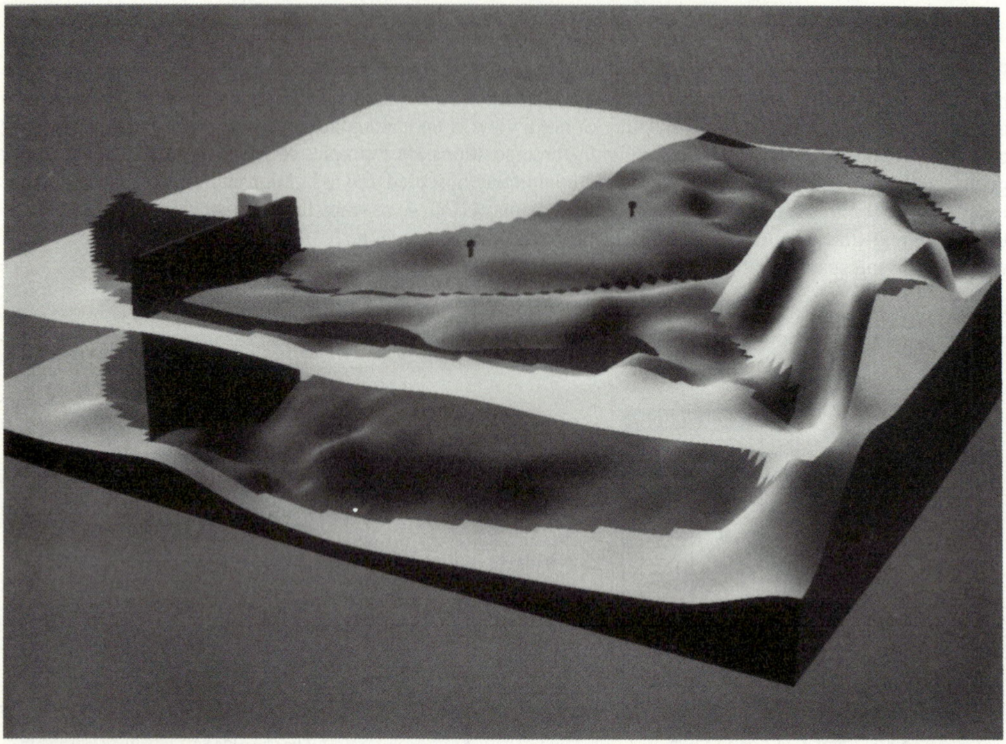

POST-PROCESSING GROUND-WATER SIMULATIONS

Efficient handling of an increasing number of ground-water problems requires more and more computer aided techniques. Many ground-water flow models used for problem solving simulate aquifers in three dimensions. Model input and output consist of large data sets that describe the geometry of an aquifer, the water table, flow rates, and more. For visualizing the simulation results, contour maps, velocity vector maps, and hydrographs are commonly plotted which provide detailed information of single parameters; e.g. ground-water-level at a given time.

3-D computer graphics, particularly shaded display techniques, can provide synoptic views of relevant simulation parameters. This is an improvement in many respects. First, interpreting 3-D data sets using contour maps requires considerable imagination and one tends to focus on features that one already knows or which one expects. 3-D views present spatial relationships without stressing the viewer's imagination. Thus, new observations for an unbiased interpretation are facilitated, and errors are more easily detected. For quantitative analysis, however, maps and graphs must be consulted or perhaps even the raw data sets inspected. Second, regions where two surfaces intersect are difficult to detect in contour maps, but they are obvious in shaded views. As a consequence, flooded areas are easily detected (Fig. 7). Third, while by classical methods flow dynamics of transient simulations can only be visualized in either the space domain or the time domain (contour maps and hydrographs, respectively), real-time 3-D computer graphics can picture the behavior of parameters such as water-table changing through space and time. Finally, 3-D views are easy to understand by non-experts. This is important when communicating with managers, experts from other fields, or the public.

Figures 7 through 10 show ground-water simulations in shaded views produced by Geo3View. In the simulated scenarios, two rivers (Fig. 9, dark grey) border an aquifer with several wells (dark spots). The swampy area (medium grey) between the rivers is drained by a brook. High water in the river to the east floods the mouth of the brook, and increases ground-water heads by infiltration into the aquifer. Together, these risk contaminating the well water with surface water. A dam (horizontal dark bar) is planned to prevent this while preserving the natural swamp north of the dam. A subsurface seal extends under the dam down to the aquifer bottom (Fig. 7). Brook water from the south must be pumped over the dam.

A finite-difference ground-water flow model (MODFLOW; McDonald & Harbaugh, 1984) was used to compute required pumping rates and ground-water tables (dark grey) in a number of both steady state and transient simulations (Stärk, 1990). Pumping rates of various scenarios involving different lateral extensions of the seal (opaque U-pipe) are related to the maximum pumping rate when no seal is present (translucent U-pipe).

Figure 7 illustrates a situation where the seal extends under the full width of the dam. Viewed from the south-west, the flooded area is restricted to regions close to the western river. The pumping rate is low. Figure 8, in contrast, shows a scenario with only half of the dam sealed. As a consequence, larger areas are flooded and water table gradients around the pump are steeper. Additionally, a slightly higher pumping rate is required which is still small compared to the maximum pumping rate needed without any seal. Topography outside the modeled area (white) is rendered translucently. Figure 10 shows the water table viewed from the north-east. Draw-down effects of both the brook and the two wells are clearly visible (the water table north of the dam is computed as if it were in the subsurface, whereas in reality the area is flooded).

A grid masking technique is used when transforming simulation result grids to triangle meshes (Fig. 11). While data such as topographic elevation or aquifer heads are stored in one grid, a second grid encodes the extension of the modeled area much as a geologic map encodes the extension of geologic units at the surface.

Fig.7. Ground-water simulation scenario with fully sealed dam; view from south-west

Fig.8. Ground-water simulation scenario with eastern half of the dam sealed

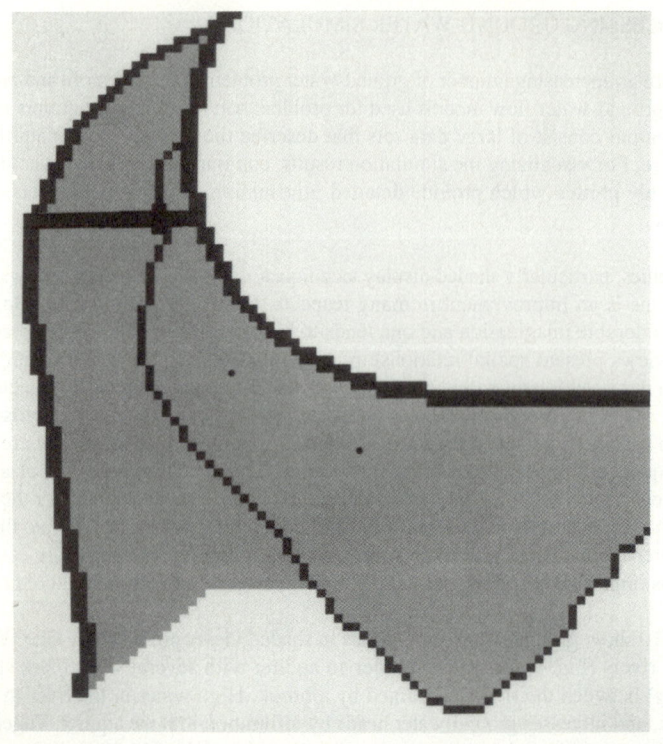

Information of the second grid is used to split up model data before generating triangle meshes. When rendering the geometric model, each triangle mesh can be treated separately, e.g. be colored differently or be made visible or invisible.

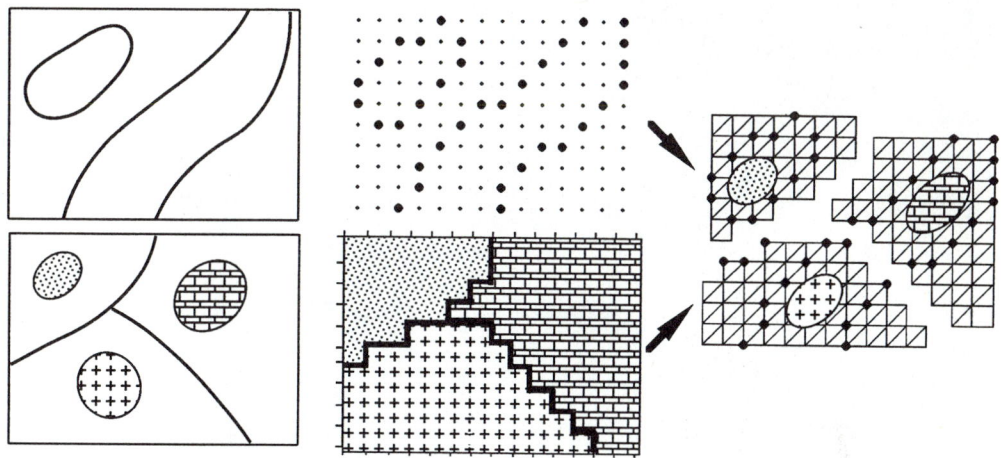

Fig.11. Creating triangle meshes from maps or from grid data

VISUALIZING SEDIMENTARY BASINS

Numerical models can simulate the physical processes that govern transportation and deposition of clastic sediments. SEDSIM (SEDimentary process SIMulation program, Tetzlaff & Harbaugh 1989), for example, generates volumetric data sets which represent sedimentary basins. These data sets need to be graphically displayed to evaluate results of simulation experiments. SedView (Ramshorn, Ottolini & Klein, in press), an interactive viewing program with the same modular design and the same user interface as Geo3View, was developed to accomplish this.

In SEDSIM, deposited and eroded material is accounted for on a grid of rectangular cells of fixed size in X and Y, and of variable size in Z. SEDSIM output essentially consists of a five-dimensional array of numbers representing the amount of pebble, sand, silt, and clay deposited in each cell, recorded at user-defined intervals ("snapshots"). SedView visualizes these volumetric data by isochron sediment surfaces, or by a lattice of vertical sections (Fig. 12), or both. Basement topography and water body can optionally be displayed. It is possible to interactively show or hide any portion of the basin, to rotate, resize, and vertically exaggerate the basin, to highlight a sediment type, to show basin evolution through time, and to switch sediment classification methods.

The sediment surface and the fence diagram are color-coded to represent sediment composition in a number of alternative ways, such as classified by sediment type (pebble, sand, silt, clay), or by sediment age (Fig. 13).

Fig.9. Ground water simulation; top view on modeled area (medium grey) bounded by rivers (dark grey)

Fig.10. Water table (half sealed dam) viewed from the north-east

SedView
SedView Comman
Light 1
View parameters
SedView
Credits...
Fences (every 2nd)
Surface (0.5)
Loop (no)
Classify (Grain size)
Highlight (none)
Water (on)
Basement (on)
Examples (Wav
Exit
*** Initial settin
Examples
Wave
Delta 1
Delta 2
Delta 3
Delta 4
Test
Tilt
Rotate
Hor shift
Ver shift
Exaggerate
Sed thick
Time step
Zoom
quit
reset
Light 2
Light Intensities
Light1
Light2
Enhance weight
Pebble
Sand
Silt
Clay
[XY] Fences
X-fence
Y-fence
Size: 7.50 x 10.00 km
Years simulated: 10000
Current state year: 5000
Sediment classes: Medium Fine

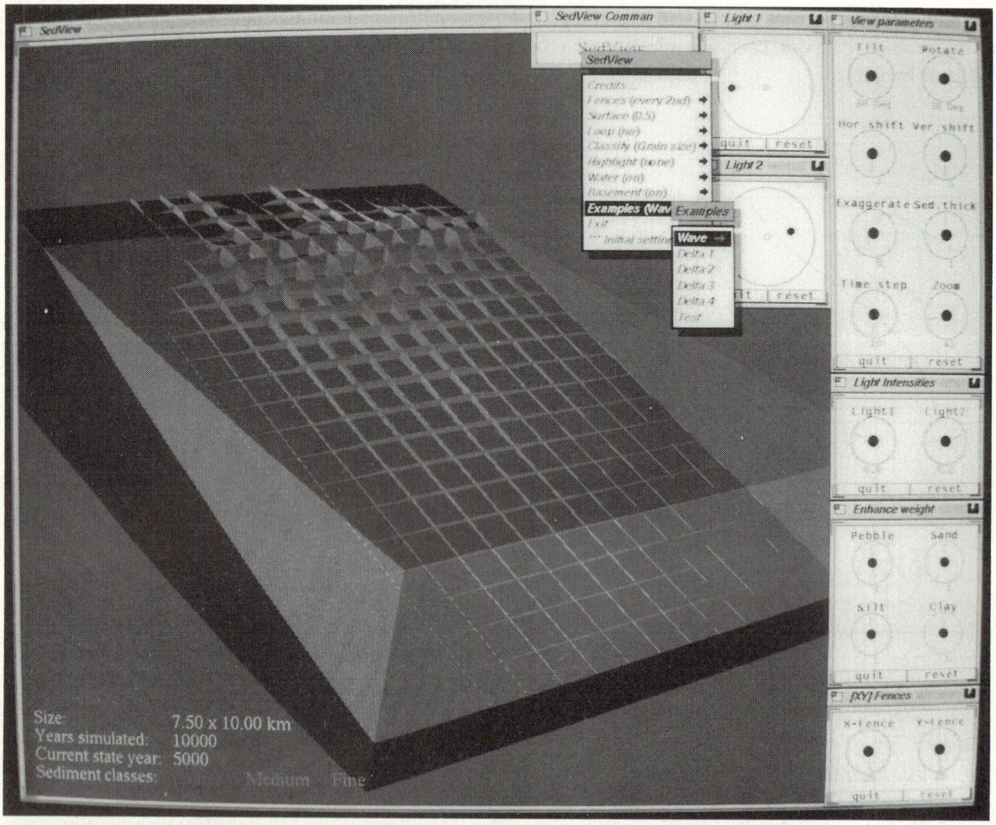

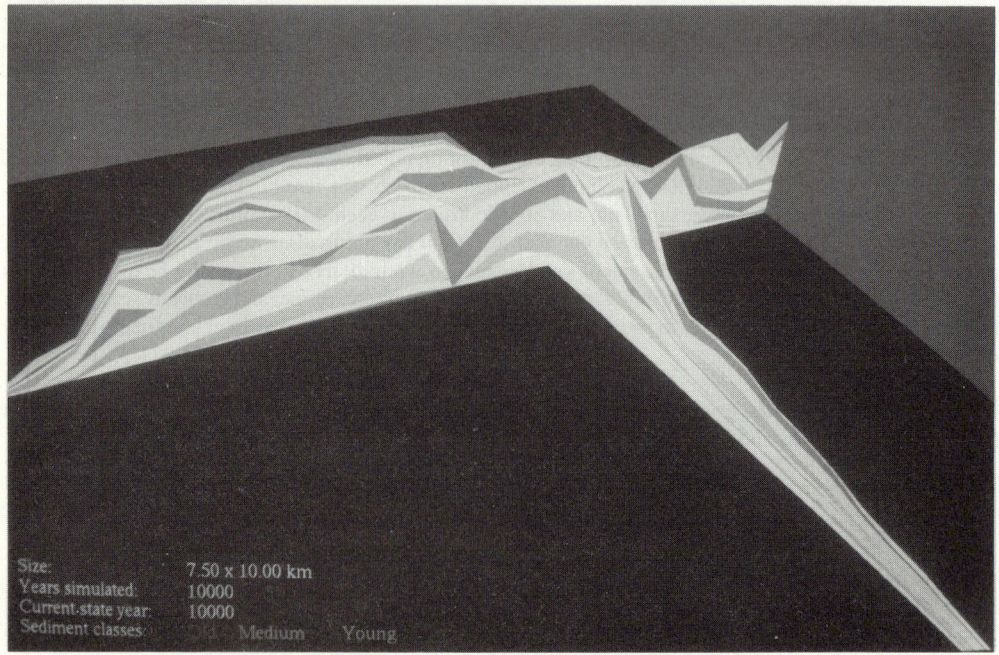

Size: 7.50 x 10.00 km
Years simulated: 10000
Current state year: 10000
Sediment classes: Old Medium Young

A translucent box outlines the water body above the sediment (Fig. 12). When displaying both the fence diagram and water, the water can be prevented from "flooding" the space between fences that reach above the water surface by rendering the sediment surface completely transparent before drawing the water. Water then is only displayed where the sediment surface drops below the water surface. Basin evolution through time is visualized by animating a sequence of snapshots.

ACKNOWLEDGEMENTS

The major part of work presented here was supported by grant Pf 65/24 of the Deutsche Forschungsgemeinschaft, while current research is sponsored by Prakla Seismos AG, Germany. IBM Germany gave financial support for developing a 3-D viewing program, and Silicon Graphics, Munich, loaned an IRIS workstation. We also wish to thank John W. Harbaugh, Stanford, for his support and we are grateful to the SEDSIM group at Stanford for discussion. We particularly appreciate contributions to SEDSIM visualization by Rick Ottolini who then worked at Stanford's Geophysics department.

REFERENCES

Christiansen, H.N & Sederberg, T.W. (1978). Conversion of contour line definitions into polygonal element mosaics. *Computer Graphics, 12,* 187-192.

Heckendorn, W. (1974) . Zur Tektonik der Vellerat-Antiklinale (Berner Jura). *Beiträge zur Geologischen Karte der Schweiz, N.F., 147.*

Klein, H., Pflug, R. & Ramshorn, Ch. (1989). Shaded perspective views by computer: A new tool for geologists. *Geobyte, 4,* 16-24.

McDonald, M.G. & Harbaugh, A.W. (1984). A modular three-dimensional finite-difference ground-water flow model. *U.S. Geological Survey Open-File Report,* 83-875.

Ramshorn, Ch., Klein, H. & Pflug, R. (in press). Dynamic display for better understanding shaded views of geologic structures. *Geologisches Jahrbuch.*

Ramshorn, Ch., Ottolini, R. & Klein, H. (in press). Interactive three-dimensional display of simulated sedimentary basins. In: *Proceedings of the Eurographics Workshop on Visualization in Scientific Computing in Clamart (France), April 1990;* Springer (Heidelberg).

Stärk, A. (1990). Grundwassermodell Ulm "Rote Wand" - Wirksamkeit und Auswirkung baulicher Veränderungen zum Hochwasserschutz. Diplomarbeit, University of Freiburg (unpubl.).

Tetzlaff, D. & Harbaugh, J.W. (1989). *Simulating clastic sedimentation.*Van Nostrand Reinhold (New York).

Tipper, J. (1977). A method and FORTRAN progam for the computerized reconstruction of three-dimensional objects from serial sections. *Computers and Geosciences, 3,* 579-599.

Fig.12. SedView fence diagram and water column. Note same user interface as used in Geo3View

Fig.13. Selected pair of fences showing sediment age

AN ATLAS OF THREE DIMENSIONAL FUNCTIONS

Jonathan F. Raper

Dept. of Geography, Birkbeck College,
7-15 Gresse St., London W1P 1PA

ABSTRACT

The rapid rate of growth in 3D modelling in the geosciences has led to an explosion of new models in fields from crystallography to crustal dynamics. However it is suggested in this paper that at present visualisation has developed faster than spatial structuring, and that the range of spatial functions supported is still dominated by the need to view and subjectively "size up" the model. This paper reviews spatial functions in terms of alternative forms of conceptualisation and representation and argues that the next phase of development should be driven by the need to form more sophisticated spatial structurings.

KEYWORDS

Data modelling; spatial structuring; spatial functions; geosciences

INTRODUCTION

All geoscientific models are rooted in a conceptualisation of reality. This process of conceptualisation therefore underlies all attempts to represent the physical world and to carry out operations upon the resulting models: it is usually known as data modelling. However, as geoscientists begin to use digital representations to carry out 3D modelling and visualisation these processes of conceptualisation need to be re-examined to ensure that the correct representation is used to obtain access to the appropriate set of 3D spatial functions. This paper aims to trace the process of conceptualisation of geoscientific phenomena from the discretisation of identity through the collection of descriptive data to the spatial structuring and choice of representation for the resulting model.

Although this process is relatively well known for a limited number of domains eg hydrocarbon exploration, the availability of a range of new 3D representations has prompted wide interest in the construction of new models to solve previously indeterminate problems. There has also been an increase in the speed of at least several orders of magnitude in the speed with which it is possible to create a 3D model and evaluate its accuracy. In this context is seems worthwhile to illustrate the relationship between conceptualisation and representation in 3D modelling.

CONCEPTUALISATION OF GEOSCIENTIFIC PHENOMENA

A convenient way to describe the infinite variability of the physical world is to consider that it is made up an infinite number of tuples X,Y,Z, P_1, P_2...P_n (Goodchild 1990), where X,Y,Z are the 3 geometric dimensions and P_1, P_2...P_n are any number of parameters describing properties at that point. Using this scheme it is possible to see the process of conceptualisation of geoscientific phenomena as simply selecting the appropriate tuples needed for a representation. This involves the discretisation of reality under controlled conditions.

The process of discretisation of a 3D reality takes place under a variety of controls. Underlying all procedures is the need to identify a structure in the target domain, here defined as 3D in scope and usually geographically bound by cadastral or topographic constraints. In a number of disciplines such as geomorphology the domain is "seen" and it may be possible to determine a functional structure by observing processes or delineating landforms. However, a typical constraint on 3D geoscientific modelling is that the domain is "unseen" and the structure must be determined by employing a discretisation technique.

The process is usually informal– geoscientists often use guestimates as the basis of a hypothesis by which a discretisation is made (Kelk 1991). Frank and Buyong (1991) have discussed this process by applying some of the techniques of cognitive science to evaluate the procedures geoscientists use.
Thus, classic cognitive science uses the concept of 'categories' to organise systems of understanding: each category is defined by an exemplar or prototype. In geology it seems reasonable to consider the stratum as an example of a category which is widely used as a means of discretising a mass of information on geological sequences. The prototype stratum might almost be traced to the work of the geological pioneers such as Hutton at classic outcrops such Salisbury Crags in Edinburgh. This concept enforces certain spatial characteristics on the observations such as lateral continuity and upper and lower bounding surfaces.

However, the concept of categories has been shown to be limited in certain respects and not all phenomena can be understood in this way (Lakoff 1987). An alternative formalisation suggests that human reasoning is guided by a range metaphors, many of which are associated with human body and the basic motor functions. Many of the metaphors used in cognition in general and geoscience in particular are therefore explicitly spatial: examples would be paths and containers. Thus, it would seem likely that geoscience categories such as strata, and spatial metaphors such as containment are important (if underlying) controls on the discretisation of physical reality.

Another source of spatial criteria for the discretisation of observed reality lies in language itself which structures space comprehensively (Talmy 1988). Linguistic structuring of space can be seen in the use of spatial prepositions such as 'on', 'in' and 'among' and in the viewpoint of an observer as expressed in the scope of attention ('local' or 'global'). The distribution of attention of an observer changes depending on circumstances: hence the observer notes different factors when actually passing through a terrain as opposed to seeing images of it.

Hence the process of discretisation can begin with the identification of structure in reality. This step is normally followed by measurement of the phenomena over a number of parameters associated with the 'structure'. The observations on the 'suggestive' parameters are usually made with reference to a 3D coordinate system which acts as a geometric frame for the located tuples X,Y,Z,P. Perhaps two main kinds of approaches can be identified:

1) Device exploratory– in this case the measurement technology defines the geometric arrangement of the observed tuples, and there is no search for an a priori object. Thus, boreholes impose a linear structure on the measurements, characterising a (non?) regular sequence of points downhole, and a photogrammetric survey of a terrain or section will usually generate a grid of measurements over the visual field. However, a survey of the positions of tracer pebbles over a channel bar will be governed by hydraulic processes and recovery factors and will generate a spatially non-regular set of observed tuples. In this process the tuples recorded only need have the means of collection or selection in common.

2) Object exploratory– in this case the search for an a priori object defines the geometric arrangement of the observed tuples. The located tuples identified by the combination of 'suggestive' parameters form a spatial cluster of arbitrary configuration. In this process the tuples recorded may have a distinct spatial structure.

When the 'object' can be identified by a single key parameter and is known to exist in a discrete form from knowledge of the domain (eg. mine access, destructive examination) then the spatial object can be termed 'sampling-limited' (Raper 1989). In this case the discretisation proceeds by using the sampling theorem to create a parsimonious description of the object from selective observations. An example would be a perched acquifer or a salt dome.

When the object is transient or part of a continuum (eg a temperature field), or exists only as spatially clustered set of observed tuples defined by a group of 'suggestive' parameters then the spatial object can be described as 'definition-limited (Raper 1989). In this case the discretisation procedes by assembling a set of 'suggestive' parameters and searching for locations matching this a priori description. An example of a definition-limited object would be plume of pollutants in the atmosphere or ocean defined by a physical threshold.

The result of this process of conceptualisation and discretisation is a set of tuples X,Y,Z,P with a particular character, not necessarily with a uniform description but generated by a particular process of domain exploration.

STRUCTURING OF SPATIAL OBSERVATIONS

Once a set of tuples are assembled according to one of the above-mentioned processes, the key step required in all 3D modelling is their spatial structuring. The structuring employed will control the form of representation and the range of functions which can be supported for the dataset. As such it is almost the single most important decision in 3D modelling: at present some forms of structuring selected for a geoscience project preclude the use of the appropriate spatial functions later in the project. The structuring chosen therefore acts as an index to the atlas of spatial functions.

Although the set of tuples selected during the data collection stage can be structured in a variety of ways, there are probably only two major strategies:

1) Domain partition– use of the selected tuples to subdivide the whole 3D domain into regular or non-regular constituent units;

2) Entity construction– use of the selected tuples to define entities within 3D space using basic geometric elements.

These two structuring strategies control the set of spatial functions available for the analysis of the model: they do not necessarily define the form of representation which should be chosen.

Regular domain partition ('tessellation') is the most straightforward of the structuring strategies. When the set of tuples produced by a data collection exercise forms a regular 3D sample of observed reality then the structuring can be defined by the X,Y,Z sample intervals. In this case the domain is partitioned into regular units (usually cuboids) which are mutually exclusive and fill the domain space exactly once. This structuring scenario is not commonly adopted in the geosciences since it normally implies the destructive examination of the feature under study: this is not always feasible except in small scale studies (Bristow and Raper 1991). Typically the tuples selected during measurement need not define the spatial configuration of the boundaries between each unit: the space around a regularly placed observation can be partitioned by defining bisector planes between sample locations and intersecting them to form cuboids. In this case the spatial structuring is defined by a constant piecewise regular subdivision of the domain.

Non-regular domain partition is more difficult to achieve in the geosciences. It may chiefly be employed in domains with some a priori knowledge of the boundary configurations such the modelling of certain fossil plants and animals. In these cases the selected tuples are grouped so as to define partitions of space where the boundaries configuration can be predicted eg walls of a chamber forming a growth unit. The main difficulty is knowing that the sampling is detailed enough to define a complete model of the phenomenon. The spatial structuring is therefore defined by a constant piecewise irregular subdivision of the domain.

However, another case of the irregular subdivision of a domain arises in the 3D equivalent of contouring– the production of isosurfaces. Here the selected tuples for an irregularly spaced set can be considered as centres of minimum variance for the measured parameter or parameter set. Using spherical search strategies the domain could be partitioned by assigning each location in the 3D domain to exactly one tuple location based on proximity, and locating the boundaries. Irregular partitions of space using selected tuples can also be handled by interpolation to a regular 3D grid partition. In this case an arbitrary grid with an interval ideally within a standard deviation of the actual tuple spacing is imposed to regularly partition the space. A mapping from the irregularly spaced tuples to the grid completes the structuring.

The main alternative form of structuring is based on the construction of entities within the study domain. This is the main method of structuring used in 3D modelling in the geosciences. Typically the procedure uses basic geometric units to outline the configuration of the entities, and it is implicit that the entities have a discrete form. This may not always be the case, in which case the actual object realised can be considered as a form associated with error conditions eg the 'maximum' or 'minimum' extent definable. The basic geometric units used are normally hierarchically organised built on the zero dimensional form of the selected tuple. Hence units equivalent to points (0D), lines (1D), areas (2D) and volumes (3D) are used in the construction of objects within the domain. Note that such objects can co-exist in the same space.

There are a number of strategies in use to assemble objects from these units. Fastening points together to form bounding edges and solids may allow the construction of a set of objects– in certain cases the data collection process will determine their configuration, for example in boreholes. In such cases the boreholes may form cross sections in the X or Y direction: the objects 'seen' in these X,Z or Y,Z planes can then be fastened together to form solids. Tuples which are spatially clustered can also be 'gathered' together to define an object eg in the determination of the epicentre of an earthquake. Some devices collect very large numbers of tuples of information, for example seismic surveys used in hydrocarbon exploration. In this situation the X,Z or Y,Z planes may be examined for discrete boundaries which can be used as bounding surfaces for sedimentary strata.

This is only a brief survey of some of the main procedures of structuring which are used in 3D modelling. However, the specific procedure used to identify the tuples described during data collection and to organise the tuples spatially conditions the scope for subsequent analysis. At present the full exploitation of this procedure for the creation of structurings is the main obstacle hindering wider use of 3D modelling tools and more research is urgently needed into the conceptual tools available for each field.

AN ATLAS OF SPATIAL FUNCTIONS

The representation chosen for the spatial structure is the final stage in the modelling and ought to be guided by the need to have an efficient implementation of the model, optimised for the spatial functions which are most important to the study. But the scheme of structuring forms the key step: no representation can compensate for the shortcomings of the structuring. Hence, the basis for spatial functions can be found in the characteristics of the structuring. A diagrammatic illustration can be found in figure 1.

Spatial functions can be subdivided by the characteristics of the space in which the structuring is embedded. Thus, space filling structurings are single valued with each location occupied by one and only one identity. Alternatively, when the space is filled with objects of varying dimensions the entities identified can be co-penetrant. Hence, the cardinal subdivision of spatial functions is whether relationships between entities can be entertained: in subdivided domains the units of subdivision are in a distinct relationship with each other defined by the structuring and such interrelationships can not be supported.

The complete set of inter-relationships between valid objects in a 3D domain have not yet been comprehensively determined. Pullar and Egenhofer (1988) and Herring (1991) have examined the 2D case and determined that there are a finite number of relationships between the interior, boundary and closure (union of boundary and interior) of two geometric figures. The figures must share common items in the sets defining their closure, ie the determination is an entirely topological one computed irrespective of metric distance. In this case the conditions of topological inter-relationship were found to be characterised by the following eight configurations (named and defined by Pullar and Egenhofer 1988):–

> Disjoint
> Overlap
> Meet
> Equal
> Covering*
> Containment*

* Two mirror image configurations

The overlap relations of the two figures are governed by the Boolean set conditions AND, OR, NOT and XOR which can used in selection. When used for selections involving more than two figures the selection relations are commutative and their order of execution must be specified.

Other forms of inter-relationships can be defined based on metric characteristics and describing qualitative aspects. These include the direction or orientation of one object relative to another. Hernandez (1990) has proposed a form of description combining topological and directional information called a Relative Projection and Orientation Node (RPON) which has the following states:

<u>Topological</u>
Inclusion
Tangency
Overlap
Disjointness

<u>Direction</u>
Front
Left-front
Left
Left-back
Back
Right-back
Right
Right-front

A 3D equivalent could be envisaged introducing 'near' and 'far' conditions to expand the set.

This set of inter-relationships can be defined for any structuring made up of objects defined by the union of geometric units. The application of these spatial functions is, however, made more complex by their heterogeneous dimensional character, requiring each operation to be carried out at successively higher dimensions. The operations can also be carried out upon probability distributions to allow for the definition of fuzzy objects, although this has not been carried out for 3D objects.

The spatial functions involved in characterisation of a structuring are applicable to all forms both in the subdivision of a domain and the construction of objects in space. The applicable 3D forms of characterisation include:

Volume
Surface area
Centre of mass
Orientation (of X,Y,Z and longest axis)

While these functions can be applied to regular partitions of a domain, they will normally be found to be trivial. In an irregular partition of a domain or arbitrary object defined in space the characterisation of each unit may generate new information.

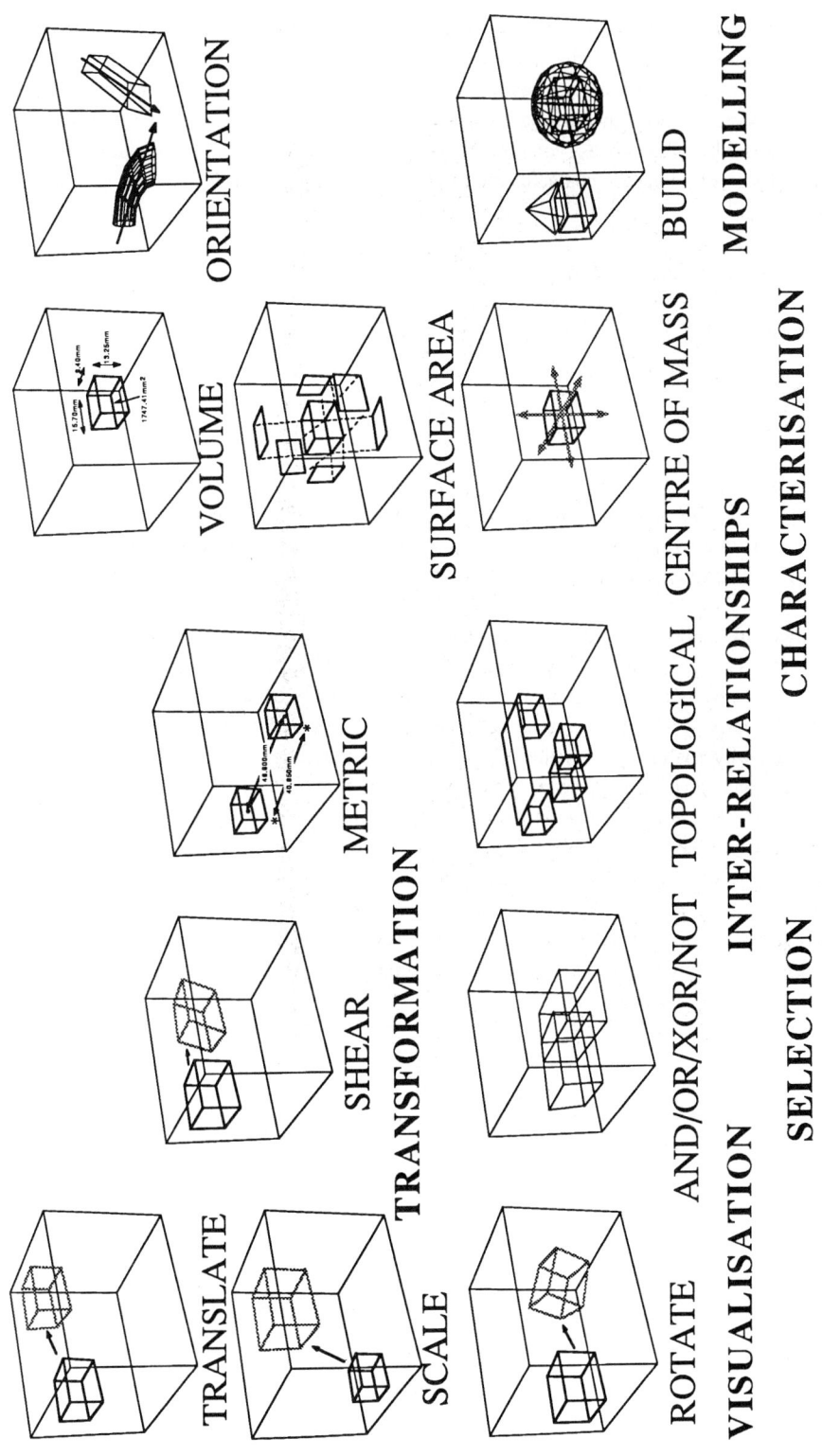

ORIENTATION

BUILD MODELLING

VOLUME

SURFACE AREA

METRIC

TRANSFORMATION

SHEAR

TRANSLATE

SCALE

ROTATE AND/OR/XOR/NOT TOPOLOGICAL CENTRE OF MASS

INTER-RELATIONSHIPS CHARACTERISATION

VISUALISATION SELECTION

3D SPATIAL FUNCTIONS

Another form of spatial function related to the characterisation of a geometric unit is the ability to expand or 'build' a new object on some selected characteristic of the first: this can be termed modelling. Hence the minimum bounding sphere for a 3D unit can easily defined. This process can also be used to define new units based on discrete aspects of the original using a constructive solid geometry eg a pyramid of base equal to one face of the original solid. This process can also be used to 'gather' 0,1 or 2 dimensional objects into an association based on some geometric characteristic eg en echelon arrangement for features such as weather fronts or faults.

Hence the spatial functions applicable to a spatial structurings belong to 4 classes: Characterisation, Inter-relationships, Selection and Modelling. These can be complemented by sets of Visualisation manipulations and complex Transformations which are used to alter the view of the structuring.

CONCLUSIONS

This paper seeks to increase the awareness of the geoscience community about the relationship of the spatial structurings available to the spatial functionality which can be realised. The challenge for developers or modelling software is to provide better 3D editors so that the structurings can be experimented with, and so that the optimum configurations can emerge.

These structuring configurations can then be represented in a variety of ways, each of which can offer access to spatial functions at a certain speed and to a certain accuracy. Typically, the regular partition of a domain will be represented by a raster representation scheme since the mapping is trivial and direct. However, other structurings can be represented either by raster or vector schemes, or by fitting functions depending on the spatial functionality which is most important. Thus, the irregular partition of a domain based on spherical intersection of 3D zones of influence can be represented either by voxels or in a boundary representation scheme by a volygon. In the former representation the spatial functions associated with space filling (eg. volume) are optimised; in the latter the spatial functions associated with space bounding (eg surface area) are optimised– although the functions can be realised in each.

The entire process discussed in this paper can be described as data modelling. Three dimensional modelling must adopt better data modelling procedures to progress beyond the simple digital reproduction of the tradition analogue models which have become so familiar.

REFERENCES

Bristow, C.S. and Raper, J.F. (1991) Modelling 3D reservoir geometry: a new approach using IVM. In Geological Society of London Special Publication "*Advances in Petroleum Geology*".

Frank, A and Buyong, T.B. (1991) Geometry for 3D GIS in geoscientific applications. In Turner, A.K. (ed.) *Three dimensional modelling with geoscientific information systems*. Dordrecht: Kluwer.

Goodchild, M.F. (1990) Geographical data modelling. *Proc. GIS Design Models Meeting*, Leicester 21-22/3/90.

Hernandez, D (1990) Relative representation of spatial knowledge: the 2D case. *Technische Universitat Munchen, Institut fur Informatik*, Report FKI-135-90.

Herring, JR (1991) The mathematical modeling of spatial and non-spatial information in GIS. *NATO ASI on Cognitive and Linguistic Aspects of Geographic Space.* Las Navas del Marqués, Spain, July 8-20, 1990.

Kelk, B. (1991) 3D modelling in geoscientific information systems: the problem. In Turner, A.K. (ed.) *Three dimensional modelling with geoscientific information systems.* Dordrecht: Kluwer.

Lakoff, G (1987) Cognitive models and prototype theory. In Neisser, U (ed.) *Concepts and conceptual development: ecological and intellectual factors in categorisation.* Cambridge: Cambridge University Press, pp 63-100.

Pullar, D.V and Egenhofer, M (1988) Towards formal definitions of topological relations amongst spatial objects. Proc. Spatial Data Handling Symposium, Sydney, 1988, pp 225-41.

Raper, J.F., (ed), 1989,*Three dimensional applications in geographical Information Systems* (London: Taylor and Francis).

Talmy, L (1988) How languages structures space. Mark, DM (ed.) (1988) *Cognitive and linguistic aspects of geographical space.* Santa Barbara, CA. National Centre for Geographical Information and Analysis.

SOLID COMPUTATIONAL MODELS
OF GEOLOGICAL STRUCTURES IN BOREHOLES

STEFAN M. LUTHI

Schlumberger-Doll Research, Old Quarry Road, Ridgefield, CT 06877, U.S.A.

ABSTRACT

Geological structures of both tectonic and sedimentary origin have been classified based on outcrop exposures. These structures may assume a very different appearance on core surfaces or borehole images because of the cylindrical shape of boreholes. A modular object-oriented solid-modeling system is used to simulate common geological structures on such curved surfaces. Starting from simple geometric primitives, geological layers are stacked and arbitrarily deformed in a number of steps specified by the user. The final object is calculated directly from the surface normals of the original object and a transformation matrix. Subsequent intersection of the body with a cylinder results in a core-like object, which may be viewed from any angle. Such models are presented for a number of sedimentary and tectonic structures with considerable changes in layer geometry over vertical distances ranging from centimeters to meters. Diagnostic features for some structures such as folds and cross-beds are found to differ from the outcrop. Because of the curved boorehole surface, dips and dip azimuths of key geological surfaces can be accurately measured. The models presented here help to properly recognize geological structures in boreholes and thus provide an important basis for improved reservoir models.

KEYWORDS

Geological structures, solid computational modeling, interactive system, modular system, geometric primitives, vector transformation rules.

INTRODUCTION

Geological structures are macroscopic geometric arrangements of rock layers. One class, termed sedimentary structures, may form during or shortly after deposition, while another group, termed tectonic structures, is formed well after deposition, normally through deformation by tectonic forces. Some structures fall between these categories, such as, for example, the viscous deformation of layers due to overburden pressure leading to structures like convoluted bedding or growth-faulting (Allen, 1984; Collinson & Thompson, 1982). These various classes of geological structures are generally based on outcrop observations, and geologists are therefore used to identify them from their appearance on a relatively planar surface.

Curved surfaces are present on the outside of cores and on the borehole wall. The habit of slabbing cores from the wellbore has, among other benefits, the effect of creating a planar surface and thus of providing the geologist with a view familiar from outcrops. On the outer curved surface of the cores, however, common geological structures may take a very different view. The same is true for high-resolution borehole images, which are obtained by scanning the borehole wall with ultrasonic and electrical sensors (Zemanek et al., 1969; Ekstrom et al., 1987). Thus, a need has been identified to recognize and understand geological structures on cylindrical surfaces.

We have used an experimental computer-aided design system to generate such geological structures. The highly modular and interactive system was developed by Kurt Fleischer and Andy Witkin at the former Schlumberger/Fairchild Palo Alto Research Center (Fleischer, 1987). It is based on a method developed by Barr (1984), which allows transposing and deforming solid geometric primitives, i.e. basic geometric objects, using simple vector transformation rules. The geometric functions used to describe the object include a surface position function to define the object in space, a surface normal function to obtain surface orientation and illumination, and an implicit function that describes the volume enclosed by the object. In addition, a material function defines object texture and light reflectance. Since this work is concerned with rocks, the primitive objects used are mostly rectangular blocks with an originally vertical layering pattern determined by a fractal curve. Subsequently, the shape of the block is changed using simple parametric deformation or transposition modules such as bending, tapering, twisting, rotating etc. These steps can be combined in a modular and interactive way, illustrated in Fig. 1 for the case of inclined isoclinal folds. A pop-up menu appears when clicking on a module and allows entry of parameter settings for that particular operation (Fig. 1). Layers may be stacked on top of each other with or without removal of the underlying layer (i.e. erosion) at their contact. In a Boolean operation the resulting structure is then intersected with a cylinder in order to obtain a core. Finally, graphic rendering is done by polygonal interpolation of the surface texture between nodes on the object surface. In the early stages of model construction the number of nodes is kept low (typically about 6 in each dimension), but once successful results are obtained, final rendering is done with a very high nodal density (typically 50) in order to achieve smooth layers and surfaces. Because of this and the relatively slow speed of our processor, rendering on the high-resolution graphic monitor took in some cases several hours.

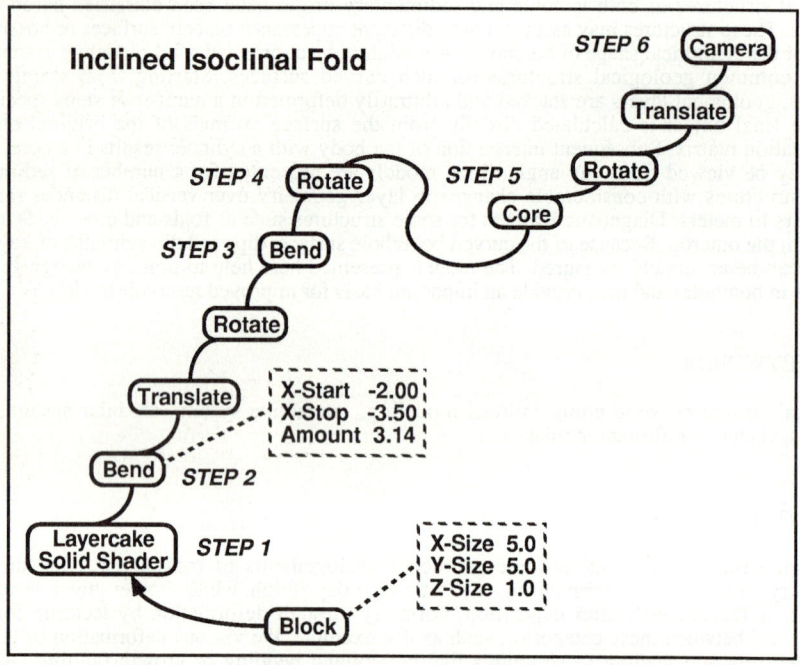

Fig. 1. Schematic path leading to the model of inclined isoclinal folds in Figures 2 and 10. The modules along the path represent successive operations on the solid object. Each operation requires input of a number of user-adjustable parameters through a pop-up menu, illustrated for the initial geometric primitive and the first bend (dashed boxes). Boolean subtraction and intersection are used to simulate the coring process in Step 5. The camera performs the final graphic rendering using parameters such as viewing angle, light source, camera focal length, background color etc.

Figure 2 shows intermediate results at selected steps in the model construction chain shown in Fig. 1. In order to view these intermediate steps, the model construction was stopped at the modules marked as Steps 1 through 6 and sent directly to the graphic rendering module designated as "camera". The more complex geological structures often required numerous iterations in order to get the desired results. Final presentation is such that the core is seen from four positions rotated by 90° and viewed from a slightly elevated position (Figs. 3-13). This allows comparison with core surfaces as well as unwrapped borehole images, since the cores are positioned close enough to one another to give an idea of how a continuous image of the borehole wall would look like.

Our cores are 6 times longer than their diameter, which for a 20 cm (8.5") borehole diameter means that they are 1.2 meters (4 feet) long. The modeled geological structures include only those in which noticeable changes in the layer geometry occur over the length of the core. Large-scale bedding dip changes as caused for example by gentle folds are, therefore, not shown. Such structures are better analyzed from seismic sections or dipmeter results. In many cases structures had to be parametrized for which no rigorous quantitative description is available. In those cases a match was sought between block diagrams used to describe the geological structure and our model prior to the "coring" process. Special attention was paid to the scale of the modeled structures, since, as mentioned above, the cores have an approximate size. This was particularly important for the sedimentary structures where size is an integral part of the classification. The selection of structures presented herein cannot be systematic nor exhaustive, since there is considerable variability of these structures in nature, and since the choice of structures which can be modeled is limited by the system's capabilities of handling complex geometric objects.

As in Rubin's (1987) geometric block models of cross-bedding types, the major limitation of these models is that they are not necessarily physically meaningful. Objects can be positioned and deformed in almost any way. For example, layer erosion is possible either by the upper layer eroding the underlying layer, which is geologically meaningful, or vice versa, which is not. The bending operation shown as step 2 in Figs. 1 and 2 is a laterally compressible, parallel fold where the layer density in the fold core has to increase relative to the outer layers. In real rocks, this is not observed; rather, the inner layers increase in thickness or rupturing, crumpling and layer-parallel shearing occurs (Billings, 1980). The modeling system also has no problems accommodating two or more solid objects at one location. The usefulness of the computing tool is therefore largely determined by the user's appropriate application of it.

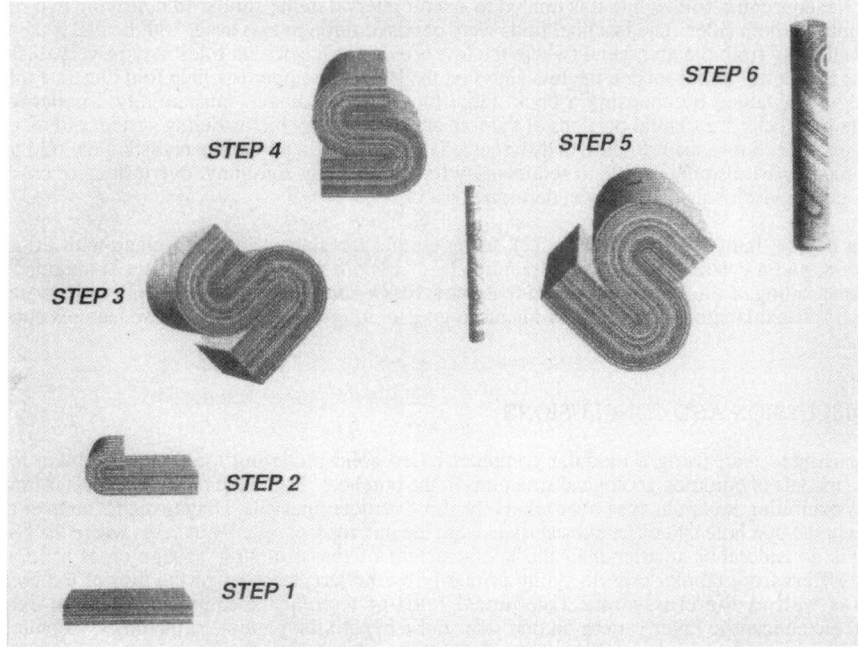

Fig. 2. Intermediate and final construction stages of the model for inclined
isoclinal folds (Steps 1 through 6 in Fig. 1).

MODELS OF GEOLOGICAL STRUCTURES

Selected models of geological structures in boreholes are shown in Figs. 3 to 13. Comparative block diagrams, sketches and photographs of the same structures can be found in Reineck & Singh (1980), Harms *et al.* (1982) and Billings (1980). The structures are grouped into sedimentary and tectonic structures, with the latter comprising folds, faults and fractures. Table 1 lists the characteristics of each structure in the borehole as well as the measurements which can be made.

Sedimentary structures

The sedimentary structures include parallel bedding with an angular unconformity (Fig. 3), tabular and trough cross-bedding (Figs. 4 and 5; McKee and Weir, 1953), hummocky cross-stratification (Fig. 6) and lenticular bedding (Fig. 7). All sedimentary structures were generated by stacking layered units of a certain geometry on top of one another in the same way as they would have been deposited, i.e. the younger layers come to lie on top of the older layers, sometimes eroding parts of the underlying material. Although trough and tabular cross-bedding seem to look very similar in boreholes, they can be distinguished based on the azimuth scatter of foresets and bounding surfaces (Luthi *et al.*, 1990). Hummocky cross-stratification has a very subtle expression in the borehole, largely due to the small vertical dip variations and because the limited lateral extent prevents identification of the typical undulating layer geometry. Lenticular bedding, on the other hand, has a typical dimension smaller than the borehole diameter and its identification in the borehole is therefore largely unambiguous.

Tectonic structures

Among the modeled folds are a recumbent concentric fold (Fig. 8), a recumbent similar fold (Fig. 9), an inclined isoclinal fold (Fig. 10) and a relatively complex flow fold (Fig. 11). The bending operation common to all fold models is a laterally compressible, parallel fold. No layer-parallel slippages or shear movements parallel to the fold axis are possible. While this is not a very satisfactory model from a physical point of view, it does not represent a serious drawback for the purely geometric models shown here, since mass-balances and other aspects are not considered. The bending operation takes place over the entire block for the concentric fold, while it is limited to a short interval in the similar fold, leaving two straight-crested limbs on both sides. The isoclinal folds were obtained through two exact 180° bends. If the well is located well away from the arch-bend (where the layers overturn), isoclinal folds may be very difficult to identify in boreholes. In our model, the two views on the left of the upper isoclinal fold illustrate this. The flow fold was obtained by choosing a thick initial block and folding it substantially. Similar to what happens in real rocks, the central portions of the fold are squeezed by the modeling system and its graphic rendering module. Subsequent folding in different axial planes produces a more realistic flow fold pattern. The fold models are also applicable to sedimentary folds caused by slumping, overturning of cross-beds and other postdepositional soft-sediment deformations.

Finally, a normal fault with drag (Fig. 12), using simple translations along a plane with prior layer deformation, and a system of cemented fractures (Fig. 13) are presented. The latter is obtained by an appropriate shading of thin, steeply inclined features across a homogeneous block. Both models are fairly simple and self-explanatory; they merely intend to give an impression for how these features appear on boreholes.

DISCUSSION AND CONCLUSIONS

We demonstrated that, using a modular computer-based solid-modeling tool, we can obtain realistic geometric models of common geological structures in the borehole. The range of these models is limited to structures exhibiting layer changes over relatively short vertical intervals. Diagnostic features of certain structures in the borehole can differ substantially from the outcrops, especially in cases where the borehole diameter is considerably smaller than the lateral extent of the structure. Trough cross-bedding, for example, differs from tabular cross-bedding primarily by the larger scatter of the dips of the bounding surfaces as well as the cross-strata. Overturned folds of tectonic or sedimentary origin exhibit a characteristic concentric layer pattern on one side, and a hyperbolic pattern on the other. Isoclinal folds may be entirely missed if the well is offset from the arch-bend, unless layer reversal is observed. Structures with subtle changes in layer geometry can be very difficult to recognize. An example is hummocky cross-stratification, where boreholes do not provide sufficient lateral extent to recognize the typically undulating layer geometries. On the other hand, structures such as lenticular bedding, whose lateral extent is smaller than the typical borehole diameter, can often be readily identified in the borehole.

In the borehole, important geological surfaces such as bedding, bounding surfaces, unconformities, fault planes or fractures can be accurately measured for their dips and dip azimuths because the curved surface provides access to the thrid dimension. This is not always the case with outcrop exposures, especially when they lack surface roughness. Measurements of these geological surfaces from borehole images or from oriented cores can be valuable in determining the local or regional subsurface geometry. In some cases, this may allow statistical modeling of reservoir compartmentalization due to cross-bedding, faulting or fracturing (Luthi and Banavar, 1988; Plumb and Luthi, 1989; Luthi, 1990).

ACKNOWLEDGEMENTS

Kurt Fleischer (formerly at SPAR) developed the geometric modeling system, written in Zeta-Lisp on a Symbolics 3640 computer. He and Andy Witkin (also formerly at SPAR, now at Carnegie-Mellon) developed the interface, adapted the system for the application presented herein and supported it. Jim Howard and Darwin Ellis critically reviewed the manuscript of this paper.

REFERENCES

Allen, J.R.L. (1984). *Sedimentary Structures, Their Character and Physical Basis.* Developments in Sedimentology, Elsevier, Amsterdam.

Barr, A.H. (1984). Global and local deformation of solid primitives. *Computer Graphics,* 18/3, 21-30.

Billings, M.P. (1980). *Structural Geology.* Prentice-Hall, Englewood Cliffs, N.J.

Collinson, J.D. and D.B.Thompson (1982). *Sedimentary Structures.* Allen and Unwin, London.

Ekstrom, M.P., C.A.Dahan, M.-Y.Chen, P.M.Lloyd and D.J.Rossi (1987). Formation imaging with microelectrical scanning arrays. *The Log Analyst,* 28, 294-306.

Fleischer, K. (1987). Implementation of a modeling testbed. In: *Object-oriented geometric modeling and rendering.* Siggraph 1987 Short Course Notes.

Harms, J.C., J.B.Southard and R.G.Walker (1982) *Structures and Sequences in Clastic Rocks.* Soc. Econ. Paleont. Min., Lecture Notes Short Course 9.

Luthi, S.M. (1990). Sedimentary structures of clastic rocks identified from electrical borehole images. In: *Geological Applications of Wireline Logs* (A.Hurst, M.A.Lovell and A.C.Morton, eds.), Geological Society Special Publications, 48, 3-10.

Luthi, S.M. and J.R.Banavar (1988) Application of borehole images to three-dimensional modeling of eolian sandstone reservoirs, Permian Rotliegende, North Sea. *Am. Ass. Petrol. Geol. Bull.,* 72, 1074-1089.

Luthi, S.M., J.R.Banavar and U.Bayer (1990). Models to interpret bedform geometries from cross-bed data. *J. of Geology,* 98, 171-187.

McKee, E.D. and G.W.Weir (1953). Terminology for stratification and cross-stratification in sedimentary rocks. *Geol. Soc. Am. Bull.,* 64, 381-390.

Plumb R.A. and S.M.Luthi (1989) Analysis of borehole images and their application to geologic modeling of an eolian reservoir. *Soc. Petr. Eng., Formation Evaluation,* 4/4, 505-514.

Reineck, H.-E. and I.B.Singh (1980). *Depositional Sedimentary Environments.* Springer, Heidelberg.

Rubin, D.M. (1987). *Cross-bedding, Bedforms, and Paleocurrents.* Soc. Econ. Paleont. Min., Concepts in Sedimentology and Paleontology, Tulsa.

Zemanek, J., R.L. Caldwell, E.E.Glenn, S.W.Holcomb, L.J. Norton and A.J.D.Strauss (1969). The Borehole Televiewer - A new logging concept for fracture location and other types of borehole inspection. *J. Petr. Tech.,* 21, 762-774.

Structure	Figure	Characterization in Borehole	Measurements
Parallel bedding with angular unconformity	3	Dipping beds have sinusoidal traces on an azimuth/depth projection. Beds below unconformity are truncated by unconformity, while beds above are roughly parallel to it.	Dip and dip azimuth of beds above and below unconformity as well as of unconformity surface itself.
Tabular cross-bedding	4	Inclined cross-strata separated by subhorizontal bounding surfaces. Relatively small scatter of dips and dip azimuths of both cross-beds and bounding surfaces.	Dip and dip azimuth of cross-strata and bounding surfaces. Thickness of cross-bedded sets.
Trough cross-bedding	5	Inclined cross-strata separated by subhorizontal bounding surfaces. Relatively large scatter of dips and dip azimuths of both cross-beds and bounding surfaces.	Dip and dip azimuth of cross-strata and bounding surfaces. Thickness of cross-bedded sets.
Hummocky cross-stratification	6	Cross-bedded sequence with low dips and wide scatter of dip azimuths of cross-beds as well as bounding surfaces. Layers often parallel to underlying bounding surfaces.	Dip and dip azimuth of cross-strata and bounding surfaces. Thickness of cross-bedded sets.
Lenticular bedding	7	Single ripple sets in shale-rich sequences. Various types of lenticular bedding dependent on shape and connectivity of ripple sets.	Thickness of sand and shale. In favorable situations dip and dip azimuth of ripple cross-laminations.
Recumbent concentric fold	8	"Hyperbolic" layer pattern on convex side of fold; elliptical pattern on concave side. Dip azimuth flips by 180° where fold axis is crossed. Layer squeezing common in fold center.	Strike of fold axis. Axial plane may be constructed from layer geometry.
Recumbent similar fold	9	Like concentric fold, but layers above and below fold axis have relatively constant dips. Generally no layer squeezing.	Strike of fold axis. Axial plane may be constructed from layer geometry.
Inclined isoclinal fold	10	Similar folds with layer dips of fold limbs and axial fold planes all parallel to each other. If arch-bend is far from borehole, layer reversal may be only indication for fold.	Dip and dip azimuth of fold limbs and axial planes.
Complex flow fold	11	Irregular changes in layer dips and layer thicknesses.	Approximate strike of fold axes.
Normal fault with drag	12	Plane across which layers are offset, sometimes with fault breccia and associated fracture network. Layers around fault assume dip azimuth close to the fault plane.	Dip and dip azimuth of fault plane and layers.
Cemented fractures	13	Planar or subplanar thin streaks of mineral precipitate dissecting rock layers without apparent displacement.	Dip and dip azimuth of fractures.

Tab. 1. Characteristics in the borehole of the structures in Figs. 3-13. A vertical borehole and zero structural dip is assumed.

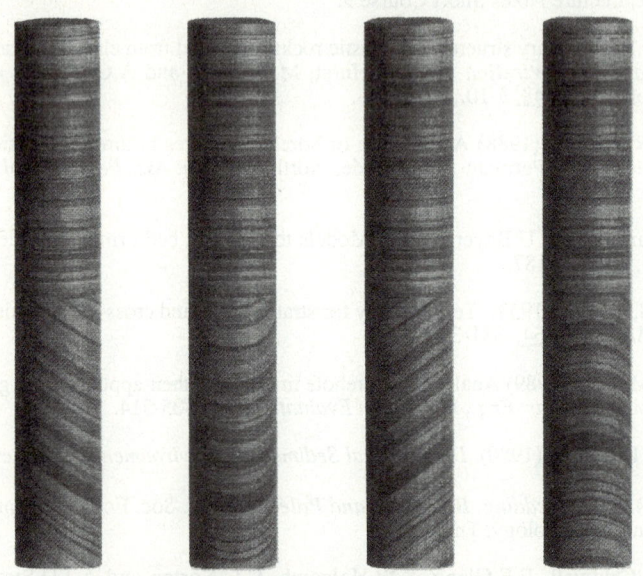

Fig. 3. Parallel bedding with angular unconformity.

Fig. 4. Tabular cross-bedding.

Fig. 5. Trough cross-bedding.

58

Fig. 6. Hummocky cross-stratification.

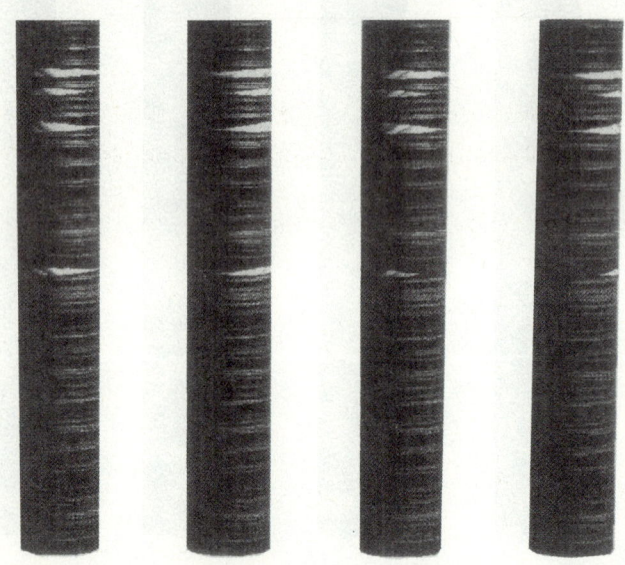

Fig. 7. Lenticular bedding.

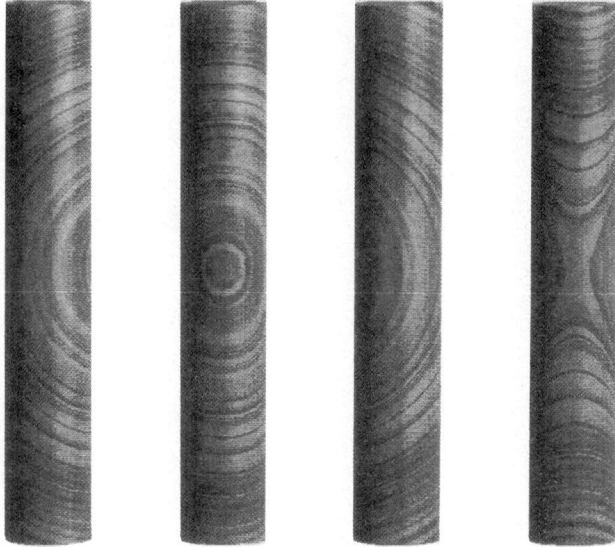

Fig. 8. Recumbent concentric fold.

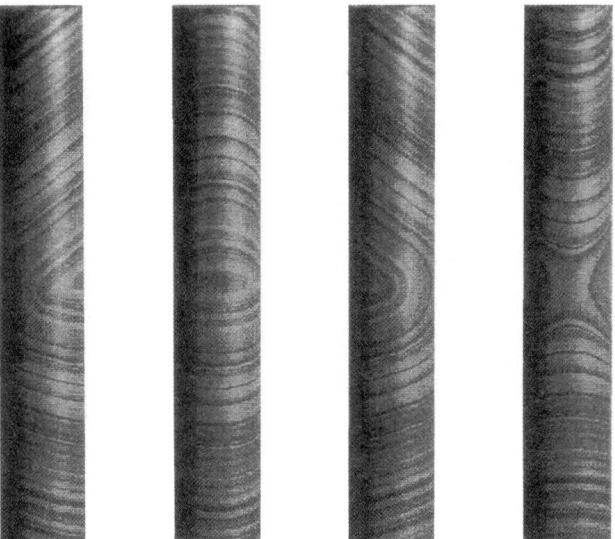

Fig. 9. Recumbent similar fold.

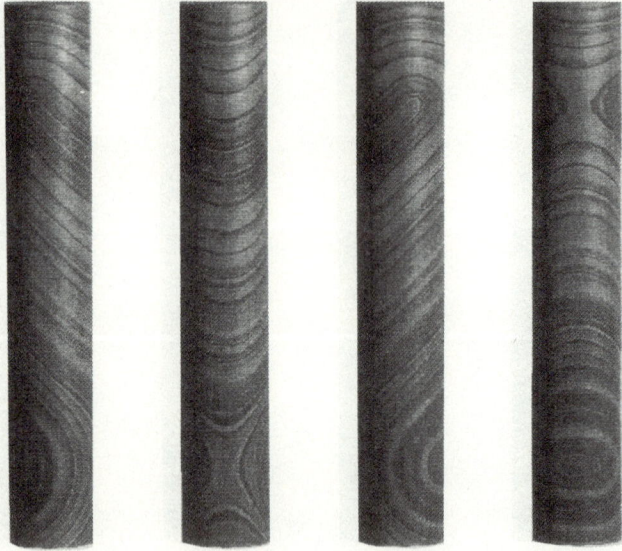

Fig. 10. Inclined isoclinal fold.

Fig. 11. Complex flow fold.

Fig. 12. Normal fault with drag.

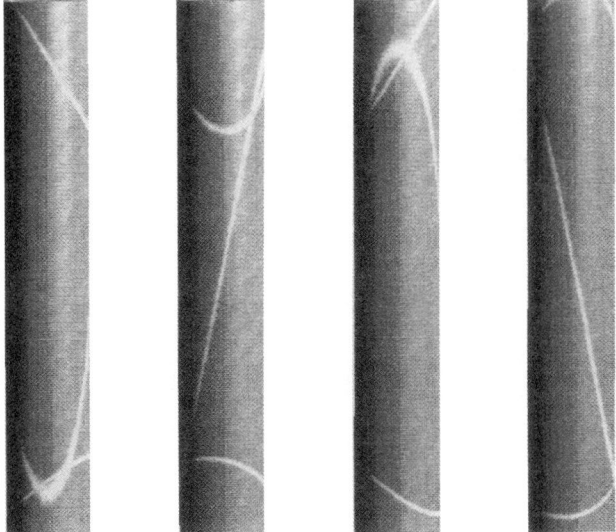

Fig. 13. Cemented fractures.

THREE-DIMENSIONAL RECONSTRUCTION
OF PORE GEOMETRY FROM SERIAL SECTIONS
- IMAGE ALGEBRAIC APPROACH

YOSHIKAZU OHASHI

ARCO Oil and Gas Company, Plano Research Center,
2300 West Plano Parkway, Plano, Texas 75075, U.S.A.

ABSTRACT

A new pixel-based technique permits the reconstruction of three-dimensional structure from serial section images. Image-algebraic operations such as erosion, dilation, skeleton and pruning are defined in terms of hit-or-miss transforms and procedures are coded in Fortran. A sequence of image algebraic operations is designed to interpolate the middle layer configuration from two serial sections. A conditional skeleton operation plays a key role to outline the middle layer. The method was applied to reconstruct the three-dimensional pore geometry of rocks.

KEYWORDS

image algebra; mathematical morphology; pore geometry; serial sections.

3-D RECONSTRUCTION FROM SERIAL SECTIONS

INTRODUCTION

Analyses of the microstructures of rocks are often based on two-dimensional cross-sections. Some bulk properties such as the porosity, grain size, etc. may be properly determined from the two-dimensional images if anisotropy is not too great. Information on truly three dimensional features such as the connectivity of pore bodies, however, can not be obtained from a single section nor from a few of oriented sections. The purpose of this paper is to describe how to reconstruct the volumetric representation of a 3-D object given a set of parallel cross sections by filling space between sections.

Interpolation of Serial Sections

Triangular Patch Methods. Keppel (1975) formulated the problem of approximating the shape of a 3-D object from a series of sections as a combinatorial problem of graphs. His approach is stated as: Given polygonal contours on two sections, find a set of triangular patches that yield the optimal shape between two sections. A criterion for the optimal shape varies from the maximum volume (Keppel, 1975) to the minimum surface (Fuchs et al, 1977) and locally minimized paths (Christiansen & Sederberg, 1978). One of the earliest applications to geological problems was for a shape reconstruction of brachiopoda (Tipper, 1976, 1977), although his algorithm was general for any triangular patch problems.

As pointed out by Ganapathy and Dennehy (1982), the number of possible triangular patches increases as a function of factorials of nodes in two polygons. Thus the strategy of the more recent work is, rather than approaching the problem from a general combinatorial aspect, to decompose contours using various criteria such as the parity of radial intersections (Dowd, 1985), tolerance sets (Zyda et al, 1987), and span pairs (Sinclair et al, 1989).

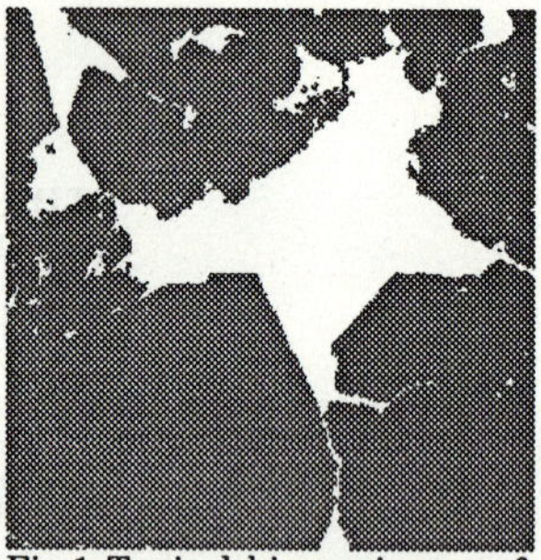

Need for algorithm based on pixel data. We have tried to reconstruct the 3-D pore space using the MOSAIC program, a part of the MOVIE.BYU system of Brigham Young University, which is based on the Christiansen and Sederberg (1978) algorithm. In order for the triangular patch algorithm to be practical, the number of vertices of a polygonal contour must not be too large (typically not more than a couple of hundred). As obvious from the typical image shown in Fig. 1, the outlines has to be overly simplified for this method.

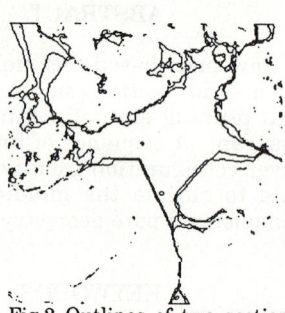

Fig.1 Typical binary image of a rock cross section. White areas represent pores, dark areas mineral grains.

Because a microscopic section of a rock specimen contains very detailed information on the boundary shape, ideally the resolution of boundaries should be maintained for interpolation without further simplification of the shape. Thus we have explored other techniques based on pixel (picture element) data of a digitized image, but not the polygonal contour approximation.

ANALYSIS OF THE PROBLEM

Interpolation strategy

In this section we will sketch our overall strategy to interpolate serial sections which is based on image algebraic approach of Serra (1982). Figure 2 shows two adjacent

Fig.2 Outlines of two sections are superimposed. The problem is to find the intermediate configuration.

sections overlaid on each other. In this superimposed view, a region on one section in general will have the corresponding region on the other. Depending on how rapidly the object changes its shape, several cases can be identified.

Case 1 - the complete inclusion. If a region A of one section is totally contained in the corresponding region B of the adjacent section, the expected boundary M of the middle layer will be somewhere between the two outlines of A and B (Fig. 3). Without additional information on the object, it is natural to assume that the outline of the region on the bisecting layer are at the equal distance from two boundaries A and B.

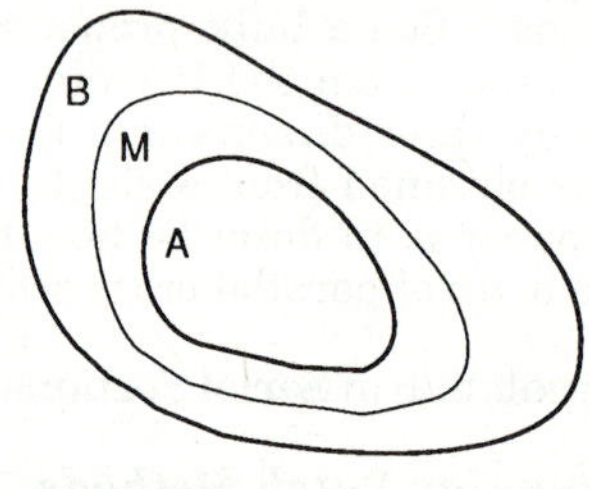

As a special case of this, a section can consist of multiple regions as shown in Fig. 4. The object forms branches between the two sections. We assume that branching occurs at the bisecting level. Then there is a special boundary of the singularity (denoted S) that is degenerated into a single curve inside the region M. As

Fig. 3. Region A is totally contained inside region B. The outline of the interpolated middle layer M should be equally distant from A and B.

is in the previous case, points that are at the equal distance from region boundaries form the outline of the interpolated region.

Case 2 - a small offset with intersections. Next, consider a case in which two outlines A and B intersect in the projection but the overall dissimilarity is relatively small (Fig. 5). In more rigorous terms, the number of points in common between in A and B is much larger than the number of points that are only in A or only in B. In this case we may still regard points at the equal distance from two outlines A and B as the solution. By definition (with the zero distance), points (P1 and P2) at which outlines A and B intersect are also on the outline of the middle layer.

Case 3 - No overlap or a large offset. Suppose the corresponding regions A and B are further displaced to the extent that fewer points are in common between A and B than those only in A or only in B (Fig. 6). In the extreme case, the regions A and B can be totally dissociated (i.e. no overlap in the projection). In this case we may still use the same approach with an additional region Z that is an area swept by an imaginary action of bringing the region A to the region B. The generation of the zone Z can be done in several ways: 1) common tangent lines between the two regions, 2) manual filling of the gaps or pits, and 3) the image algebraic CLOSE operations with a

relatively large kernel or structuring element . Furthermore, the zone Z can be subdivided to Z_A and Z_B which are for the regions A and B, respectively.

Considering all above cases, we can generalize that the outline of the middle layer is included in the skeleton of the region that is in A but out of B (or vice versa) plus Z, provided that some points such as P1 and P2 are specified, if necessary, as seeds for the skeleton. The region Z can be empty (cases 1 and 2) or a list of special points P1, P2, etc. can be empty (case 1).

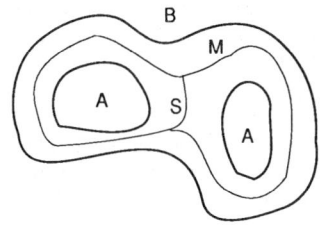

Fig 4. A case of branching. The boundary S is degenerated to a single curve.

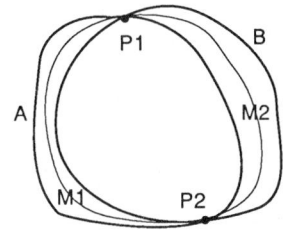

Fig 5. A case in which two regions are slightly offset.

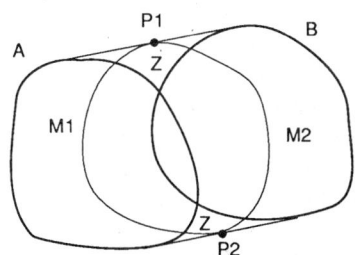

Fig 6. A case in which two regions are offset greatly.

Preparation of Digital Images

Types of digitization grids. Two frequently used patterns to digitize an image are a hexagonal grid and a square grid (Fig.7). Although the hexagonal grid has an apparent advantage of all direct neighbor points being at the equal distance, its implementation on general purpose graphics terminal is more involved than for the case of the square grid. Many of dedicated image analysis systems adopt the hexagonal grid. All discussions in this study are based on the square grid using no special hardware.

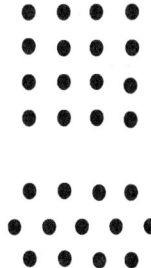

Fig 7. Square grid and hexagonal grid.

Binary Images. The original gray-scale images of microphotographs are transformed to a binary array using a threshold value appropriate to separate opaque regions (mineral grains) and transparent regions (pores). Pixels falling in pores are assigned the value 1 and those for mineral grains the value 0 because the pore geometry is of main interest (a reverse assignment will yield a negative of the same image). For this study an array of 256 rows and 256 columns was created this way for each section. All image-algebraic operations described below are performed with a positive logic with the value of 1 representing true and 0 false.

Connection modes of cells in the square grid. In the square grid, the surrounding cells are not all at the equal distance from a cell of interest. Depending on which neighbors are considered as connected to the center cell, one can distinguish two cases (Fig. 8). The connectivity of cells in the square grid can be defined in two ways: 1) a four-connected mode in which only the immediate neighbors with the shortest distance are considered as connected or 2) an eight-connected mode where both first and second neighbors make connections. A selection of the connection mode is totally arbitrary, but once decided, it should be used consistently for all subsequent operations. A different set of structuring elements (i.e. predefined template patterns) is required for image-algebraic operations such as erosion, dilation, and thinning. In this study the eight-connected mode is used throughout. The descriptions of algorithms and also their software implementations, however, include both modes.

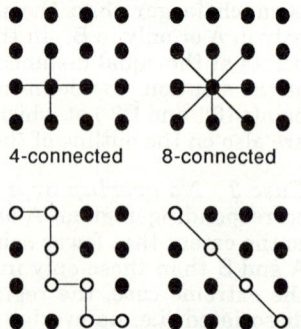

4-connected   8-connected

Fig 8. Connection mode in the square grid.

DEFINITIONS OF BASIC OPERATIONS

There are several books (Serra, 1982; Dougherty & Giardina, 1987; Giardina & Dougherty, 1988) and reviews (Maragos & Schafer, 1986; Haralick, et al, 1987; Heijmans & Ronse, 1990; Ritter, et al, 1990) on image algebra or mathematical morphology. The definitions, however, are not always consistent among the literature. The minimum set of definitions necessary for describing this study is summarized here.

Definitions of symbols for image algebra. We mostly follow the nomenclature used by Serra (1982). Upper case letters designate sets and lower case letters elements of a set. For example, a set X is a collection of points x the values of which are true or 1, i.e. $X = \{x \; ; x=1\}$. If a set X does not include any element, it is an empty set, $X = \emptyset$. A complement of X, expressed as X^C, is a set of points x' that do not belong to a set X, i.e. $X^C = \{x' \; ; x' \notin X\}$. Two fundamental set theoretical operations between two sets X and Y are the union ($\cup$) and the intersections ($\cap$).

$$X \cup Y: \text{a set of points belonging to X } \textbf{or } Y \tag{1}$$

$$X \cap Y: \text{a set of points belonging to X } \textbf{and } Y \tag{2}$$

Obviously by definition,

$$X \cup X^C = \text{the entire image} \tag{3}$$

$$X \cap X^C = \emptyset \tag{4}$$

Also the set difference is defined as,

$$X \, / \, Y: \text{a set of points belonging to X but not Y , i.e. } X \cap Y^C. \tag{5}$$

Hit-or-miss transformation. This is a key concept for many image algebraic operations used in this study. An intuitive definition of a hit-or-miss transform is given as follows: Superimpose a template (a structuring element) on a given image and do cell-by-cell comparison between the structuring element and the image. If all comparisons conclude true, this is a hit. If any of comparison turns out false, that is a miss. Record the hit position, normally by the location of the center cell of the structuring element with respect to the image grid. Translate a template and repeat the same procedure. For example, a structuring element $\begin{smallmatrix} 1 & 1 & 1 \\ \bullet & 1 & \bullet \\ 0 & 0 & 0 \end{smallmatrix}$ can be used to detect a boundary cell. A hit with this structuring element means that the center cell, which is in the object (value 1), is at such a position that the top three cells belong to the object (1) and the bottom three cells to the background (0). Cells with a symbol $\bullet$ on the left and right of the center can be either in the object or the background.

The original definition of the hit-or-miss transform is (Serra, 1982, p.39): A point x belongs to the hit-or-miss transform, HIT-OR-MISS(X,B), if and only if B^1_x is included in X and B^2_x is included in X^c, the complement of X, i.e.

$$\text{HIT-OR-MISS}(X,B) = X \otimes B = \{x \; ; \; B^1_x \subset X, \; B^2_x \subset X^c\} \tag{6}$$

where B^1 and B^2 are disjoint subsets of B and B_x means B positioned at x. For the above example of $B = \begin{smallmatrix} 1 & 1 & 1 \\ \bullet & 1 & \bullet \\ 0 & 0 & 0 \end{smallmatrix}$, disjoint sets are $B^1 = \begin{smallmatrix} 1 & 1 & 1 \\ \bullet & 1 & \bullet \end{smallmatrix}$ and $B^2 = \begin{smallmatrix} \bullet & \bullet & \bullet \\ \bullet & \bullet & \bullet \\ 1 & 1 & 1 \end{smallmatrix}$. Note that 0 elements in B are complemented to 1 in B^2 because $B^2_x \subset X^c$ is for a *complement* of X where original zeros are changed to ones.

Erosion and dilation. Erosion and dilation, two fundamental operations of image algebra, are usually derived from the Minkowski addition and subtraction (Serra, 1982, p.44; Giardina & Dougherty, 1988, p.6). A hit-or-miss transform can also be used to define elegantly these operations as its special case. If the structuring element does not contain elements 0 or a subset B^2 is empty, a hit-or-miss transform will give the erosion. For example the structuring elements in

Table 1: Structuring elements B*
for unit erosion/dilation in square
grids

4-connected	8-connected
1 1 1	$\bullet$ 1 $\bullet$
1 1 1	1 1 1
1 1 1	$\bullet$ 1 $\bullet$

* A symbol $\bullet$ denotes either 0 or 1.

Table 1 are for the unit erosion by a unit translation, i.e. the region is eroded by one pixel deep from the boundary. We use a symbol X^e for the unit erosion and X^d for its complement, the unit dilation.

$$X^e = \text{HIT-OR-MISS}(X,B \text{ in Table 1}) = X \otimes (B \text{ in Table 1}) \quad : \text{Unit erosion} \tag{7}$$

The erosion contracts the set and the dilation expands the set. The dilation operation can be defined as the erosion of the complement of the set by the duality theorem (Serra, 1982, p.588).

$$(X^d)^c = (X^c)^e \tag{8}$$

or taking the complement of the both sides, we have,

$$X^d = ((X^c)^e)^c \quad : \text{Unit dilation} \tag{9}$$

Boundary. A boundary of a region does not have any thickness in a *continuous* image but has a finite thickness of one pixel in the *digital* image. One could define the boundary as a transition *between* neighboring grid points. We prefer, however, the boundary is also represented by a connected set of pixels so that image algebraic operations are also valid for the boundary. There are two ways to define the boundary: one by the outer-most pixels of the region, the other by pixels in the background that enclose the region. The boundary of a set X is then defined as a difference between the original and the eroded sets.

$$\partial X = X / X^e \quad : \text{Inner boundary} \tag{10}$$

$$\partial X^c = X^c / (X^c)^e \quad : \text{Outer boundary} \tag{11}$$

Thinning. Thinning is the operation to reduce the size of the region, or to change cell contents selectively from 1 to 0. As for erosion, a hit-or-miss operation is used to find appropriate cells for thinning. The center cell is changed from 1 to 0, if it is a hit, for the thinning operation. For example, using a structuring element $B = \begin{smallmatrix} 1 & 1 & 1 \\ \cdot & 1 & \cdot \\ 0 & 0 & 0 \end{smallmatrix}$ for a hit-or-miss transform, *bottom* boundary cells can be detected and it is changed to 0, that is, one cell at the bottom boundary is removed or the object is thinned.

Skeletonization. In essence the skeleton operation is sequential thinning operations to peel a "skin", or the outer-most elements, of the object layer by layer until only one cell wide of the object is left. What is left after these repeated peeling operations is a backbone of the object. With the above-mentioned structuring element $B = \begin{smallmatrix} 1 & 1 & 1 \\ \cdot & 1 & \cdot \\ 0 & 0 & 0 \end{smallmatrix}$, pixels at the bottom boundary can be removed. Another hit-or-miss transformation $B = \begin{smallmatrix} 1 & \cdot & 0 \\ 1 & 1 & 0 \\ 1 & \cdot & 0 \end{smallmatrix}$, which is derived by rotating the first one by 90 degrees counter-clockwise, is followed in order to remove boundary cells from right. Thinning from the top and left directions will be done in the same way using the structure element that is rotated 90 degrees each time. In addition to the these vertical and horizontal removals, diagonal elements, e.g. $B = \begin{smallmatrix} \cdot & 1 & \cdot \\ 0 & 1 & 1 \\ 0 & 0 & \cdot \end{smallmatrix}$ and its rotations are also used alternately to complete thinning from eight directions 45 degrees apart. In total eight structuring elements are used repeatedly until no new hit is found.

$$\text{SKELETON (X)} = \text{REPEAT}(...(X O L_1) O L_2)...) O L_8 \text{ UNTIL NO HIT} \qquad (12)$$

where a symbol O denotes a thinning operation and L_i (i=1 to 8) are structuring elements corresponding to eight directions 45 degrees apart. As given in Table 2, diagonal structuring elements are different for the eight-connected and the four-connected mode.

Pruning. A pruning operation is a type of sequential thinning especially for removals of terminal cells of chains. The pruning operation can eliminate non-boundary superfluous branches of chains. Similar to the skeleton operation, the pruning operation is defined as

$$\text{PRUNE (X)} = \text{REPEAT}(...(X O L_1) O L_2)...) O L_N \text{ UNTIL NO HIT} \qquad (13)$$

where L_i is structuring elements listed in Table 2 and N=4 for the 4-connected and N=16 for the 8-connected mode.

Conditional skeletonization and pruning. The skeleton and prune operations, or in general any thinning or thickening operations, are relatively limited in their usefulness to real-world problems when applied unconditionally. In more realistic situations, the image is to be simplified by thinning operations while some of the original structure should be maintained. In other words, some pixels in the image are to be protected from being thinned. This type of image-algebraic operations is called conditional operations (Serra, 1982, p.393) and operationally these protected pixels are reactivated after every thinning cycle.

$$\text{CONDITIONAL-SKELETON (X)} = \text{SKELETON (X ; Y)} =$$
$$\text{REPEAT}(...((X O L_1) \cup Y) O L_2) \cup Y)...) O L_8 \cup Y) \text{ UNTIL NO HIT} \qquad (14)$$

$$\text{CONDITIONAL-PRUNE (X)} = \text{PRUNE (X ; Y)} =$$
$$\text{REPEAT}(...((X O L_1) \cup Y) O L_2) \cup Y)...) O L_N \cup Y) \text{ UNTIL NO HIT} \qquad (15)$$

where Y is a set of protected points and N=4 for the 4-connected and N=16 for the 8-connected mode (See Table 2). For image-algebraic operations that increase the region size such as dilation and thickening, these special pixels are prohibited from being turned on.

Table 2: Structuring elements L* for skeleton
and prune operations in square grids

	4-connected		8-connected			
Skeleton	1 1 1 • 1 • 0 0 0	1 1 • 1 1 0 • 0 •	1 1 1 • 1 • 0 0 0	• 1 • 1 1 0 • 0 0		
Prune	• • • 0 1 0 0 0 0		0 • 0 0 1 0 0 0 0	• 0 0 0 1 0 0 0 0	• 1 • 0 1 0 0 0 0	1 • 0 • 1 0 0 0 0

* A symbol • means either 0 or 1. Other orientations with 90°, 180° and 270° rotations of the above should also be included for a hit-or-miss transform.

Thus conditional operations for these operations include the intersection ($\cap$) operation instead of the union ($\cup$).

As shown later, the conditional thinning operations are critically important for image interpolation for the bisecting layer.

Curve intersection in the eight-connected mode. When a curve is regarded as continuous, the intersection of two curves is by definition points that belongs to both curves. Mathematically,

intersection of curve-A and curve-B = (curve A) $\cap$ (curve B)   (16)

Fig. 9 Possible curve intersections for the 8-connected grid.

When a curve is represented by a set of discrete points, however, the above set theoretical operation may not be extended to find intersection points (See Fig.9). Let us define a chain as a set of connected points.

intersection of chain-A and chain-B = (chain A) $\cap$ (chain B)   for 4-connected mode   (17)

intersection of chain-A and chain-B =[chain-A $\cap$ chain-B] $\cup$
$\quad$[(chain-A $\otimes$ T^1) $\cap$ (chain-B $\otimes$ T^2)] $\cup$
$\quad$[(chain-A $\otimes$ T^2) $\cap$ (chain-B $\otimes$ T^1)]   for 8-connected mode   (18)

where $T^1 = \begin{bmatrix} 1 & • \\ • & 1 \end{bmatrix}$ and $T^2 = \begin{bmatrix} • & 1 \\ 1 & • \end{bmatrix}$. The second and third terms are hit-or-miss transforms

TABLE 3. Summary of image algebraic operations used in this study

$X \cup Y$	union	a set of points in X or in Y
$X \cap Y$	intersection	a set of points in X and in Y
X / Y	difference	a set of points in X but not in Y
X^c	complement	a set of points not in X
X^e	(unit) erosion	a set eroded by a unit translation
X^d	(unit) dilation	a set dilated by a unit translation
∂X	boundary	a set of outer-most points of X $\partial X = X / X^e$
B_x	translation	a set B translated by x
$X \otimes B$	hit-or-miss transform	a set of points x at which an exact match of B_x is found in X
$X \circ B$	thinning	a set of points in X but not in $X \otimes B$ $X \circ B = X / (X \otimes B)$

which return logical true when chains A and B crossover at non-grid points such as $\begin{bmatrix} a & b \\ b & a \end{bmatrix}$

and $\begin{bmatrix} b & a \\ a & b \end{bmatrix}$.

RECONSTRUCTION ALGORITHMS

Image Algebraic Processes for Section Interpolation

In this section, we describe a series of image algebraic operations to generate a middle layer configuration M from two regions A and B. The algorithm can be split into two: one for the region inside A but outside B, that is

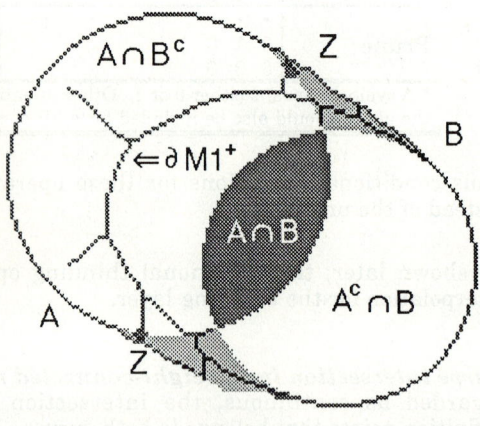

Fig. 10. Two disk-shaped regions A and B overlap. Concave portions of outlines are filled to form a region Z. A skeleton
$$\partial M1^+ = \text{SKELETON} ((A \cup Z) \cap (B^e)^c ; C)$$
includes several extra branches.

$$A \cap B^c , \tag{19}$$

and the other for the region outside A but inside B, $A^c \cap B$. Note that B^c is a set not in B, i.e. outside B. Because the second case can be obtained by exchanging A and B, only the first case will be discussed. If boundaries of A and B coincide in a certain area, however, the above operation will leave no elements along the coincided boundary. To ensure such a boundary to be included, we erode one layer before the complement operation,

$$A \cap (B^e)^c. \tag{20}$$

If an additional region Z needs to be added,

$$(A \cup Z_A) \cap (B^e)^c. \tag{21}$$

When the conditional skeleton operation is performed, the result (denoted as $\partial M1^+$) contains a half of the outline (denoted as $\partial M1$) of the region M. Note that the result of the above operation may still include extra points which are not a part of the boundary, thus in general $\partial M1^+ \subseteq \partial M1$.

$$\partial M1^+ = \text{SKELETON} ((A \cup Z_A) \cap (B^e)^c ; C) \tag{22}$$

where C is a set for conditional points. The other half of the boundary, M2, can be found by exchanging A and B.

$$\partial M2^+ = \text{SKELETON} ((B \cup Z_B) \cap (A^e)^c ; C') \tag{23}$$

Add these two skeletal structures and remove branches with the conditional prune operation:

$$\partial M^+ = \text{PRUNE} (\partial M1^+ \cup \partial M2^+; C'') \tag{24}$$

where C" is a set for conditional points for pruning. Finally the region M will be

$$M = \text{FILL} (\partial M+) \tag{25}$$

The FILL operation is to fill inside the boundary from seed cells. This is done by turning on cells, if they are not on the boundary, around seed cells recursively until there are no more cells to be transformed. The contour-filling-by-connection algorithm is given by Pavlidis (1982, p.180).

The difference set S^+ between ∂M^+ and a true boundary ∂M will include a set of singularity points if branching occurs.

$$S^+ = \partial M^+ / \partial M = \partial M^+ \cap (\partial M)^c \tag{26}$$

The above procedure is schematically summarized in Fig. 11.

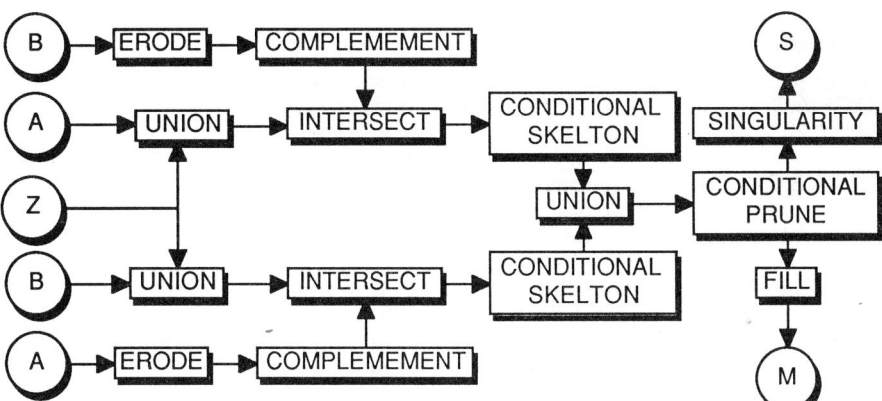

Fig.11. A flow chart of the algorithm to interpolate the middle layer M from two images A and B and optionally Z.

Computer Implementation

This section describes software code development of the above mentioned functions for an Apple® Macintosh® computer. Algorithms were coded in Fortran 77 and compiled with Language Systems Fortran™ for MPW (Macintosh Programmer's Workshop™).

To examine and edit images of individual sections, we extensively used the NIH Image program written by Wayne Rasband. This software was found also extremely valuable to locate causes of the software problems when the algorithm did not work as we expected. By examining intermediate images pixel by pixel, many subtle cases were found and the algorithm was modified to make it more robust.

Data representation - Augmented image array. One of sources of the algorithm complexity is how to treat rows and columns at the four edges of the image array which have valid entries only on one side. When a 3×3 structuring element is superimposed on an m×n image, for example, operations can be defined only for cells of which index ranges 2 to m-1 (or n-1). Repeated operations will end up with smaller and smaller images. Instead, we define an augmented array that has one extra row or column on each side of the array.

An array index can be assigned from 0 to m+1 and 0 to n+1. The image still occupies the center portion of the array but a hit-or-miss transform can be now defined at cell (1,1) by using cells on the margin. When a 3×3 structuring element is overlaid on the image array, the advantage of using the augmented array is 1) no special treatment is necessary for the first and the last rows (or columns) of the image, 2) an input array and an output array have the same dimension, and 3) thinning from the array boundaries can be prevented by setting margin cells all true. The margin cells, e.g. the 0-th row, can be set all true, or all false, or identical to the first row or to the m-th row.

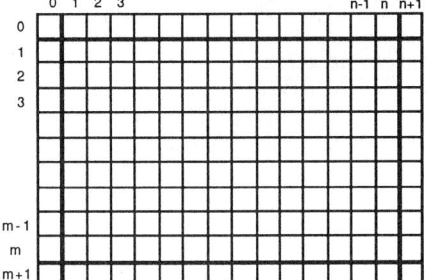

Fig. 12. Augmented image array.

Below is a short list of functions written in Fortran. When two or more arguments are listed, the output array may be the same as the input. For example, NOT_IMAGE(A,A) will replace A with its complement.

Primary image algebraic operations. This group of functions includes:

CLEAR_IMAGE(A) ....Reset all entries of image array A to 0.
COPY_IMAGE(A,C) ....Copy image array A to another array C.
NOT_IMAGE(A,C) ....Create complement image C of A
                or change entries 0 to 1 and 1 to 0.
OR_IMAGES(A,B,C) ....Create union image C of images A and B
                or pixel-wise OR operation.
AND_IMAGES(A,B,C) ....Create intersection image C of images A and B
                or pixel-wise AND operation.
DEDUCT_IMAGES(A,B,C)...Create subtraction image C with images A less B
                or pixels in A but not in B.
XOR_IMAGES(A,B,C) ....Create difference image C between images A and B
                or pixel-wise XOR operation.

Erosion and dilation.
ERODE_IMAGE(A,C) ....Create unit erosion image C of image A.
DILATE_IMAGE(A,C) ....Create unit dilation image C of image A.

Skeleton and prune. Template patterns listed in Table 2 are stored as template data and the SET_TEMPLATE function creates a target structuring element for the specified pattern and orientation.

SET_TEMPLATE(pattern,orientation)
              ....Internally called function to set a structuring element for a
               given pattern for a given orientation.
THIN_IMAGE(A,C) ....Create thinned image C of image A using structuring
               element set by SET_TEMPLATE function.
SKELETON_IMAGE(A,C).....Create skeletonized image C of image A.
PRUNE_IMAGE(A,C) ....Create pruned image C of image A.

Margin functions. In order to initialize margin cells of the augmented array, four functions were implemented. Before any thinning operations, the TURN_MARGIN_ON function is called to prevent cells from being thinned at the image edges.

TURN_MARGIN_ON(A) ....Set all margin elements of array A to 1.
TURN_MARGIN_OFF(A) ......Reset all margin elements of array A to 0.
CYCLE_MARGIN(A) ....Copy bottom row to top margin, left column to right margin
               etc. or arrange margins in cyclic fashion.
REPEAT_MARGIN(A)............Copy top row to top margin, right column to right margin
               etc. or arrange margins as repeat of adjacent row/column.
Boundary functions.
IN_BOUNDARY(A,C) ....Create inner-boundary set C of array A, or C=∂A.
OUT_BOUNDARY(A,C) ....Create outer-boundary set C of array A, or C=∂A^c.
CROSS_BOUNDARY(A,B,C)
              ....Create a set C of intersecting points of boundaries A and B.
               (non-grid intersections are treated correctly for a 8-
               connected mode).

TEST RESULTS

Fig. 13-a and -b are 256x256 binary images of two adjacent sections. Inter-layer spacing is approximately 20 pixel units of the sections. These two represent input arrays A and B.

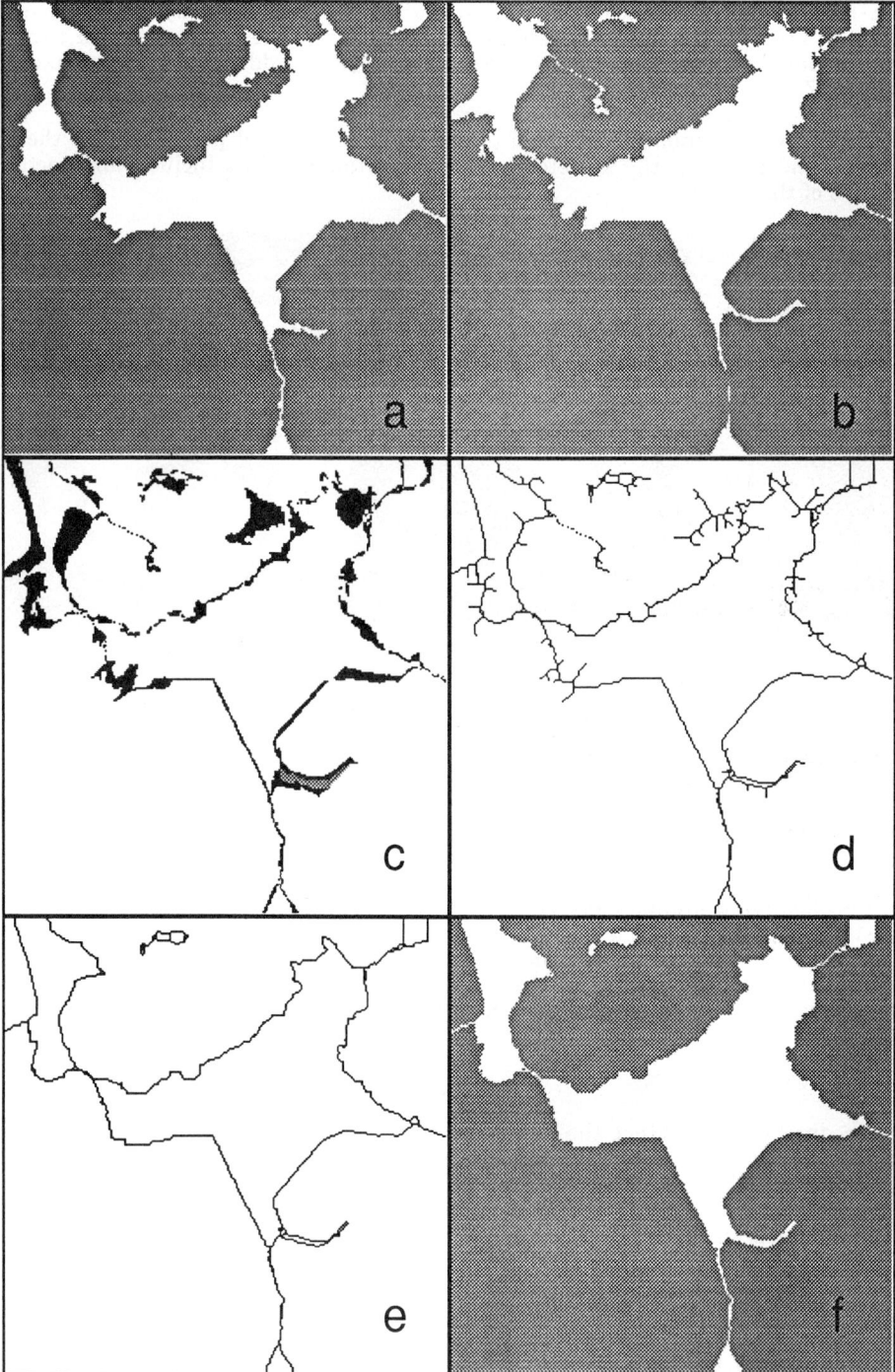

Fig. 13 (a) and (b) input images A and B . (c) Set difference A and B. A set Z, small gray area in lower-right quadrant, was created. (d) After the skeleton operation. (e) After the pruning operation. (f) A filled image for the middle layer.

With a help of the difference image (shown Fig. 13-c) obtained by the exclusive-OR operation of the two, the gap between two non-overlapping narrow regions in the lower-right quadrant was filled (gray area in Fig. 13-c). These additional pixels make up an input array Z.

Midpoints of the Z boundary are added to the set of conditional points so that these will become the end points of the skeleton. Other conditional points include intersections of boundaries of the regions.

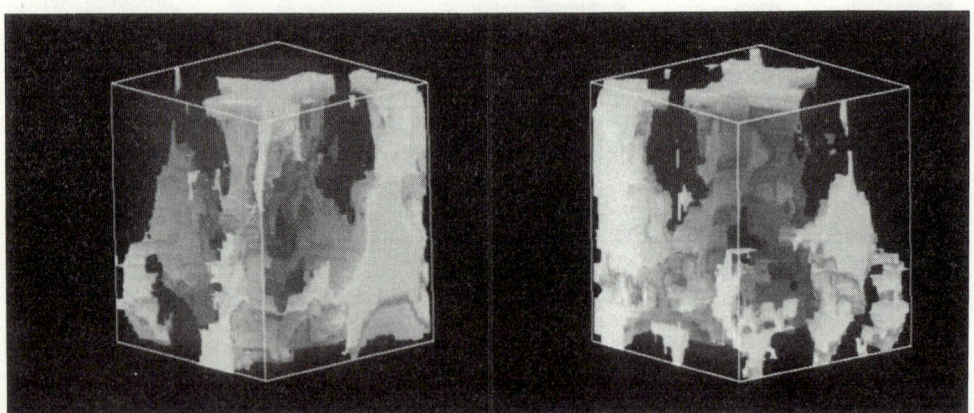

Fig. 14   Front and rear perspective views of reconstructed pore space from serial sections. The top of the cube is the section given in Fig.13-a. This rendering was done on a SUN SPARK™ station with SUN VISION™ software.

After a series of image-algebraic operations — erosion, complement, union, intersect, and conditional skeleton — corresponding to the left two-thirds of the flow chart (Fig. 11), Fig.13-d was obtained. Although the general feature is acceptable, there are dangling branches. These branches were successfully removed by the conditional prune operation (Fig. 13-e). By filling the contours, the middle layer configuration is obtained in Fig. 13-f.

Similar procedures were applied to 16 serial sections, yielding 15 bisecting layers. Using these 31 slices, the three-dimensional pore structure was constructed and rendered (Fig. 14). The height of the cube is approximately 310 pixel units and the side edges are 256 pixel units long. In the three-dimensional perspective views, shown white is the pore and all volume elements in mineral grains are set to be fully transparent.

CONCLUSIONS

The procedure to reconstruct the 3-D voxel array can be defined as a series of image algebraic operations. The complexity of the problem then can be transformed to higher level manipulations of the images. Since each image algebraic operation is simple and clearly defined, computer implementation is straightforward. Furthermore, each operation is fairly parallel in nature and the process can be designed as a pipeline for a high speed hardware implementation.

The current implementation level, however, is not totally automatic. Images are still examined visually and, if necessary, an image of additional points is created by editing a bit-mapped display on a computer monitor. It was also found that some preprocessing, such as a removal of small noise specks, resulted in less ambiguous outlines.

ACKNOWLEDGMENTS

Dr. Charles L. Vavra provided original image data of serial sections and Ms. Mary J. Cole helped with volume rendering of the reconstructed pore geometry. Critical reviews by Dr. Kishore K. Mohanty and Dr. Charles L. Vavra improved the manuscript. I thank Arco Research and Technical Services for permission to publish this paper.

REFERENCES

Christiansen, H. N., & Sederberg, T. W. (1978). Conversion of complex contour line definitions into polygonal element mosaics. *Computer Graphics, ACM-SIGGRAPH,* 12(3), 187-192.

Dougherty, E. R., & Diardina, C. R. (1987). *Matrix structured image processing.* Prentice-Hall, Englewood, N.J.

Dowd, P. A. (1985). Algorithms for three-dimensional interpolation between planar slices. In R. A. Earnshaw (Ed.), *Fundamental Algorithms for Computer Graphics* (pp. 531-554). Springer-Verlag, Berlin.

Fuchs, H., Kedem, Z. M., & Uselton, S. P. (1977). Optimal surface reconstruction from planar contours. *Communications of the ACM,* 20, 693-702.

Ganapathy, S., & Dennehy, T. G. (1982). A new general triangularization method for planar contours. *Computer Graphics, ACM-SIGGRAPH,* 16(3), 69-75.

Giardina, C. R., & Dougherty, E. R. (1988). *Morphological methods in image and signal processing* . Prentice-Hall, Englewood, N.J.

Haralick, R. M., Sternberg, S., & Zhuang, X. (1987). Image analysis using mathematical morphology. *IEEE Transactions on Pattern Analysis and Machine Intelligence,* 9(4), 532-550.

Heijmans, H. J. A. M., & Ronse, C. (1990). The algebraic basis of mathematical morphology. *Computer Vision, Graphics, and Image Processing,* 50, 245-295.

Keppel, E. (1975). Approximating complex surfaces by triangularization of contour lines. *IBM Journal of Research and Developments,* 19, 2-11.

Maragos, P., & Schafer, R. W. (1986). Morphological skeleton representation and coding of binary images. *IEEE Transactions on Acoustics, Speech, and Signal Processing,* 34(5), 1228-1244.

Pavlidis, T. (1982). *Algorithms for graphics and image processing.* Computer Science Press, Rockville, Maryland.

Ritter, G. X., Wilson, J. N., & Davidson, J. L. (1990). Image Algebra: An overview. *Computer Vision, Graphics, and Image Processing,* 49, 297-331.

Serra, J. (1982). *Image analysis and mathematical morphology.* Academic Press, Orlando, Florida.

Sinclair, B., Hannam, A. G., Lowe, A. A., & Wood, W. W. (1989). Complex contour organization for surface reconstruction. *Computers & Graphics,* 13(3), 311-319.

Tipper, J. C. (1976). The study of geological objects in three dimensions by the computerized reconstruction of serial sections. *Journal of Geology,* 84, 476-484.

Tipper, J. C. (1977). A method and Fortran program for the computerized reconstruction of three-dimensional objects from serial sections. *Computers and Geosciences*, 3, 579-599.

Zyda, M. J., Jones, A. R., & Hogan, P. G. (1987). Surface construction from planar contours. *Computers & Graphics*, 11(4), 393-408.

COMPUTER GRAPHICS TECHNIQUES TO REPRESENT
VOLCANIC HAZARD AND RISK

M.T. PARESCHI

IBM Scientific Center, Via Santa Maria 67, I-56100 Pisa, Italy.

ABSTRACT

Physical behavior of volcanic eruptions can be numerically simulated to build hazard and risk maps. To fully exploit model results and adequately use them to hazard and risk presentation, computer graphics techniques can be profitably utilized. Computer-built shaded images and satellite scenes are used as background for numerical result representation. Moreover both the cases of static results and time-dependent data have been considered. In the last case, dynamic animation has been adopted to show system evolution. Applications are presented for tephra fallout risk, lava flow hazard and lahar flowing.

KEYWORDS

Volcanic eruptions; hazard and risk maps; digital terrain models; model simulation.

INTRODUCTION

Concurrent with the rapid progress in the realms of workstations, expansive efforts are also being made worldwide to develop visual methods of present computer scientific outputs. Initially there was the simple method to projecting the mesh covering the surface of simple three dimensional objects, so-called wire-frame modelling, followed by the surface or solid modelling which expresses more realistic images of objects. Presentation of outputs of numerical models simulating volcanic eruptions and compilation of hazard and risk maps based on numerical approach can gain significant advantages from the actual computer graphics state of art.

Volcanic eruptions are among the most spectacular and widely feared natural disasters. In this century, they have killed more than 70,000 people and caused billion dollars in property damage. Damages are not only related to the size or violence of the eruption, but also to the proximity to the volcano of towns and productive resources. Even a small eruption in a highly populated region may cause the death of thousands of people. The evaluation of risk related to volcanic eruptions is an important goal of modern volcanology. From a quantitative point of view, risk is defined as the product of value (the number of human lives at stake, capital value, or productive capacity), vulnerability (the fraction of the value which is likely to be lost as a result of a given event) and hazard (the probability that a particular area will be affected by a destructive eruption in a given time interval). The compilation of an hazard and then of a risk map is based on the study of the past behavior of a volcano, on the

Pareschi, M.T. & Macedonio, G., (1990).   An algorithm for the triangulation of
arbitrarily distributed points.   Applications to volume estimate and terrain
fitting, submitted to *Computer & Geoscience*.

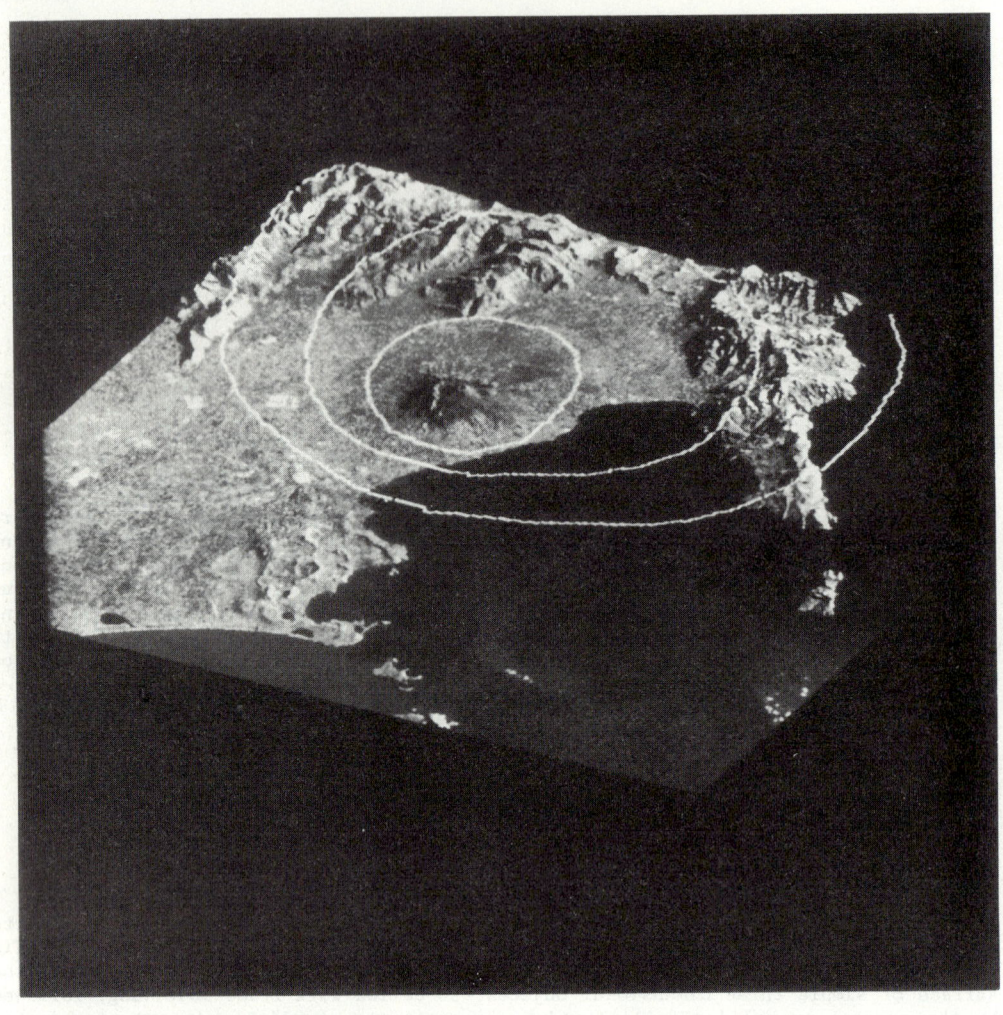

Fig.1 Perspective view of Mt.Vesuvio (Italy), obtained
superimposing a Landsat image to a digitalized
topography. Yellow lines refer to different hazard
values for an hypothetical tephra fallout eruption.

development of mathematical models able to simulate eruptible typologies and on the upgrading of a data base about the actual conditions of the volcano and its surrounding areas. The final goal, in addition and parallel to an improvement of the scientific knowledge, is to provide safety responsible officials with the necessary information during impending volcanic crises and for plans of urban settlement developments. Being these the aims, hazard and risk maps should be very legible, that is quite different information, such as buildings, power plants, factories, agricultural lands, etc., exposed to some consequences of an eruption, and a quantification of the hazard have to be kept together and simply and efficaciously displayed.

RESULT PRESENTATION

The fist step toward the quantification of hazard and risk is the modelling of volcanic eruptions. To quantify volcanic hazard, the approach is based on the attempt to formulate equations (balance of mass, of momentum, of energy, laws of state) simulating eruption behavior. The system of equations is then solved via numerical methods. Such an approach, if correctly performed, guarantees <u>rapidity</u> and <u>flexibility</u> (by changing the input parameters according to new data from volcanic surveillance, the techniques of mapping adequately provide new information). As expected, the reliability of the results depends on the capability of the models to adequately describe the physical and chemical behavior of the volcanic phenomena and also on the correct choice of the input parameters. This can be investigated by testing and validating models on past eruptions. The second step is the analysis of volcanic areas under investigation and the combination of this information with hazard from numerical models. Information needed for computation and presentation includes:
1) remotely sensed images in RGB and IHS;
2) classified remotely sensed images (in order to outline inhabited areas, agricultural lands, etc.);
3) elevation and moments of the distribution of elevation (DEM) of volcanic areas, computed by triangulation techniques or other methods (Pareschi, 1987; Macedonio and Pareschi, 1990).
DEMs are indispensable to physical numerical models simulating topography controlled phenomena. Moreover they allow to display multiple images using 3D oblique views. Landsat data, shading images, segmented images, etc. are used as color source and DEM matrix is used to support a 3D projection processing (Pareschi and Bernstein, 1989). By varying the zenith and the azimuth angles, different perspective views of the same area can be computed and displayed; by changing the vertical scaling factor, an exaggeration of the elevation can be obtained in order to enhance ground features (Fig. 1).

To analyze and exploit the large amount of output data from models, it is very important to display the results in a visual (or graphical) manner. The fastest and easiest way to achieve a visualization goal is to use satellite image data. From the model results, a set of lines of equal deposition values can be derived and then displayed and superimposed onto a satellite image (Fig. 1). This method provides a realistic presentation of complex model results. The relative positions of inhabited areas and zones covered by volcanic products can be rapidly shown. Another possibility is to display hazard values (for example with different colors) only in given regions (as resulting from segmentation of Landsat images). For example, to quantify hazard related to roof collapse, the probabilities of accumulation of tephra fallout can be depicted only in the areas of interest (inhabited ones) as resulting from segmentation of Landsat images.

Compilation of hazard maps has been realized for tephra fallout at Vesuvius (Barberi *et al.*, 1990a) and lava flows at Mt.Etna (Barberi *et al.*, 1990b).

For Vesuvius, on the basis of ten years wind statistics, the different occurrences of accumulation of tephra on ground have been computed according to a model for transport-dispersion of volcanic particles (Armienti *et al.*, 1988; Macedonio *et al.*, 1988). The size of the hypothesized eruption has been inferred from the past behavior of

Fig. 2 Risk map for roof collapse at Vesuvio as obtained
by an hazard map for concentrations on ground greater
than $200 kg/m^2$ (after Barberi et al., 1990a). Grey
corresponds to zero risk, black to 1-5, yellow to
5-10, red to 10-20, violet to 20-100.

Fig.3 Hazard map for lava invasion from a subsummital
crater at Mt.Etna (Sicily).

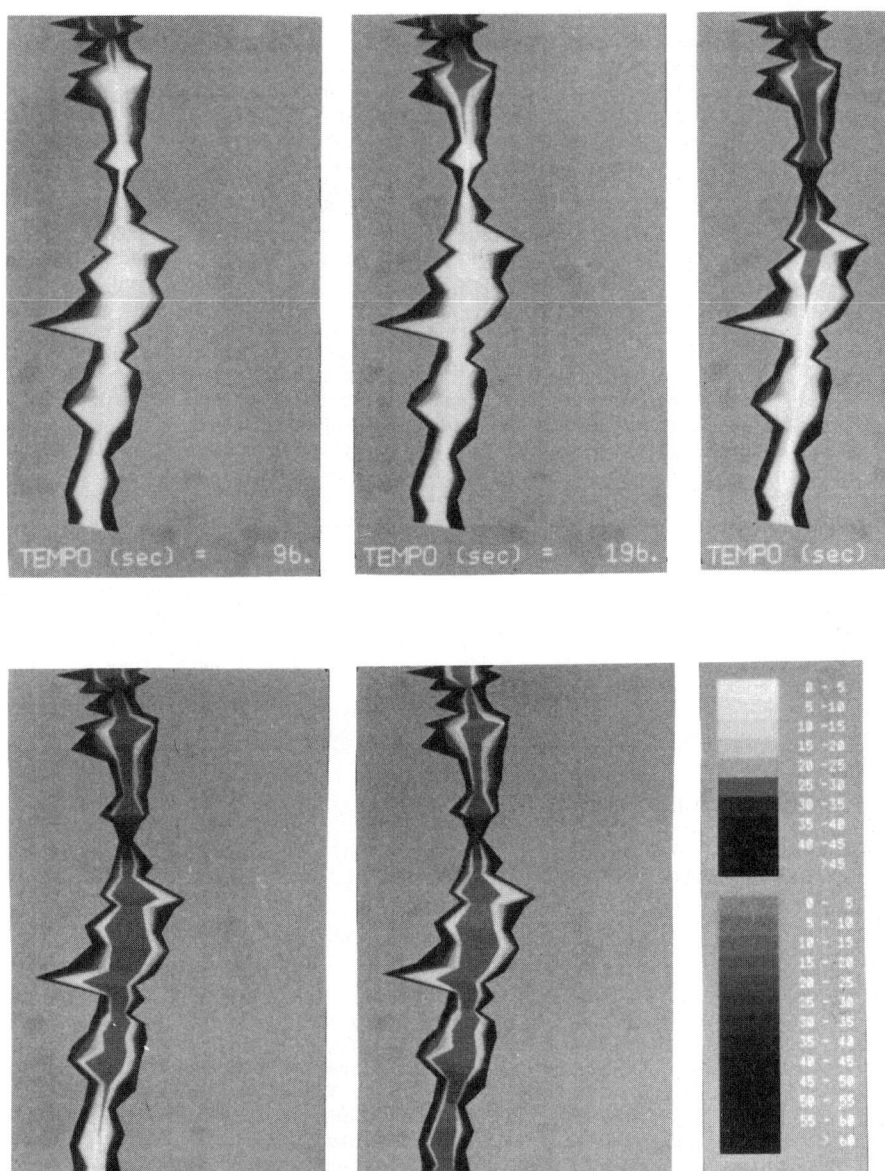

Fig.4 Lahar evolution along a meandering valley.
Coordinates across channel are magnified 10 times
respect to that along flow.

Vesuvius, and from its actual status. From hazard map, an example of risk map for roof collapse under the load of fall deposit has been then derived. The density of urban settlement (moving average of inhabited areas on 600x600 squares from a Landsat image) has been assumed as representative of the value and a weight greater than 200 kg/m^2 has been considered enough to provoke roof collapse (vulnerability=1). In each point, the product of the value times probability of having more than $200 kg/m^2$ of tephra fall deposit allows the establishment of risk maps (Fig. 2).

For Mt.Etna a digitalized topography has been obtained by the Delaunay triangulation of a set of about 100,000 points speeded along isolines and covering a surface of 2300 km^2. This is the basis for a roughly evaluation of lava flow path, based on maximum slope. Perturbation of elevations (simulating uncertainties on quotes and lava self adjusting topography) and vent location (according to a distribution deduced by past vent positions) allow the reconstruction of preferred paths and then of an hazard map. In Fig.3 the invasion area of lava flowing from a sub-summital crater is present. Green area corresponds to probabilities till 1%, yellow till 10% and red in the range 10 -100%. The display has been obtained by a package (Pickover, 1990) allowing control of lighting, surface shading and object rotation. The input to figure is a convenient computer graphics metafile which specifies the necessary information needed to produce a figure. For example in the case of Fig.3 the input could be a regular matrix in ZRGB format (elevation plus RGB color information) or a set of triangles characterized by the spatial coordinates of their vertexes and by a color.

Time dependent results can be visualized on topographic background as shown in Fig. 4, where the propagation of a lahar along a meandering channel is depicted (Macedonio *et al.*, 1990c). Different GRAPHIGS structures (IBM, 1989) corresponding to mud elevation along the valley are built and successively destroyed as mouse is clicked, producing animation. From numerical model results (muddy water elevation at each cross section), a set of triangles and quadrilaterals approximating terrain surface and flow invasion are computed. Color is used to give information on elevation. Flow behavior (initial crack, steep front propagation, spillways or gates effects, etc.) can be easily followed during time (Fig. 4).

The software runs on an IBM RISC 6000.

This research has been implemented in the scope of a joint project with the Dipartimento di Scienze della Terra of the University of Pisa.

REFERENCES

Armienti, P., Macedonio, G. & Pareschi, M.T., (1988). A numerical model for the simulation of tephra transport and deposition: applications to May 18 1980 Mt. St. Helens eruption, *Journal of Geophysical Research, 96, B6,* 6463-6476.

Barberi, F., Macedonio, G., Pareschi, M.T. & Santacroce, R., (1990a). Mapping the tephra fallout risk: an example from Vesuvius, Italy, *Nature, 344,* 142-144.

Barberi, F., Macedonio, G., Pareschi, M.T. & Santacroce, R., (1990b). Simple model for lava flow hazard assessment: Mt.Etna, *IAVCEI, Abstract, Mainz 1990.*

Barberi, F., Macedonio, G., Pareschi (1990c). Floods and lahars: equation of motion and numerical simulation, *IAVCEI, Abstracts, Mainz 1990.*

Pickover A. ., (1990). GALAXY, a versatile visualization program, *User manual, IBM T.J.Watson Research Center, Yorktown Heights, NY.*

IBM (1989). The Graphigs Programming Interface, Subroutine Reference, Version 2, Release 1.0, *SC33-8194-0.*

Macedonio, G., Pareschi, M.T. & Santacroce, R., (1988). A Numerical simulation of the plinian fall phase of 79 A.D. eruption of Vesuvius, *Journal of Geophysical Research, 93, B12,* 14817-14827.

Pareschi, M.T., (1987). Earth imaging and data processing for mapping and analysis, In: *Digital Signal Processing* (Cappellini, V. & Costantinides, A.G., Eds.), pp.916-921. Elsevier Science Publishers, North Holland.

Pareschi, M.T. & Bernstein, R. (1989). Modeling and image processing for visualization of volcanic mapping, *IBM Journal of Research and Development, 33, 4,* 1989.

3-D MODELING OF A COMPLEX FAULT PATTERN ON AN ENTRY LEVEL 2-D WORKSTATION

Mark VERSCHUREN

Renard Centre of Marine Geology, Gent University
Krijgslaan 281-S8, B-9000 GENT, BELGIUM

ABSTRACT

A programme is introduced that allows to model horizons cut by a complex system of normal faults, on an entry level 2-D workstation, but with sophisticated 3-D control and visualization.

First, polygonal approximation reduces continuously digitized reflector lines to more homogeneously distributed reflector points. Next, an automatic Delaunay triangulation of these most relevant points is edited interactively. Colour code contoured and shaded triangles are redrawn as soon as they are altered, or selected as scarp triangles. A dense grid is initialised with the triangulation and iteratively smoothed to a minimum tension surface constrained by isolated reflector points, except for a narrow zone along fault scarps, where a high tension is applied. Thus, 'overshooting' can be avoided completely. This combination of tools allows effective geologic surface modeling and fault correlation.

Full visual 3-D inspection proceeds through hidden surface removal, with optionally colour coded and line contoured depth intervals, and Phong shading. Thus, any discontinuity can be realistically modeled and visualized in 3-D as an integral part of an otherwise smooth surface, on a platform that consists of nothing more than a 2-D user interface and an 8-bit frame buffer.

KEYWORDS

Horizon modeling; fault modeling; polygonal approximation; smoothing; splines in tension; 3-D visualization.

INTRODUCTION

High resolution reflection seismic profiling in the Southern Bight of the North Sea has revealed large-scale compaction faults confined to the Ypresian Clay (fig. 1a; Henriet *et.al.*, 1990). A small case-study area was covered with a dense pseudo-3-D seismic network, in order to build a 3-D model of the fault pattern, depicted in fig. 1b.

POLYGONAL APPROXIMATION

Continuous digitization of seismic profiles yields a vast amount of points along interpreted horizons and faults, far more than useful for surface reconstruction purposes (fig. 3). A much lower data redundancy follows from an often preferred, discrete, point-by-point digitization. However, this method is based on subjective and ever shifting selection criteria. Moreover, the required level of detail depends entirely on the eventual scale. We therefore developed a geometrical filter algorithm that selects points out of a polygonal line, based on a user-defineable condition of minimum local significance. Its principle is explained in fig. 2.

Depending on the filter setting and line smoothness, continuously digitized lines can be reduced to 5-10%, without losing essential detail. Such a reduction can dramatically speed up drawing, interactive modeling as well as gridding. Quite as expected, it turned out that routine reduction to 50% never entailed loss of meaningful detail for any purpose, so that data storage efficiency increases as well.

Compared with a host of other, published polygonal approximation algorithms, ours can be qualified as simple and fast. Moreover, extension to filtering of spacefilling polygonal lines (e.g. seismic reflectors along a curved profile) is straightforward, and increases processing time only marginally. The optimized algorithm will be published elsewhere.

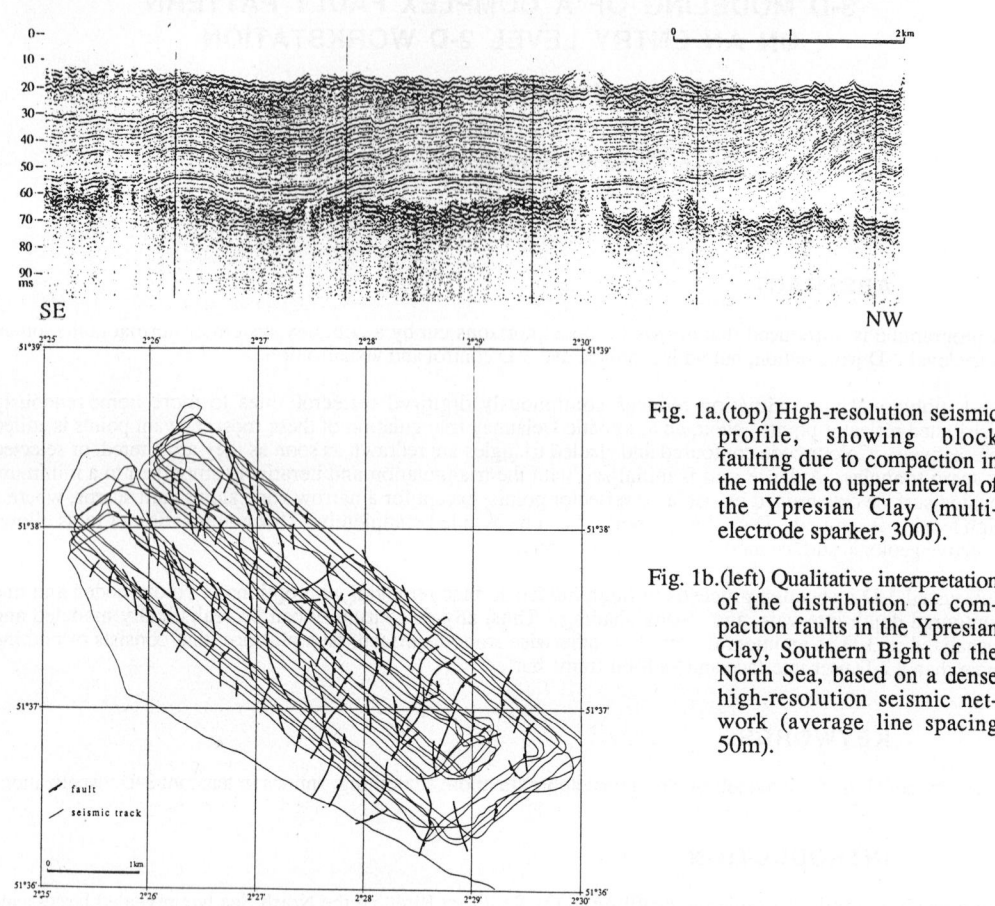

Fig. 1a.(top) High-resolution seismic profile, showing block faulting due to compaction in the middle to upper interval of the Ypresian Clay (multi-electrode sparker, 300J).

Fig. 1b.(left) Qualitative interpretation of the distribution of compaction faults in the Ypresian Clay, Southern Bight of the North Sea, based on a dense high-resolution seismic network (average line spacing 50m).

SURFACE MODELING

Modeling by Triangulation

Modeling and visualization of faults in 3-D is fairly tricky on a simple 2-D workstation. First, a surface needs to be constructed that does not smooth out every discontinuity. For most general surface interpolation techniques, discontinuities such as normal faults or escarpments are merely a nuisance. Pure triangulation can preserve discontinuities. Hidden surface removal on a large, irregular triangulation is not practical on a basic 2-D workstation, however. Moreover, projected triangles are hard to interpret three-dimensionally, since they lack perspective. A combination of triangulation, offering discontinuity preservation, and gridding, offering perspective and straightforward hidden surface removal with a painter's algorithm, is therefore the preferred combination.

In practice, the automatic triangulation (e.g. Akima, 1978a, b) needs to be edited near the discontinuities by pointing to the sides of triangles that need to be swapped. Points along fault scarps are thereby effectively connected. Certain edges are not allowed to be swapped, such as the outer edges of boundary triangles and the common edge of two triangles that form a concave polygon. In the latter case, a swap would otherwise result in a locally multivalued triangulation. During this surface modeling phase, some triangles are interactively identified as 'scarp'-triangles, if their edges delineate any relevant surface discontinuity (fig. 3). Thin boundary triangles or badly constrained internal triangles can be deleted as well.

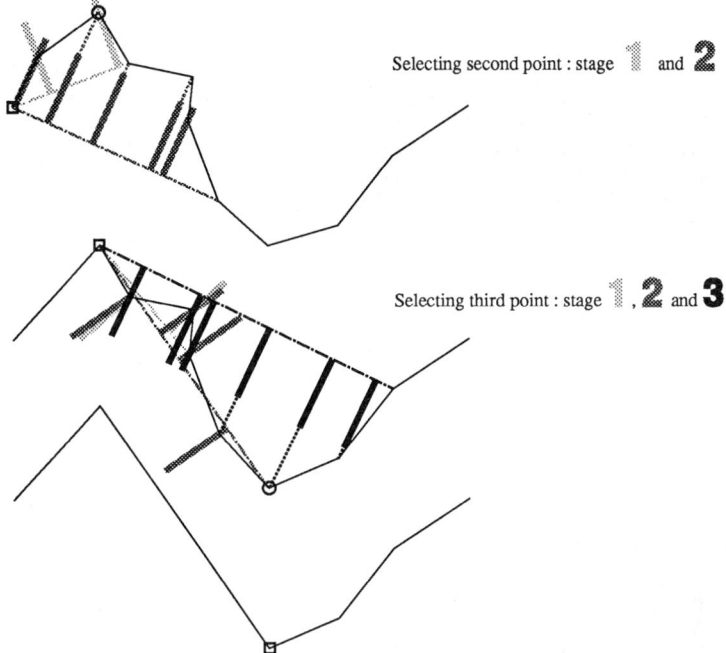

Selecting second point : stage 1 and **2**

Selecting third point : stage 1 , **2** and **3**

Fig. 2. Principle of polygonal approximation algorithm.
A polygonal approximation of a line (thin black solid) is found by selecting points on a user-defineable condition of minimum local significance. Starting at one of the ends (square in uppermost drawing), a local reference line (thin dash-dotted) reaches in stages farther and farther along the polygonal line. Each time, the distances to all of the intermediate points are calculated. As soon as the maximum distance is found to exceed a user-defineable measure (thick solid line), the corresponding vertex is considered to be locally significant (circle) and is selected as the next starting point.

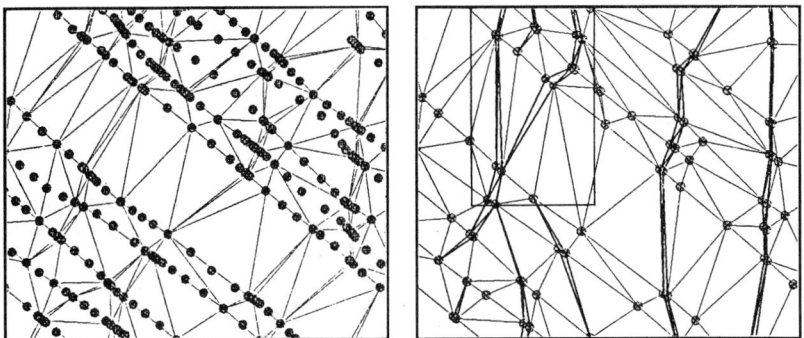

Fig. 3.   (left) Part of the automatic Delaunay triangulation of reflector points, selected by polygonal approximation of digitized reflector lines along the seismic network, shown in Fig. 1.
(right) Same part of interactively edited triangulation. Several triangle edges needed to be swapped, fault triangles have been identified (thick lines) and badly constrained triangles have been deleted. Sector (box) delineates part of the triangulation used to initialize the grids of Fig. 4.

Gridding with splines in variable tension

In order to preserve small scarps, the resulting triangulation is densely gridded. Gridnodes not covered by triangles are initialised with a distance weighted average of nearby data points. This also provides a more general initialisation procedure for fast contouring of sparse seismic networks or well data. Subsequent smoothing iteratively reduces curvature in the surface, so that the underlying triangulation does not show up as contour breaks. The popular 'minimum-curvature' method proposed by Briggs (1974) is applied to this end. The scope of Briggs' difference equations has been found to be insufficient to reduce tension completely in badly constrained parts of a dense grid, however. The difference equation for the normal gridnode away from the edges, has therefore been extended to include 33 instead of 13 nodes. With $u_{i,j}$ the value of a grid node in row i and column j, it becomes

$$468u_{i,j} - 152(u_{i+1,j}+u_{i-1,j}+u_{i,j+1}+u_{i,j-1})+ 32(u_{i+1,j+1}+u_{i+1,j-1}+u_{i-1,j+1}+u_{i-1,j-1})- 24(u_{i+2,j}+u_{i-2,j}+u_{i,j+2}+u_{i,j-2})$$
$$+ 8(u_{i+1,j+2}+u_{i+1,j-2}+u_{i-1,j+2}+u_{i-1,j-2}+u_{i+2,j+1}+u_{i+2,j-1}+u_{i-2,j+1}+u_{i-2,j-1})$$
$$+ 2(u_{i+2,j+2}+u_{i+2,j-2}+u_{i-2,j+2}+u_{i-2,j-2}) + 8(u_{i+3,j}+u_{i-3,j}+u_{i,j+3}+u_{i,j-3}) + u_{i+4,j}+u_{i-4,j}+u_{i,j+4}+u_{i,j-4} = 0.$$

In combination with a multiple grid strategy, successive overrelaxation and convergence detection (Smith and Wessel, 1990), the more elaborate equation actually entails faster convergence to a smoother surface.

Smith and Wessel (1990) introduced gridding with splines in tension, a generalization of the minimum curvature method, in order to reduce unwanted oscillations near steep dip changes in badly constrained parts of the interpolant. In order to avoid 'overshooting', tension near scarps can be chosen to be maximal (T=1 in Smith and Wessel's notation), while at the same time global tension can be minimal (i.e. T=0). During each stage of grid refinement, a gridnode is considered to be near a scarp, if the difference equation would incorporate nodes from across the scarp. Grid nodes coincident with scarp-triangle edges or isolated data points, remain untouched and constrain the smoothing calculations.

The *single* resulting grid is therefore only locally discontinuous, and at the same time smoothly honours the data points. Furthermore, this smoothing scheme with variable tension provides for faster convergence than with either high or low tension applied to the whole grid : low tension smoothing spends time to converge on 'overshoots', while high tension smoothing tends to draw the interpolant away from fixed data points, which end up peaking above or below the interpolated surface (fig. 4).

Discussion

Zoraster and Ebisch (1990) described another method to incorporate fault geometry into a horizon model. Their method requires the fault geometry (pattern, local dip and throw) to be interpreted and digitized separately for each horizon in a specific way. Any realistic normal fault geometry is bound to be based on continuous (seismic) profiles, on which both the up- and downthrown horizon ends can be located precisely. Since these two points are situated simultaneously on a fault and a horizon, it is possible and more convenient to model the fault system and the affected horizon *together*. This is precisely what the programme presented in this paper allows the geologist to do. If need be, the fault geometry can still be extracted as the set of all fault triangles. Zoraster and Ebisch's method to produce cross-fault structural continuation can also be incorporated in a straightforward way.

3-D VISUALIZATION

The quality of geological surface models critically depends on the detection of flaws and a spacial understanding of the structures involved. It was felt that high 3-D visual realism would greatly improve both. Figure 5 should also make the point that solid, colour- and line contoured, and above all realistically shaded surfaces can more directly convey far more structural information than a conventional line-drawn map or 3-D

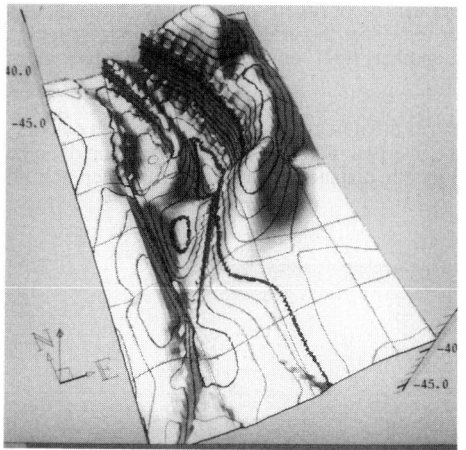

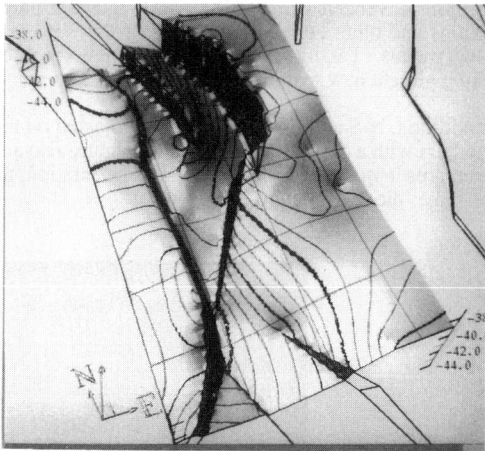

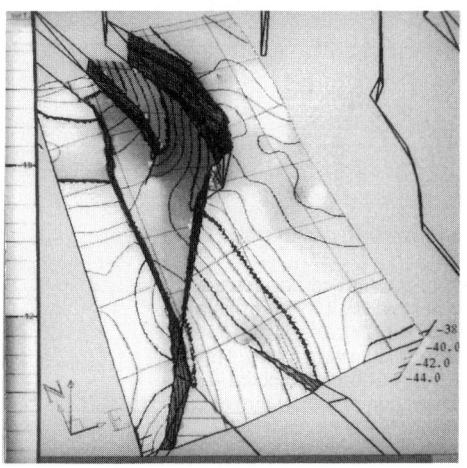

Fig. 4. Three iterative smoothing methods on the same grid, initialized with part of the triangulation shown in fig. 3, and with nodes fixed at isolated data points and at fault scarps. (top left) Global tension T=0 (equivalent with Briggs, 1974) leads to large 'overshoots' near scarps; (top right) With global tension T=0.75 (advocated by Smith and Wessel (1990) for topographic data) overshoots are far less prominent, but the interpolant peaks near isolated data; (left) With global tension T=0, and T=1 near scarps, the interpolant shows no un-wanted oscillations and smoothly honours the data at the same time.

picture. 3-D visual realism comes about when an image offers one or more of the following depth cues (Foley and Van Dam, 1982; Flynn, 1990; ordered with increasing soft- and hardware sophistication):

- perspective;
- solid surfaces through hidden surface removal (e.g. with a painter's algorithm);
- colour and line contours emphasizing height;
- smooth shading (e.g. using Phong's shading model), except for selected discontinuities;
- texture (e.g. 'smoothness' in Phong's shading model);
- stereopsis;
- shadows;
- transparencies;
- real-time movement.

The first five of these depth cues have been implemented in *Geofox*, on an entry-level 2-D workstation (a Sun-386i with an 8-bit frame buffer), without the graphical hardware or software support which is standard on more advanced workstations. All relevant parameters can be changed interactively, through sliders, buttons and other convenient means of the SunView user interface, in various pop-up windows and pull-down menus. Taking another 'point of view' or throwing another 'light' on the same grid or stack of grids, may highlight new features or flaws.

In addition, N-S and E-W guide-lines projected *on* the surfaces at known intervals, together with colours and contours with a definite height or depth value, also allow the interpretation and spacial correlation of complex structures in *quantitative* terms such as orientation, throw, length, distance, area and height difference, all on the same 'nice 3-D picture'.

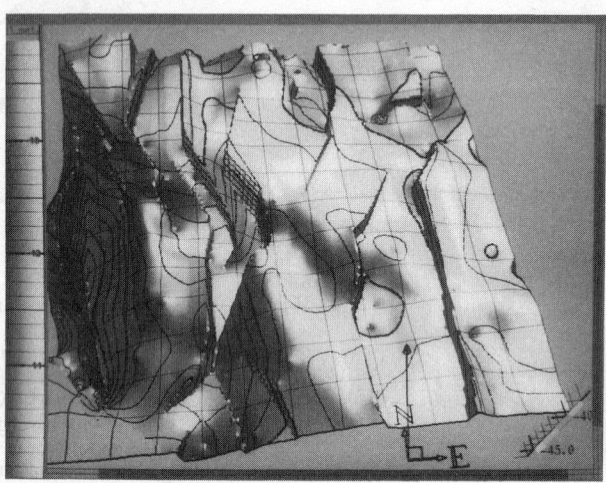

Fig. 5. Phong shaded 3-D view of one of the faulted horizons in fig. 1a. The horizon and the normal fault system have been quantitatively modeled *together*, with one triangulation and one grid. Colour intervals every 1ms TWT and NS/EW guide-lines every 100m, projected *on* the grid, allow quantitative interpretation of the 3-D image.

Acknowledgements

The 3-D analysis of clay deformations and the development of modeling and visualization software is part of project TH/06036/87 of the Commission of European Communities (Hydrocarbons). This article presents part of the author's Ph.D. work, funded by the National Fund for Scientific Research.

References

Akima, H. (1978a). A method of bivariate interpolation and smooth surface fitting for irregularly distributed data points. *ACM Trans. Math. Softw.* 4, 2, 148-159.
Akima, H. (1978b). Algorithm 526. *ACM Trans. Math. Softw.* 4, 2, 160-164.
Briggs, I. (1974). Machine contouring using minimum curvature. *Geophysics*, 39, 1, 39-48.
Flynn, J.J. (1990). 3-D Computing Geosciences Update. *Geobyte*, 5, 2, 30-36.
Foley, J.D. and A. Van Dam (1982). *Fundamentals of Interactive Computer Graphics*. Addison-Wesley, Reading, Massachusetts.
Henriet, J.P., M. De Batist, and M. Verschuren, (1990). Field evidence of early fracturation of potential source rocks. In: *Proceedings of the First EAPG Conference, 1989*, in press.
Smith, W.H.F. and P. Wessel, (1990). Gridding with continuous curvature splines in tension. *Geophysics*, 55, 3, 293-305.
Zoraster, S. and K. Ebisch, (1990). Incorporating fault geometry into geologic horizon models. *Geobyte*, 5, 2, 30-36.

COMPUTER MODELLING OF SURFACES: STRUCTURAL GEOLOGY APPLICATIONS

M.BARCHI[1], C. FEDERICO[1], F. GUZZETTI[2] and G. MINELLI[1]

1-Dipartimento di Scienze della Terra, Università degli
Studi, P.za dell'Università, 06100 Perugia, Italy
2-C.N.R. - Istituto di Ricerca per la Protezione
Idrogeologica nell'Italia centrale,
V. Madonna Alta 126, 06100 Perugia, Italy

ABSTRACT

Map analysis techniques are an important tool for a structural geologist. We explore
the possibility of using these techniques to reconstruct the geometry of complex
geological surfaces, such as thrust planes and deformed stratigraphic boundaries. We
suggest that this procedure can be used to extract more information from existing
geological maps, as well as to verify in three dimensions the geometry of structures
during the map-making process.

INTRODUCTION

Geologists work every day in a three-dimensional world. Since it is difficult to
visualize three dimensions graphically, the Earth scientists generally use two-
dimensional tools, such as maps and cross-sections, to project three dimensional
information onto a more manageable two-dimensional space (Jones and Leonard, 1990).
Several different kinds of geological, geochemical, hydrological and geophysical
data can be displayed in this way. A comprehensive review of the problem encountered
in these fields can be found in Jones et al. (1986). In the last decade increasing
computer advances have made it possible to develop applications that create, display
and operate on databases which fully describe the three dimensional geometry and
attributes of geological objects (Flynn, 1990). Different three-dimensional spatial
representation models (e.g. voxels, G-octrees, isosurfaces, nurbs, etc.) have been
proposed and tested (Fried and Leonard, 1990). Unfortunately most of these
applications, developed chiefly for mining and petroleum explorations, are very
expensive and still out of reach for 'the rest of us' in the geology and geography
departments.

Our work is concerned with the application of standard, all-purpose map analysis
techniques (Davis, 1986). We reconstruct and visualize the geometry of complex
geological surfaces, from data easily derived from existing geological maps. We
focus in particular on the solution of structural geology problems, attempting to
reconstruct the geometry of thrusts and deformed stratigraphic horizons. In general
our method requires the studied surface to have an extensive and articulated
intersection with the topography. It is therefore applicable to any low-angle
structure, stratigraphic or tectonic.

To carry out our experiment we used two software packages available at the C.N.R.
Istituto di Geologia Marina in Bologna (Italy): DIGMAP-DATUM-PLOTMAP and SURFACE II.
The first package is a series of programs that allow the digitization,
transformation of coordinates, preparation and plotting of high quality topographic
maps as well as simple geological maps (Bortoluzzi and Ligi, 1986; Ligi and

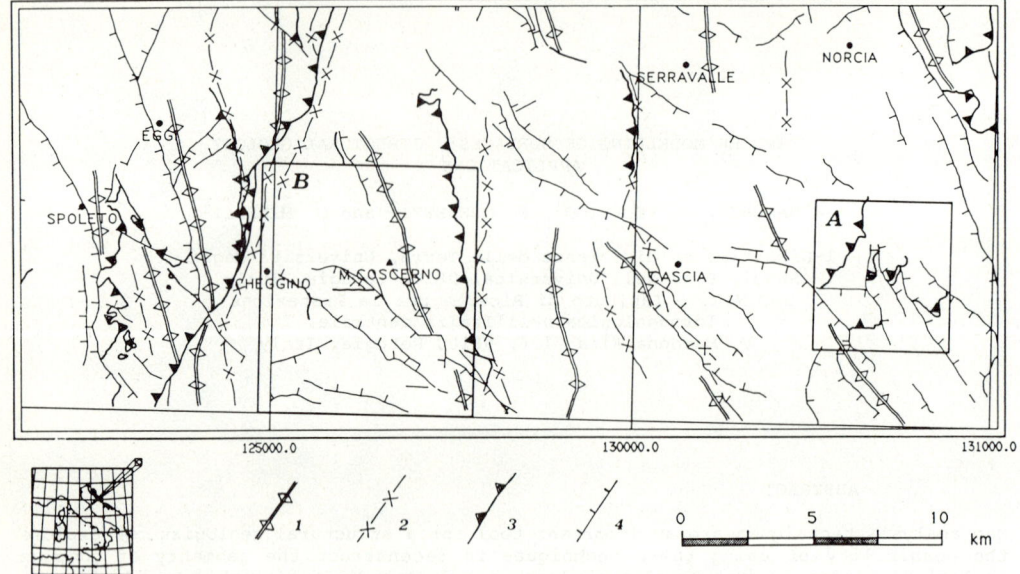

Fig.1. Location map showing the main structural features of the studied areas. A- Monte Pozzoni area; B – Monte Coscerno area; 1- anticline; 2- sinclines; 3- thrusts; 4 – normal faults.

Bortoluzzi, 1989). The second is a well known commercial package for the creation and display of spatially distributed data (Sampson, 1975).

DATA PREPARATION

The first step in the data preparation procedure consists in simplifying an existing geological map. For our experiments we use recent, detailed geological maps, at a scale of 1:25,000, that clearly portray all the structural features. These maps are to be simplified to produce schematic maps, specifically aimed for use in the study of the problem at hand. As an example, in the case of the reconstruction of a thrust surface, the schematic structural map shows only the trace of the major thrust and eventually of later faults that significantly affect its geometry. All the intermediate slices, eventually present between the hanging wall and the foot wall, are generally considered to be part of the overthrusted block. The formation boundaries and the minor faults, that are not essential for the reconstruction, do not appear on the schematic map.

During the second step of data preparation we intersect structure and topography, to obtain a population of scattered elevation points. By laying the schematic structural map on top of the topographic map, we determine the elevation of the studied structure at the intersection with each contour line. Scattered elevation points so obtained are then digitized, with the aid of a large format tablet.
It should be clear that the data preparation procedure is somewhat subjective and not at all automatic. To avoid the risk of introducing substantial biases, as well as to improve the quality and quantity of input data, the operation should be carried out by a geologist with a good knowledge of both the regional and local geology.

GRIDDING AND GRAPHICAL DISPLAY

Scattered elevation points are used to evaluate an elevation matrix (grid), i.e. a set of z-values arranged in a regular, rectangular or square, pattern. This grid will then be used to prepare conventional contour maps and perspective block diagrams of the surfaces under investigation. Several gridding algorithms are available to interpolate the value at each node of a regular grid from a variable number of scattered elevation points. The choice of the gridding algorithm is crucial to obtain correct results. In the choice of such an algorithm it is important to consider: the size of the grid matrix, i.e. the number of grid nodes or their spacing; the procedure to search for sample data points around each node; the interpolating algorithm and the weighting function to be used in the averaging process.

The size of the elevation matrix is to be kept consistent with the number of the original scattered elevation points. When graphically displayed, a grid that contains too many elements will give a misleading wiggling appearance of the studied surface. On the other hand a grid that is too coarse will produce a sharp and unrealistic image of the same surface.

The parameters that control the searching procedure specify the number, position and distance of the sample data points used to estimate the elements of the grid. It is advisable to adopt severe searching criteria to limit the unrealistic extrapolations in areas where only minimal information is available. We have generally preferred an octant search procedure, that divides the area around each grid node into eight equal sectors, forcing the estimating algorithm to use sample data points radially distributed around each grid node. Compared with other procedures, such as the nearest neighbour search, octants give a somewhat smoother surface.

Different interpolation algorithms may be applied to different geological problems. Some algorithms make use of the weighted average of data, with weights based on distance; others, more complex, calculate values through fitted, simple, low-order functions, commonly planes (Jones, 1989). We used an algorithm that estimates the elevation value at each grid node as the distance weighted average of the projection of the surface dip, computed at nearby scattered elevation points (Davis, 1986). The weighting function was assumed to be inversely proportional to the square of the distance. The choice of such an algorithm is consistent with the problem at hand, where the estimation of the dip of the geological surface is of particular interest in the reconstruction of its geometry.

It is worth pointing out that we have used all-purpose gridding algorithms, that don't have any built-in specific "knowledge" of the type of surface to be investigated (e.g. stratigraphic framework, geological history, style of deformation etc.). The elevation matrix obtained by gridding the sample elevation points can then be displayed in two dimensions by preparing contour maps (Figs. 2, 4). For a valid contour the density of elevation points must be sufficiently high, showing spatial persistence (Sharp, 1987). Contour lines are drawn using a simple spline interpolator, to avoid excessive and unrealistic angularity.

An alternative way of displaying the elevation matrix is to prepare perspective block diagrams (Figs. 3, 5). The ease with which the observation point (azimuth and elevation), the viewing distance and the vertical scale can be changed allows the production of several different synoptic views. These are very useful to gain a more effective and immediate understanding of the geometry of the studied surface. To obtain an overall view of the surface, eliminating short term variations, a smoothing algorithm may also be applied to the elevation matrix. We have successfully used an algorithm that re-evaluates the elevation value at each grid node as the distance-weighted average of the two nearby grid values (Barchi et al., 1989).

GEOLOGICAL EXAMPLES

We present the results of two experiments carried out in the Umbria-Marche Apennines of Central Italy, a thrust and fold belt, Upper Miocene-Lower Pliocene in age. In this area the deformed Mesozoic-Cenozoic sedimentary cover consists of two litho-structural units: a Lower Liassic carbonate platform, and a Middle Liassic-Lower

Miocene pelagic multilayer. The basal décollement is generally supposed to correspond to a thick sequence of Triassic Evaporites. Folds and thrusts are displaced by later normal faults, Upper Pliocene-Quaternary in age. A comprehensive overview and an extensive reference list of the Umbria-Marche geology has been recently published by Bally et al. (1986).

The two reported examples are located in the Southern part of the Umbria-Marche Apennines (Fig. 1). The first example, in the Monte Pozzoni area, concerns the reconstruction of the geometry of a low-angle thrust surface; whereas the second, in the Monte Coscerno area, regards the reconstruction of the geometry of a deformed stratigraphic boundary.

The Monte Pozzoni thrust

The Monte Pozzoni thrust consists of a series of small klippen of Liassic rocks, overthrusting a previously folded Mesozoic-Cenozoic multilayer, and displaced by a set of NW-SE trending normal faults. This structural setting produces a flat geometry of the thrust, and consequently an extensive intersection with the topographic surface.

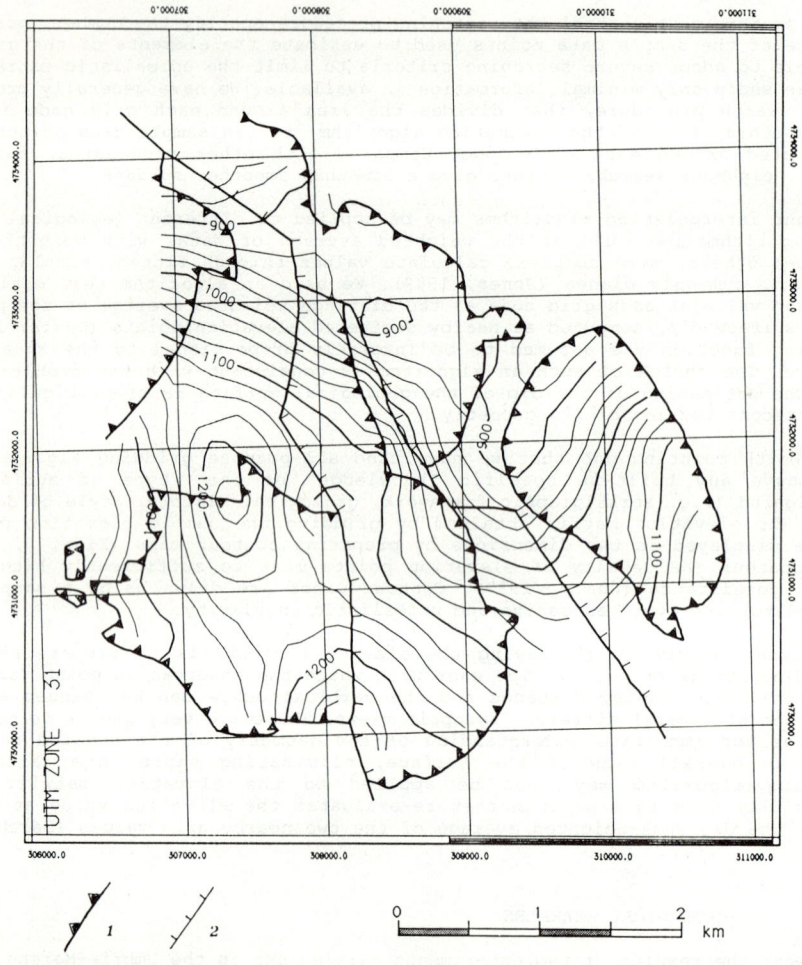

Fig.2.   Contour map of the Monte Pozzoni thrust.
         1 - thrust; 2- normal fault.

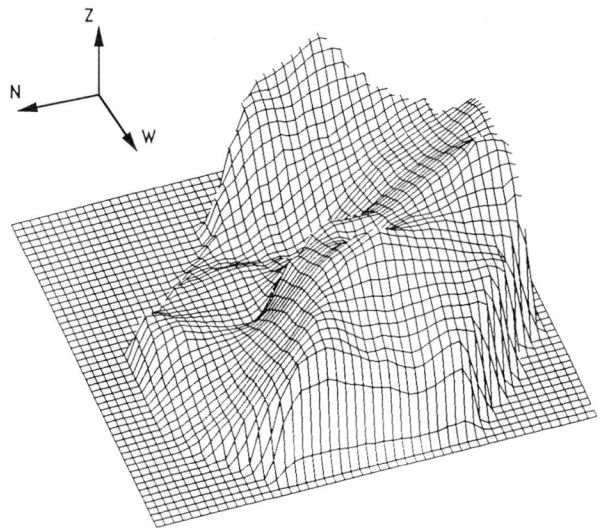

Fig.3. Perspective block diagram of the Monte Pozzoni
        thrust. The vertical exaggeration is 4, the
        azimuth of the observation point is N290°, the
        angle of the observation point is +45°.

The basis of our work was a detailed geological map published by Calamita *et al.*
(1981), from which we derived a schematic structural map, portraying two major
normal faults cutting the thrust surface. From the schematic map sample elevation
points were collected along the trace of the thrust surface, to prepare a 30x30
elevation matrix, that was then visualized through a contour map (Fig. 2) and a
perspective block diagram (Fig. 3). The modelled surface, according to the well
known deformation history of the region, was interpreted as a W dipping thrust cut
and displaced by a NE dipping normal fault. It is important to note that a similar
geometry can be interpreted also as a thrust surface folded along a NW-SE axis. In
this case the knowledge of the local geology is essential to discriminate in favour
of one of the two different interpretations.

We have successfully applied the same methodology to the reconstruction of other
low-angle thrusts in Central Italy. In particular we have analyzed the Spoleto
thrust, also shown in Fig. 1, and the Gran Sasso thrust system, located further to
the South. The obtained results and further discussion can be found in Barchi et
al., 1989.

The Val Casana graben (Monte Coscerno)

The Val Casana graben is a narrow and relatively deep extensional structure, which
cut across the N-S trending Monte Coscerno anticline. It is bounded by high-angle,
NE-SW trending normal faults, whose vertical displacement exceeds 500 m in the
central part of the graben, and quickly decreases along the strike. To understand
this rather complex structure, starting out from a detailed geological map prepared
by Barchi (1990), we reconstructed the top of the Maiolica Formation (Lower
Cretaceous), an easily traced stratigraphic marker.
The reconstruction was carried out collecting sample points in three different ways.
The elevation of the top boundary of the Maiolica Formation was measured directly on
the map. The elevation of other formation boundaries was corrected, adding or
subtracting the corresponding stratimetric distance from the studied horizon. This
very simple criterion is sufficiently accurate if bedding planes are not too steep

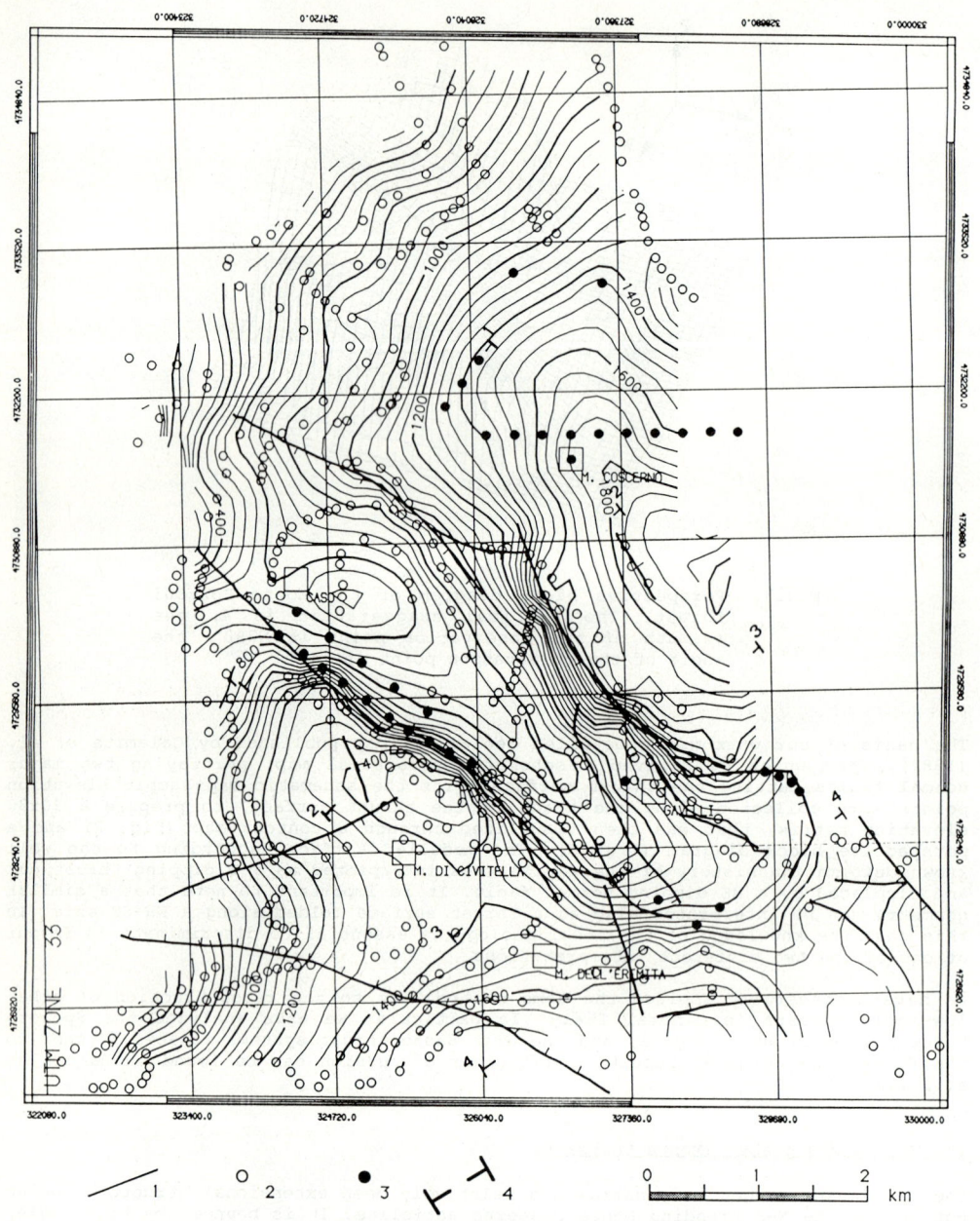

Fig.4.   Contour map of the top of the Maiolica Formation
         in the Monte Coscerno area.   1- normal fault; 2 -
         sample elevation point obtained from formation
         boundaries; 3- sample elevation points obtained
         from geological cross-sections; 4 traces of the
         sections of Fig. 6.

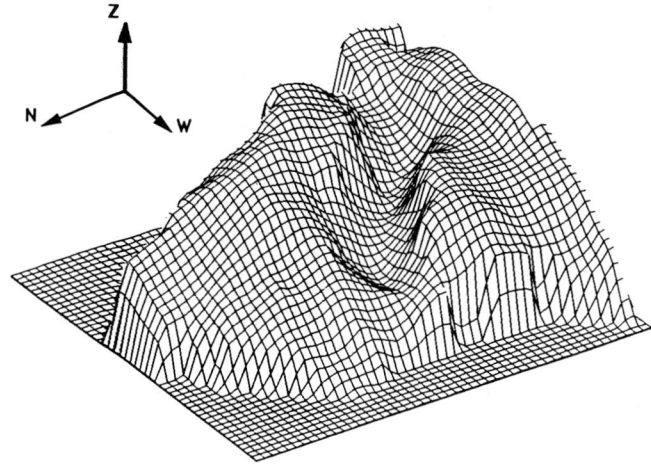

Fig.5.   Perspective block diagram of the Val Casana graben
in the Monte Coscerno area. The vertical
exaggeration is 3, the azimuth of the observation
point is N305°, the angle of the observation point
is +30°.

(if the dip is < 30° the error is < 15%). Finally, in the areas with a poor
distribution of elevation data, points were obtained from expressly prepared short
geological profiles.

The reconstructed surface is shown as a contour map in Fig. 4 and as a perspective
block diagram in Fig. 5. Both representations clearly show the Monte Coscerno
anticline, striking N-S, cut by the deep and narrow Val Casana graben. Note in Fig.
4 the good correspondence between the closely spaced contour lines and the location
of major normal faults. To check the accuracy of the results we also prepared four
profiles across the contour map and we compared them with the corresponding
geological cross-sections. The results, summarized in Fig. 6, show a good fit. The
somewhat smoother appearance of the estimated surface, in comparison with the
geological data, is due to the type of gridding algorithm and, locally, to the
distribution of sample elevation points.

In the Monte Pozzoni and Monte Coscerno areas the geological surfaces were
reconstructed using the entire set of available elevation data, not taking into
account the presence of any fault. To study the effects of these faults, a simple
method is to divide the elevation points into several sub-groups, one for each
faulted block, and treat them separately. The resulting contour maps indicate that
the extrapolation across major faults can produce local, unrealistic features that
do not relate to the actual geometry of the studied surface (Barchi et al., 1989,
Fig. 9). In the presented examples we were most interested in the reconstruction of
the overall geometry of the surfaces, and therefore we have disregarded all local
effects.

        CONCLUDING REMARKS

The starting point of this work was the observation that the information content of
a geological map is not readily and completely available. To analyze and fully
represent the geometry of complex structures, geologists usually build up two
orthogonal sets of balanced cross-sections, checking their reciprocal consistency.
This is a time consuming and rather difficult procedure. The method that we propose
is based on the possibility of rapidly extracting, from a geological map, elevation
data relative to structural surfaces such as thrusts and deformed stratigraphic

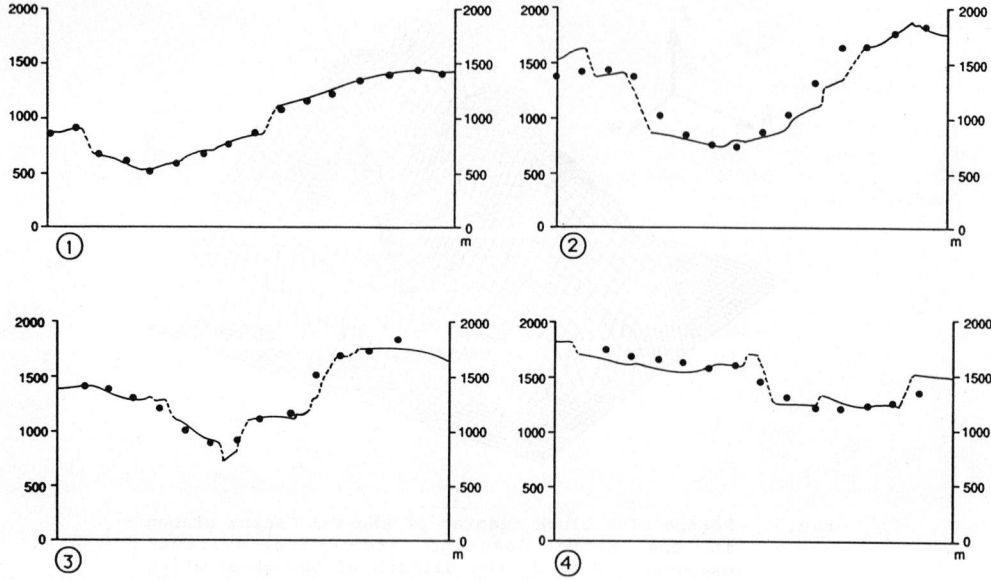

Fig.6. Schematic profiles across di Val Casana graben.
Solid line shows the trend of the top of the
Maiolica Formation derived from geological cross-
sections. Closed dots show the elevation of the
same horizon as obtained from the contour map of
Fig. 4. Horizontal scale equal to vertical scale.

boundaries. The geometry of these surfaces can then be reconstructed using simple, standard map analysis techniques.

With these techniques it is not possible to investigate surfaces that have more than one z value for any x-y location, such us recunbent folds. 3-D technology is rapidly evolving, and approaching towards the solution of these problems, but it requires laborious procedures for data acquisition, and hardware/software resources which are not widely available.

Finally, it must be stressed that modelling geological surfaces by computer should not become an automatic procedure. A good knowledge of the local and regional geology is essential to obtain realistic results.

Acknowledgements - We are particularly grateful to G. Bortoluzzi and M. Ligi of the C.N.R. Istituto per la Geologia Marina (Bologna) who supplied us with the graphical software. We also wish to thank P. Reichenbach for helping in various stages of the work. This work was supported by grants from the CNR-IRPI and CNR-GNDCI as well as from MURST (40% and 60%: U.O. Perugia, Responsabile G. Pialli) grants.

REFERENCES

Bally A.W., L. Burbi, C. Cooper and R. Ghelardoni (1986). Balanced sections and seismic reflection profiles across the Central Apennines. *Mem. Soc. Geol. It.*, 35, 257-310.
Barchi M. (1990). Analisi strutturale e cinematica del settore meridionale dell'Appennino umbro-marchigiano. - Ph.D. Thesis, University of Perugia, Italy.

Barchi, M., C. Federico and F. Guzzetti (1989). Reconstructing thrust geometry by map analysis techniques: three examples from the Central Apennines (Italy). In: *Atti del Convegno "La geologia strutturale ed i possibili contributi all'interpretazione della sismica profonda", Pisa.*

Bortoluzzi G. and M. Ligi (1986). DIGMAP: a computer program for accurate acquisition by digitizer of geographical coordinates from conformal projections. *Computers & Geosciences,* 12, 175-197.

Calamita F., G. Deiana and G. Pambianchi (1981). Considerazioni strutturali sull'area compresa tra la conca di Cascia e la valle del Tronto (Appennino Umbro-marchigiano meridionale). Problemi di raccorciamento e neotettonica. *Boll. Soc. Geol. It.,* 100, 415-422.

Davis J.C. (1986). Statistics and data analysis in geology. 2nd ed., John Wiley and Sons, New York, 646 pp.

Flynn J.J. (1990). 3-D computing geoscience update: hardware advances set the pace for software developers. *Geobytes,* 5, 33-36.

Fried C.C. and J.E. Leonard (1990) . Petroleum 3-D models come in many flavors. *Geobyte,* 5, 27-32.

Jones, T.A. (1989) - The three faces of geological computer contouring. *Mathem.Geol.,* 21, 271-283.

Jones T.A. and J.E. Leonard (1990) - Why 3-D modeling?. *Geobyte,* 5, 25-26.

Jones T.A., D.E. Hamilton and C.R. Johnson (1986). Contouring geologic surfaces with the computer. Van Nostrand Reinhold Co., New York, U.S.A., 314 pp.

Ligi M. and G. Bortoluzzi (1989). PLOTMAP: Geophysical and geological applications of good standard quality cartographic software. *Computers & Geosciences.* 15, 519-585.

Sampson R.J. (1975) - The SURFACE II graphic system., in: *Display and analysis of spatial data* (J.C. Davis and M.J. McCullagh, eds.), pp. 244-266. Wiley Intersciences, London.

Sharp W.E. (1987). Two basic rules for valid contouring. - *Geobyte,* 2, 11-15.

PRELIMINARY MASS-BALANCED 3-D RECONSTRUCTIONS OF THE ALPS AND SURROUNDING AREAS DURING THE MIOCENE

WILLIAM W. HAY[1,2], CHRISTOPHER N. WOLD[1,2] and JOHN M. HERZOG[1]

[1]Department of Geology and CIRES, Campus Box 216, University of Colorado, Boulder, CO 80309 USA
and
[2]GEOMAR, Wischhofstrasse 1-3, D-2300 Kiel 14, Germany

ABSTRACT

The masses of sediment in the Molasse Basins of Switzerland, Germany and Austria total 0.299 x 10^{21} g. In the Neogene, other sediment shed north of the Alps went to the Rhine Graben (0.03 x 10^{21} g), and on to the northern European plains and North Sea (0.81 x 10^{21} g). During the Quaternary, the drainage systems were drastically reorganized, with the Danube capturing streams that drain the German and Austrian Alps, and delivering the sediment to the Pannonian Basin (0.15 x 10^{21} g) and on to the Black Sea (0.31 x 10^{21} g). Sediment shed to the west has entered the Bresse and Rhone Grabens (0.05 x 10^{21} g) and been deposited in the Rhone Delta (0.19 x 10^{21} g) and on the floor of the Gulf of Lions (0.24 x 10^{21} g). Sediment shed from the south side of the Alps resides in the Po Basin and Adriatic Sea that have total sediment masses of 0.38 x 10^{21} g and 0.36 x 10^{21} g respectively. Mass balanced reconstructions suggest that during the Chattian and Aquitanian (Late Oligocene-Early Miocene) thrusting in Switzerland, southern Germany and western Austria (Vorarlberg-Tyrol) resulted in mountains of Himalayan height, with elevations up to 7 km. Erosion attacked these mountains, existing sediments in the Molasse basins alone are equivalent to a layer 0.5 km thick, but represent only a fraction of the material originally deposited in them. Sediments equivalent to another 1.7 km thickness occur in the more peripheral basins. Sediment supply was greatly reduced in the middle Early Miocene (Burdigalian) indicating that the elevation of the mountains had decreased more rapidly than can be explained by erosion alone, which, after isostatic adjustment would have reduced the elevations only to about 5 km. The loss of elevation might have been due to a slow (ca. 1 my) flexural response to the narrow load on the southern edge of the European continental block, or to physical collapse of the mountains.

A second phase of orogeny occurred during the Helvetian (late Early Miocene), loading additional thrust sheets onto the southern margin of the European continental block in southern Germany and throughout Austria. These mountains were not as high as those that had been produced to the west during the earlier orogenic phase, averaging about 3 km. Again, sedimentary mass-balance suggests that although a total thickness of about 0.4 km of material was eroded off these mountains before the end of the Early Miocene, their elevation declined more than can be explained by erosion alone. During the later Miocene and Pliocene, further erosion of the Alps and of the Molasse basins delivered some 0.49 x 10^{21} g of sediment to the North Sea. The Pleistocene glaciations deepened and widened the valleys in the Alps, removing another 0.33 x 10^{21} g of material, much of which was delivered to the Danube and Rhone deltas and associated deep-sea fans. This resulted in isostatic adjustment that uplifted the summits of the Alps about 700 m. The total thickness of rock eroded off the Alps is about 9.5 km.

KEYWORDS

Mass-balance, paleotopography, erosion, Alps, Molasse.

INTRODUCTION

Although there is extensive literature on both the Alps and the Molasse Basins along their northern margin, there have been only a few attempts to evaluate the alpine orogeny in terms of the topographic history of the rising mountains and their subsequent erosion (Büchi, 1975; Schaer, 1979, Guillaume and Guillaume, 1980; Hantke, 1984, 1985a, 1985b). Quantitative estimates of uplift and erosion rates were presented by Schaer (1979) who suggested that along a NW-SE section from across the St. Gotthard 20-50 km of "normal continental crust" has been eroded during the last 20 million years, and only 10-25% of this eroded material was retained in the Molasse basins. Hantke (1984) has speculated that the elevation of the Bergell Massif was in excess of 5000 meters during the late Oligocene, and that glaciers on the mountain range exerted a significant influence on the sedimentation in the surrounding basins. He suggested that the coarse detritus delivered to the Molasse basins may be in part the result of catastrophic floods from breaching of the moraine dams of glacial lakes.

Mass-balanced paleogeographic reconstructions utilize information on the mass/age distribution of sediment to reconstruct ancient topography. The assumptions and methods have been discussed by Hay et al. (1989a, b), and Shaw and Hay (1989). The method depends on the assumption that the mass of detrital sediment delivered from a source area to a basin sink is directly proportional to the volume of the source area (area x elevation) above an erosional base level. The masses of sediment in the basins are redistributed onto the existing topography, and the elevations of the resulting topography are proportionally adjusted so that the amount of sediment that would be delivered by erosion of the reconstructed topography is equal to the amount actually delivered to the basin.

Mass-balanced reconstruction of the topography of the ancient Alps is especially complex because the sediments have been delivered both to proximal foreland and other peripheral basins (the Molasse basins along the northern side of the Alps, the Rhine Graben, the Bresse and Rhone Grabens, and the Po Basin) and to more distant cratonic and ocean basins unrelated to the formation of the Alps (northern European basins, North Sea, Pannonian Basin, Black Sea, western Mediterranean Basin, Adriatic Sea). Some of these basins have also had other sediment sources.

THE DATA

To determine the masses of sediment in the different basins, we used a variety of literature sources, some of the most important were: for the Molasse basins, Büchi (1975), Malzer (1981), and Lemcke (1988), for southeastern France, Debran-Passard (1984), for the Rhine Graben, Pflug (1982), for northern Europe and the North Sea, Vinken (1988), for the Black Sea, Tugolesov (1985). The current state of knowledge is such that the spatial resolution of the data available in the published literature is about 1/2° of latitude and longitude (ca. 50 km spacing). Even at this coarse resolution, complete stratigraphic sections are not available everywhere, and in some sections the stratigraphic resolution is coarse. For the Rhone Delta, Po Basin, Adriatic Sea, and Pannonian Basin west of 19° 40', we used the mass estimates of Guillaume and Guillaume (1980, 1982). The sediments in the Gulf of Lions were estimated from the interpreted seismic section presented in Burrus and Audebert (1990). Stratigraphic terminology and usage differs markedly from one region to another, particularly for the Neogene. In some cases supposed international stage names are used for units of quite different age in different areas. Figure 1 shows the major terms used in the older and newer literature on the Molasse basins. The newer terminology corresponds more closely with stratigraphic usage in other parts of Europe.

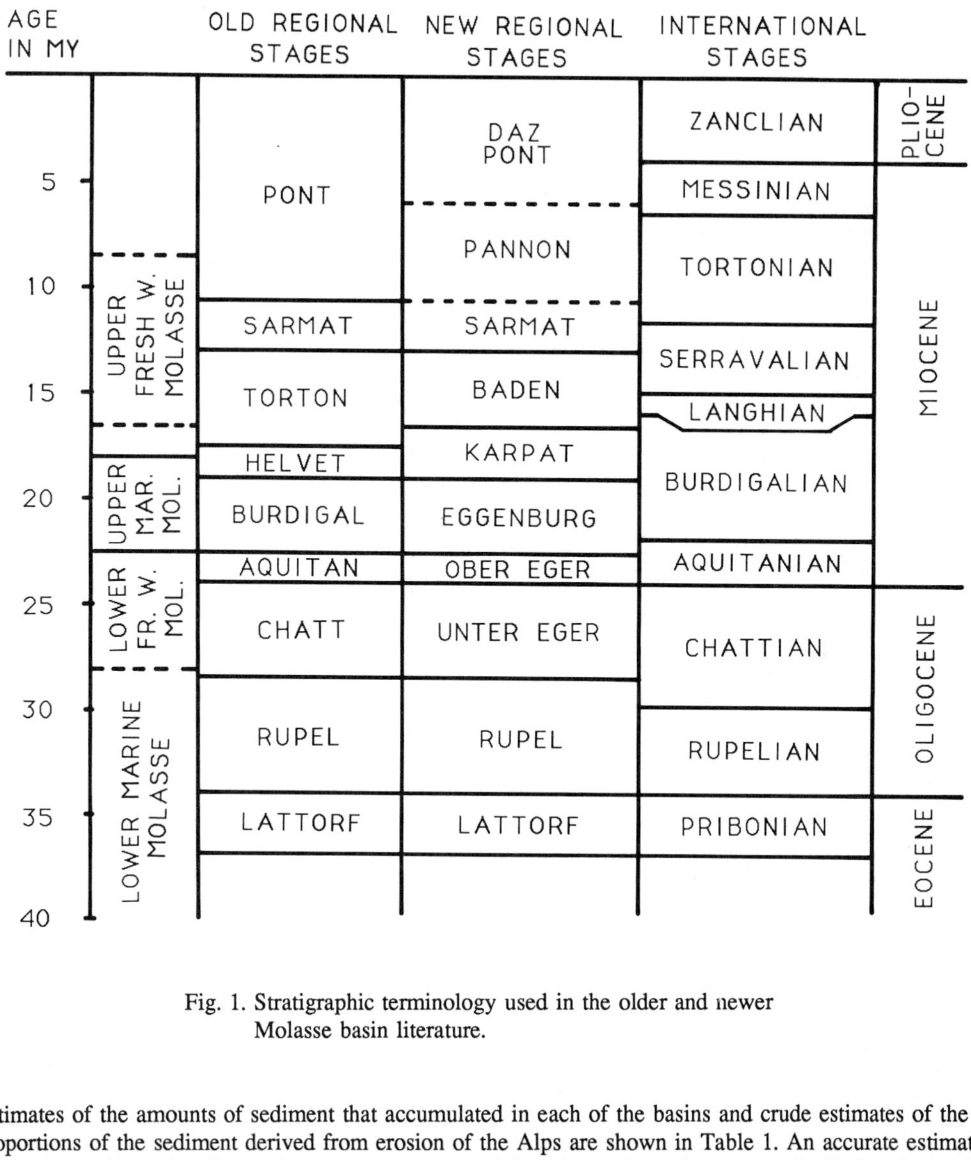

AGE IN MY	OLD REGIONAL STAGES	NEW REGIONAL STAGES	INTERNATIONAL STAGES	
		DAZ PONT	ZANCLIAN	PLIO-CENE
5	PONT		MESSINIAN	MIOCENE
10	UPPER FRESH W. MOLASSE	PANNON	TORTONIAN	
		SARMAT	SARMAT	SERRAVALIAN
15		TORTON	BADEN	LANGHIAN
	UPPER MAR. MOL.	HELVET	KARPAT	BURDIGALIAN
20		BURDIGAL	EGGENBURG	
	LOWER FR. W. MOL.	AQUITAN	OBER EGER	AQUITANIAN
25		CHATT	UNTER EGER	CHATTIAN	OLIGOCENE
30	LOWER MARINE MOLASSE	RUPEL	RUPEL	RUPELIAN
35		LATTORF	LATTORF	PRIBONIAN	EOCENE
40				

Fig. 1. Stratigraphic terminology used in the older and newer
Molasse basin literature.

Estimates of the amounts of sediment that accumulated in each of the basins and crude estimates of the proportions of the sediment derived from erosion of the Alps are shown in Table 1. An accurate estimate of the proportions of sediment derived from erosion of the Alps will require a more detailed analysis of the topography and geology of all of northwestern Europe and the offshore.

For some of the areas, the mass-age distributions of the sediments are relatively well known, for others they are only crudely known. Figure 2 shows the major sinks for material eroded from the Alps, and the mass/age distributions for Late Oligocene, Neogene and Quaternary sediment in them.

In relating the Alps as a source region to the sinks, we have divided the mountain chain into five source regions: the French Alps south of Mont Blanc Massif, the Swiss Alps and Mont Blanc Massif, the German and western Austrian (Vorarlberg-Tyrol) Alps, and the eastern Austrian

Table 1. Masses of sediment and proprotion derived from erosion
of the Alps (units of 10^{21} g)

	MASS 10^{21} g	PROPORTION FROM ALPS
West		
RHONE DELTA/GULF OF LION	0.430	95%
RHONE AND BRESSE GRABENS	0.053	95%
North		
NORTHERN EUROPE/NORTH SEA	0.810	80%
RHINE GRABEN	0.030	80%
SWISS MOLASSE	0.072	95%
GERMAN MOLASSE	0.170	95%
AUSTRIAN MOLASSE	0.057	95%
East		
PANNONIAN BASIN W OF 19°40'	0.148	100%
BLACK SEA	0.624	50%
South		
PO BASIN	0.376	80%
ADRIATIC SEA	0.355	80%

(Salzburg-Styria-Carinthia) Alps, and the southern Alps of Austria, Italy, and Switzerland. These source regions delivered material to different basins at different times as drainage patterns changed. The coarsest sediments are in the foreland basins, the Austrian, German and Swiss Molasse basins on the north side of the Alps, and in the Po Basin on the south side. Three peripheral basins received sediment from the Alps. Small amounts of material were deposited in the Bresse and Rhone Grabens, but these do not clearly reflect episodes of orogeny in the Alps. A larger mass was deposited in the Rhine Graben during the Aquitanian, apparently an overflow of excessive amounts of the Lower Fresh Water Molasse from the Swiss Molasse Basin, reflecting the climax of deformation of the Swiss and northern French Alps. Finally, a large mass of sediment accumulated in the western part of the Pannonian Basin during the Quaternary, after the Danube drainage captured the streams draining the northern flank of the Alps west to the Black Forest. Less is known of the stratigraphy of the more distal deposits. The mass of Pliocene and Quaternary sediment in the Rhone Delta has been estimated by Guillaume and Guillaume (1980), but the mass of older material that entered the Mediterranean from the Rhone and is beneath the Messinian evaporites is unknown, hence reconstruction of the southern French Alps is highly uncertain. During the Late Oligocene and during much of the Miocene, when the Fresh Water Molasse was deposited, large masses of finer grained sediments were carried to the northern European Plains and on to the North Sea. During much of the Quaternary, the ice sheets restricted sediment transport to the north, and much of the Quaternary sediment eroded off the north and western sides of the Alps was delivered to either the Rhone or Danube deltas and their associated fans. The Danube deltaic and fan complex in the Black Sea did not originate until the Quaternary, but since its origin it has been the site of deposition of much of the material eroded from the Alps, Carpathians and Balkans. The stratigraphy of the Po Basin and Adriatic region has only been superficially treated in the published literature, so that reconstruction of the southern Alps, while better than that of the southern French Alps, is also uncertain.

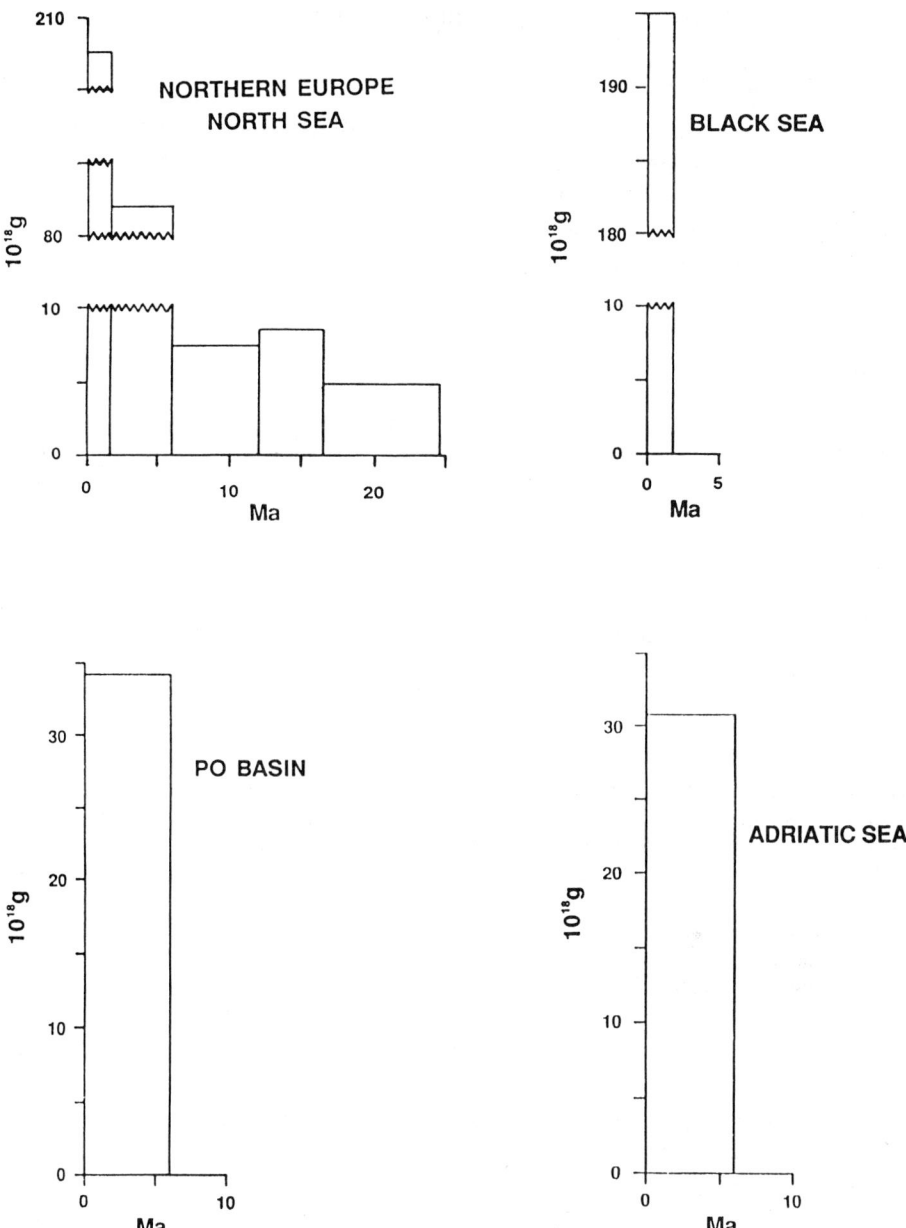

Fig. 2a. The mass/age distributions of Neogene and Quaternary
sediments existing in the major sinks for material eroded
from the Alps (Northern Europe/North Sea, Black Sea, Po
Basin and Adriatic Sea).

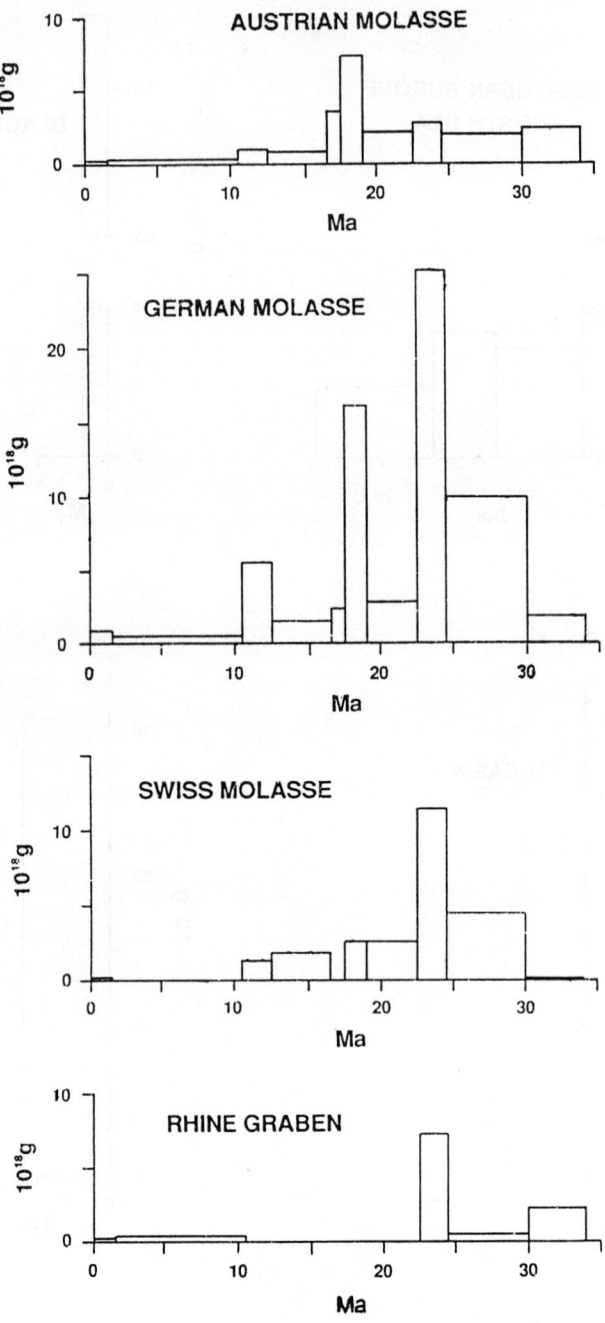

Fig. 2b. The mass/age distributions of Late Oligocene, Neogene and Quaternary sediments existing in the major sinks for material eroded from the Alps (Austrian, German and Swiss Molasse and Rhine Graben).

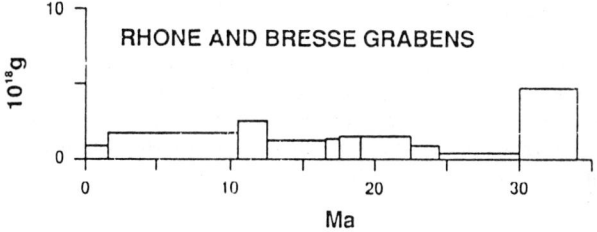

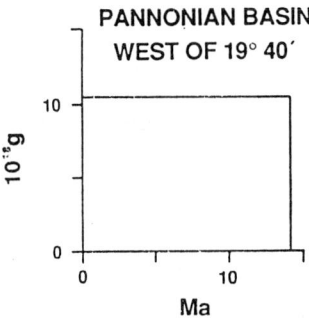

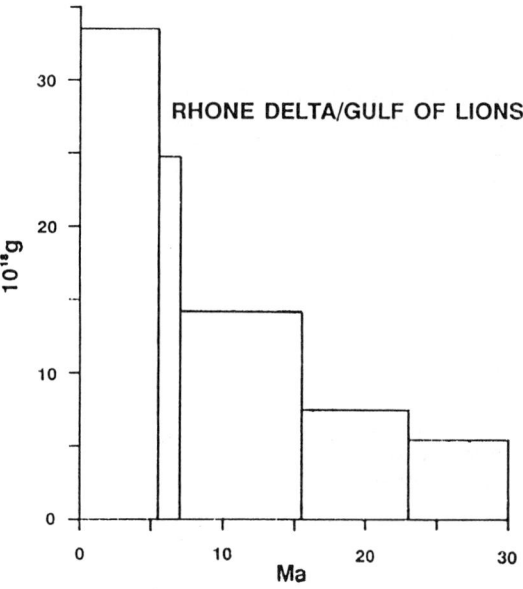

Fig. 2c. The mass/age distributions of Late Oligocene, Neogene
and Quaternary sediments existing in the major sinks for
material eroded from the Alps (Pannonian Basin west of
19°40', Rhone and Bresse Grabens and Rhone Delta/Gulf
of Lions).

RECONSTRUCTION

The masses replaced onto each of the five source areas of the Alps were determined by estimating the amount delivered to each basin from each source area at each of the stratigraphic intervals. The topography of the source area was then reconstructed directly from the detrital sediment masses by calculating the average elevation of the area required to yield the sediment masses according to the detrital erosion formula

$$\frac{dMd_e}{dt} = A \ (H - Ed_b) \ Kmd_e, \tag{1}$$

where dMd_e/dt is the rate of mechanical denudation expressed in terms of mass of sediment eroded per square meter of exposed sediment surface per meter of elevation above the detrital erosional base level per year, A is the area represented by each grid square, H is the average elevation of the grid square, Ed_b is the elevation of the erosion base for detrital material, and Kmd_e is the constant of detrital erosion expressed in terms of the mass of the solid phase eroded per incremental meter of elevation per year (for discussion see Hay et al., 1989). For the topographic models shown here we used Ed_b = 200 m and Kmd_e = 0.3 g m^{-2} m^{-1}_{elev} yr^{-1}. The conversion to thickness was made assuming a rock density of 2.1g/cm^3, and the thicknesses were then distributed over the source area grid squares proportional to their elevation above the assumed detrital erosion base of 200 meters. For comparison we also calculated the elevations using Ed_b = 600 m and Kmd_e = 0.419 g m^{-2} m^{-1}_{elev} yr^{-1}, as suggested by Pinet and Souriau (1988) for "young uplifts". The effect is to reduce the highest elevations from 7 to 6 km and to increase the elevations between 600 and 2235 meters, ie. to make a less sharply crested mountain range. Both sets of erosion constants suggest that during the episodes of thrusting, the Alps reached Himalayan heights.

The comprehensive mass-balanced paleogeographic reconstruction method described by Hay et al. (1989a) was not used for the region beyond the Alps and the Molasse basins because the stratigraphic compilations for these areas are still incomplete. Instead, the elevations of the surrounding areas were calculated assuming the inverse of simple erosional decay, using the formula

$$\frac{dTd_e}{dt} = (H - Ed_b) \ Ktd_e. \tag{2}$$

where dTd_e/dt is the rate of mechanical erosion expressed in terms of thickness of detrital sediment eroded per year, H is the average elevation of the grid square, Ed_b is the erosional base level for detrital material, and Ktde is the erosion constant expressed in terms of thickness of the solid phase eroded per meter of elevation per year. For the models presented here we used Ed_b = 200 m and Ktd_e = 0.113 x 10^{-6} m m^{-1}_{elev} yr^{-1}. The inverse of simple erosional decay restores the surrounding areas to the elevations they would have had if they had been subjected to erosion since the specified age and no uplift other than that necessary for isostatic adjustment has occurred.

The block diagrams were plotted using SURFER (Golden Software, Golden Colorado, USA).

DISCUSSION

Figures 3-6 show the development of the Alps in the Early Miocene and their present configuration at the 1/2° level of resolution that corresponds to the presently available data.

Figure 3 shows the reconstruction for the early Early Miocene ("Aquitanian" = Lower Fresh Water Molasse II = Ober Eger), the time of the climax of Late Oligocene-early Early Miocene mountain building. The highest elevations are in Switzerland and the northern French Alps. The elevation of the Bergell region is indicated to be about 7 km. The Alps of western Austria and Germany form a narrow, high (4-5 km) range, but there was only minor mountain building in eastern Austria.

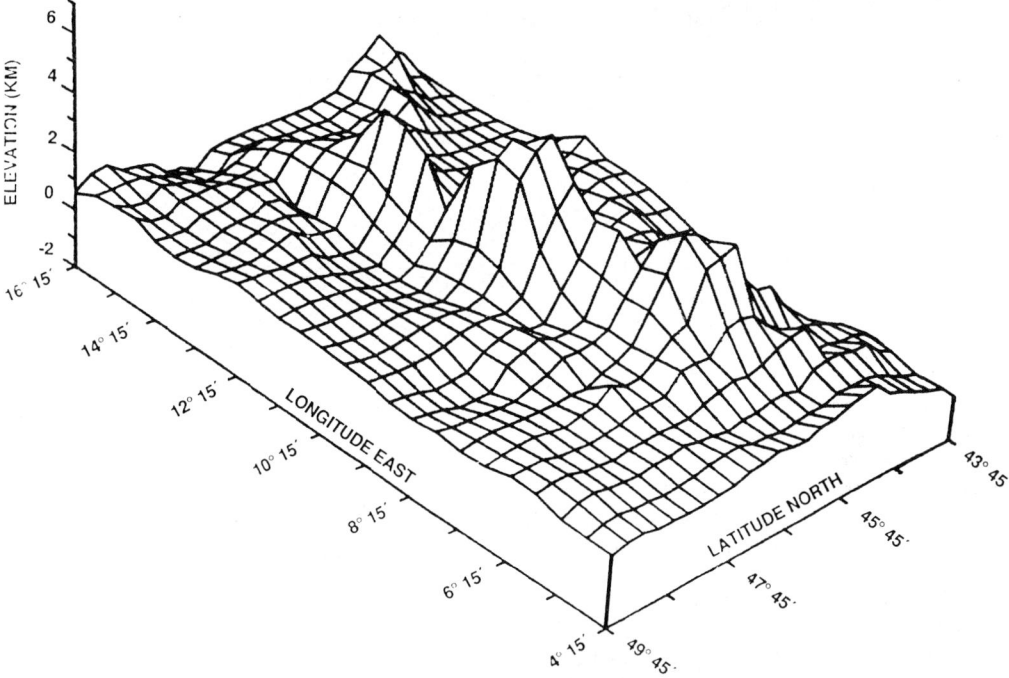

Fig. 3. Early Early Miocene ("Aquitanian = Lower Fresh Water
Molasse II = Ober Eger) reconstruction of the Alps based
on sediment mass-balance.

Figure 4 shows the reconstruction for the Alps and surrounding regions for the middle Early Miocene ("Burdigalian" = Eggenburg = Upper Marine Molasse). The entire region is much lower than in the previous reconstruction. Although the main uplift of the Swiss and northern French Alps had culminated only a few million years earlier, these mountains have subsequently been reduced to 2-3 km elevations. The low elevations shown here are required to balance the relatively low sediment output recorded in the Molasse basins. The loss of elevation is significantly greater than could be expected from normal erosion alone. Although the erosion expected from the crest of a 5 km mountain range over a 5 my period is equivalent to a layer of rock having a thickness of about 2700 m, but, taking isostasy into account, the loss of elevation is only about 800 m. The rapid loss of elevation since the climax of mountain building in Switzerland suggests either a delayed long term isostatic response to the load, or collapse of the mountains. Delayed isostatic response could only be possible if there were a long time constant associated with the sharp bending of the crustal slab required to accommodate the load of the Alps. Collapse of parts of the Alps has been postulated elsewhere (Hantke, 1985a).

Figure 5 shows the reconstruction of the Alps and surrounding regions for the late Early Miocene ("Helvetian" = Ottnang = Upper Marine Molasse). The Austrian and German Alps form a 4-5 km high narrow mountain range. A second phase of thrusting has affected the western Austrian Alps, and the

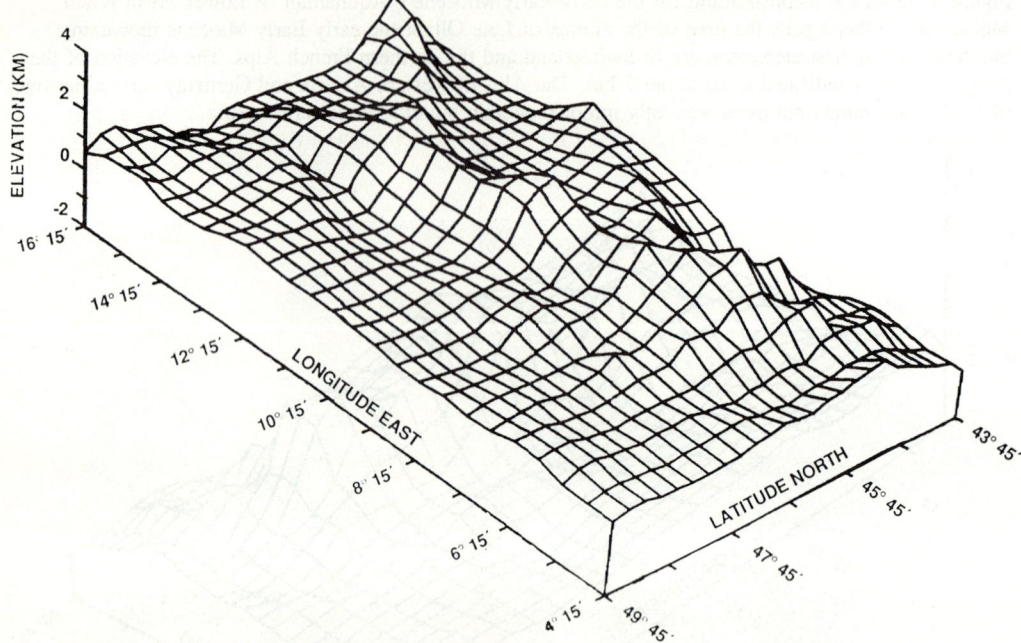

Fig. 4. Middle Early Miocene ("Burdigalian" = Eggenburg = Upper Marine Molasse) reconstruction of the Alps based on sediment mass-balance.

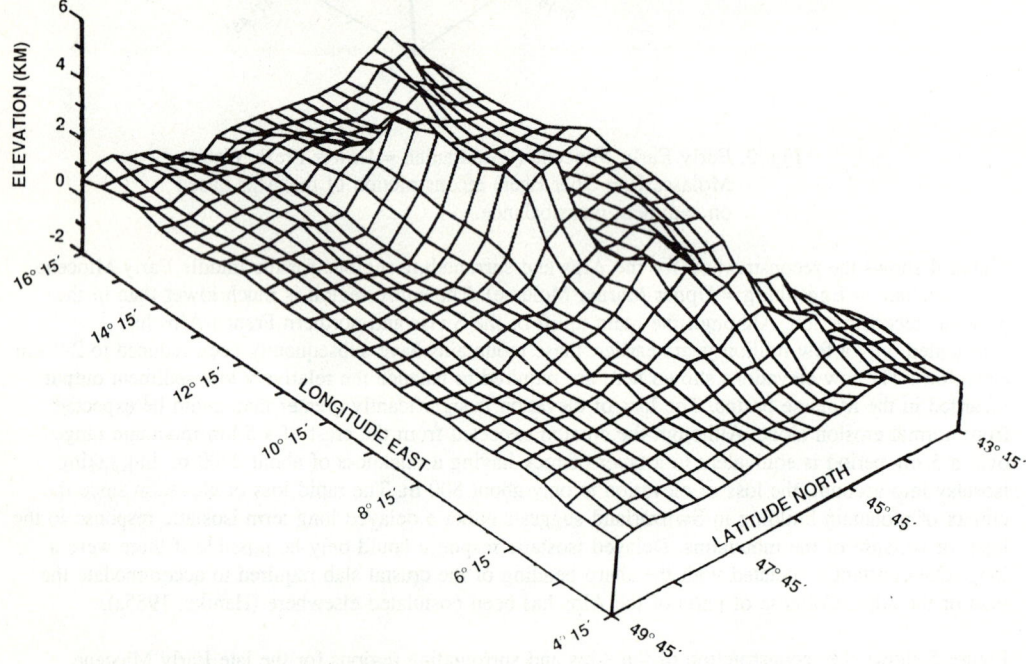

Fig. 5. Late Early Miocene ("Helvetian" = Ottnang = Upper Marine Molasse) reconstruction of the Alps based on sediment mass-balance.

eastern Austrian Alps have just been formed. Mass-balance suggests that a rapid loss of elevation by the German and Austrian Alps, similar to that suggested by the middle Early Miocene sediments (Fig. 4), followed this episode of mountain building.

Figure 6 shows the present configuration of the Alps and surrounding regions, based on the average elevations of 1/2° latitude x 1/2° longitude squares calculated from the ETOPO-5 data base (average elevations of 5' squares, available from National Geophysical Solar-Terrestrial Data Center, Boulder, Colorado, USA). At the 1/2° resolution only the general configuration is seen. The present peaks are 1-1.5 km higher than the average elevation surface shown here.

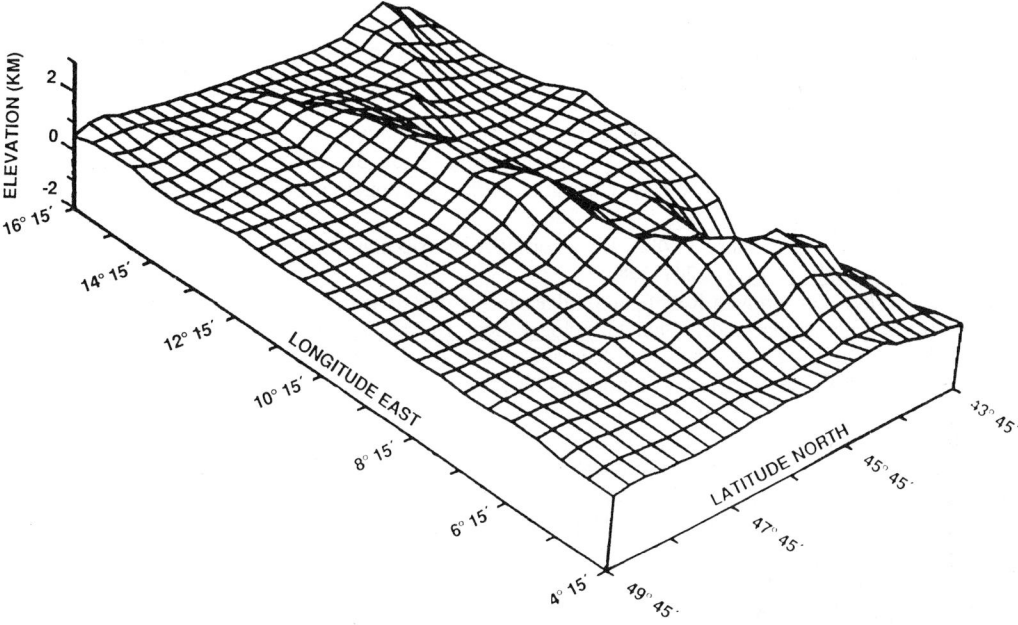

Fig. 6. The present configuration of the Alps and surrounding
regions.

These models, although in a preliminary and primitive state, support the idea that during their formation the Alps reached Himalayan heights. Assuming sea level temperatures of 17-18° C and a lapse rate of 6.5°C/km, timberline (mean annual temperature = 0° C) would have been at 2.75 km and the upper elevations would surely have been ice covered. If sufficient moisture were available, then extensive glaciers would have existed as postulated by Hantke (1984, 1985a,b). The coarse nature of the detritus and reddish staining of the rocks in the Molasse basins along the north side of the mountains suggests arid (rain shadow?) conditions (Garner, 1959; Hantke, 1985). More humid conditions may have prevailed in the south.

Three dimensional paleotopographic models are useful in visualizing the geologic history of a region and are useful in developing an understanding of the dynamic interrelationships between sources and sinks.

ACKNOWLEDGEMENTS

We are indebted to Robert DeConto for compiling data on northern Europe and the North Sea. This work was supported by grant OCE 8716408 from the US National Science Foundation, grant 19274-AC2 from the Petroleum Research Fund of the American Chemical Society, and by a gift from Texaco, Inc.

REFERENCES

Büchi, U.A. (1975). Geologie der Schweizer Molasse (Vorlesung Sommersemester 1975). Geologisches Institut der Eidgenössichen Technischen Hochschule und Universität Zürich, Zürich, 95 pp.

Burrus, J., and F. Audebert (1990). Thermal and compaction processes in a young rifted basin containing evaporites: Gulf of Lions, France. Amer. Assoc. Petrol. Geol. Bull., 74, 1420-1440.

Debran-Passard, S. (1984). Synthése Géologique du Sud-Est de la France. Mém. Bur. Rech. Géol. Min., 125, pp. 1-613, 126, Atlas.

Garner, H.F. (1959). Stratigraphic-sedimentary significance of contemporary climate and relief in four regions of the Andes mountains. Geol. Soc. Amer. Bull., 60, pp. 561-590.

Guillaume, A., and S. Guillaume (1980). L'érosion au Plio-Quaternaire dans les Alpes - Bilan quantitatif. Eclogae Geol. Helvet., 73, pp. 326-329.

Guillaume, A., and S. Guillaume (1982). L'érosion dans les Alpes au Plio-Quaternaire et au Miocène. Eclogae Geol. Helvet., 75, pp. 247-268.

Hantke, R. (1984). Zur tertiären Relief- und Talgeschichte des Bergeller Hochgebirges, der zentralen Südalpen und der angrenzenden Gebiete. Eclogae Geol. Helvet., 77, pp. 327-361.

Hantke, R. (1985a) Zur Relief-, Tal-, und Klimageschichte der zentralen und östlichen Schweizer Alpen: Teil 1: Das Geschehen vom mittleren Oligozän bis ins frühe Miozän. Vierteljahresschrift Naturf. Gesell. Zürich, 130, pp. 92-114.

Hantke, R. (1985b) Zur Relief-, Tal-, und Klimageschichte der zentralen und östlichen Schweizer Alpen: Teil 2: Das Geschehen vom mittleren Miozän bis ins frühe Pleistozän. Vierteljahresschrift Naturf. Gesell. Zürich, 130, pp. 144-156.

Hay, W.W., C.A. Shaw and C.N. Wold (1989a). Mass-balanced paleogeographic reconstructions. Geologische Rundschau, 78, pp. 207-242.

Hay, W.W., C.A. Shaw and C.N. Wold (1989b). Mass-balanced paleogeographic maps: Background and input requirements. In: Quantitative Dynamic Stratigraphy (T.A. Cross, Ed.), pp. 261-275.

Lemcke, K. (1988). Geologie von Bayern I, Das bayerische Alpenvorland vor der Eiszeit, Erdgeschichte, Bau, Bodenschätze. E. Schweizerbart'sche Verlagsbuchhandlung, 175 pp.

Malzer, O. (1981). Geologische Characteristik der wichtigsten Erdöl- und Erdgasträger der oberösterreichischen Molasse. Erdöl-Erdgas- Zeitschrift, 97, pp. 20-28.

Pflug, R. (1982). Bau und Entwicklung des Oberrheingrabens. Erträge der Forschung. Wissenschaftliche Buchgesellschaft Darmstadt, 145 pp.

Pinet, P., and M. Souriau (1988). Continental erosion and large scale relief. Tectonics, 7, pp. 563-582.

Schaer, J.-P. (1979). Mouvements verticaux, érosion dans les Alpes, aujourd'hui et au cours du Miocène. Eclogae Geol. Helvet., 72, 263-270.

Shaw, C.A., and W.W. Hay, (1989). Mass balanced paleogeographic maps: modeling program and results. In: Quantitative Dynamic Stratigraphy (T.A. Cross, Ed.), pp. 277-291.

Tugolesov, D.A. (1985). Tektonika Mezokainozoiskik Otleshchenii Chernomorskoi Vpadin'. NEDRA, Moscow, 213 pp.

Vinken, R. (1988). The Northwest European Tertiary Basin. Geol. Jahrb., Reihe A, 100, pp. 1-508.

Chapter 2

Process Simulation

STANFORD'S SEDSIM PROJECT:
DYNAMIC THREE-DIMENSIONAL SIMULATION OF GEOLOGIC
PROCESSES THAT AFFECT CLASTIC SEDIMENTS

Young-Hoon Lee and John W. Harbaugh
Department of Applied Earth Sciences, Stanford University
Stanford, CA 94305, U.S.A.

INTRODUCTION

SEDSIM focuses on simulating geologic processes in which the objective is to have the computer operate the way the earth actually operates, paralleling the processes in three-dimensional detail. SEDSIM provides an integrated experimental approach in dealing with sedimentary features. Its potential applications include hydrocarbon exploration and exploitation, where improved understanding and characterization of reservoirs are important. Major emphasis is placed on fluid movement, including flow in open bodies of water, nearshore wave activity, and expulsion of fluids as compaction of sediments take place.

SEDSIM's plan involves computing modules that represent aspects of specific geologic processes and are linked together to operate in concert with each other, adhered to four major precepts:

(1)   Process modules adhere to the physical fundamentals that govern the actual geologic processes.
(2)   Modules adhere to the conservation laws.
(3)   Modules are three dimensional.
(4)   Modules represent results graphically, using dynamic 3-D color displays.

SEDSIM represents geologic processes by breaking them down into fundamental components. For example, transportation and deposition of grains of clastic sediment, such as sand, silt, and clay, are represented by behavior of individual grains swept along by currents of water. Flow and transport in streams, or by turbidity currents, or in currents induced by waves in the surf zone, is represented by motion of large numbers of individual fluid elements and by transport of individual particles. In turn, these fundamental process components are represented by equations that are solved numerically millions of times in any specific experiment. Fast computers carry out the arithmetic.

SEDSIM treats the transport of clastic sediment grains figuratively on a "grain-by-grain" basis, but uses appropriate shortcuts to facilitate computation. Each grain of sediment cannot be represented, but the collective behavior of grains can be. Simulated deposits can evolve gradually through geologic time in an experiment. Locally, sediment may be deposited, and either preserved, or re-eroded and redeposited elsewhere. Following deposition, beds are progressively compacted and deformed, while the basin itself is broadly affected by eustatic changes in sea level, and by isostatic compensation and other tectonic processes. Like the earth itself, SEDSIM is persistent and tireless, preserving a detailed three-dimensional record through geologic time for analysis and comparison with actual deposits.

SEDSIM differs drastically from conventional forms of sedimentologic analyses because it deals with deposits in terms of geologic processes, which are represented in three-dimensional detail. SEDSIM conserves mass, energy, and momentum, and deals effectively with interdependent processes, such as sediment transport, deposition, compaction, and isostatic compensation, all of which are represented consistently through geologic time. For example, a geologist focusing on a specific sedimentary sequence, could perform experiments that analyze the consequences of assumptions regarding sediment sources, wind direction and intensity, eustatic changes in sea level, compaction, and strength of the earth's crust in response to changes in sediment load.

One of SEDSIM's hallmarks is dynamic graphic display, utilizing 3-D graphics workstations. The results of simulation experiments are shown in vivid color, with facies maps, cross-sections, shaded relief maps, and perspective diagrams. Full three-dimensional color displays permit sedimentary facies and geologic structure to be viewed simultaneously as a single image. With SEDSIM's dynamic display capabilities, changes within a sedimentary sequence can be shown in video-like form, permitting thousands of years of geologic history to be reviewed in rapid succession on a workstation's high-resolution screen (Lee & Harbaugh, 1990). Furthermore, users can generate synthetic well logs and seismic sections for comparing SEDSIM's output with actual logs and seismic sections. A modern 3-D graphics workstation is a necessity. SEDSIM currently operates on both, Silicon Graphics workstations and STARDENT Titan Graphics SuperComputers, which do both, the background "number crunching" and the 3-D graphics. Alternatively, a fast general purpose computer could be used for the number crunching, but it would have to be linked with a 3-D graphics workstation for dynamic graphic display.

Only one major publication describing an aspect of SEDSIM has appeared, namely a book-length monograph that describes SEDSIM's procedures for transporting and depositing clastic sediments (Tetzlaff & Harbaugh, 1989). Meanwhile, theses by Martinez (1987), Scott (1986), and Laudati (1988) have appeared.

SEDSIM has close ties with other universities, including Freiburg University, West Germany. SEDSIM also has ties with oil-company affiliates that include Conoco, Japan National Oil Company, Mobil, Shell, Statoil, Marathon, Amoco, Texaco, Phillips, Elf Aquitaine, and others.

SEDSIM's modular organization will permit most important geologic processes that affect sedimentation and the development of sedimentary basins overall to be incorporated in SEDSIM. The work will span a number of years. Meanwhile, certain aspects of SEDSIM have been developed and are now undergoing extensive testing. Four of these will be described at this symposium, namely (1) erosion, transport, and deposition of clastic sediments in alluvial channels or more broadly as "open channel flow", (2) reworking and modification of deposits by wave activity in nearshore environments (Martinez, 1989) and, (3) compaction of clastic sediments following deposition, and response of the crust to changes in sediment load in an attempt to achieve isostatic equilibrium (Wendebourg and Ulmer, 1990). Each of these modules can be operated by itself, which is particularly useful for the testing and calibration. In turn, however, these modules, as well as others that will be developed later, can be operated in concert with each other, such that all become mutually interdependent.

SIMULATION OF OPEN-CHANNEL FLOW

The module described in this paper provides the basic elements involving the simulation of open-channel flow, and the erosion, transport, and deposition that is coincident with fluid movement. Flow is modelled using a "marker-in-cell" method (Harlow, 1965) where flow parameters are represented with respect to the moving fluid itself as a frame of reference. Fluid elements move according to the Navier-Stokes equations for describing open channel flow. The fluid elements react to changes in topography, obeying laws of conservation and momentum. Changes in fluid flow and changes in deposits of clastic grains are simulated over intervals of geologic time spanning in the aggregate, hundreds or thousands of years, but individual time increments for updating of flow calculations span intervals of only a few seconds. SEDSIM's procedures for simulating fluid flow, and sedimentation are presented here, along with an example of SEDSIM's application to modeling a fluvial system.

Before discussing stream systems in transporting sediment and in creating fluvial deposits, we need to focus on the hydraulics of flow. Flow in a stream or open channel may be unsteady, varying in time and space. The flow is three-dimensional, with significant cross-channel and vertical components. It may become relatively unconfined during overbank flooding. The mechanics of the stream's flow and the hydraulic structures that it creates are related to the transport of sediments by flow. Even the simple seemingly process of sediment movement as bed load and as suspended load involves many interrelated factors, and analytical methods for treating flow and transport are difficult because sediment transport varies with complex characteristics of sediment and flow.

SEDSIM provides an alternative for predicting geometrical forms of sediment bodies because it directly and mathematically represents the processes that form deposits, which in the example described here consist of a hypothetical fluvial deposit. The processes are responsive to hydraulic and geologic parameters, such as discharge volume, velocity, sediment load and ratio, that combine to regulate the frequency distribution of sizes of sediment grains that are eroded, transported and deposited. The objective of open-channel experiments with SEDSIM include improved understanding of the spatial distribution of sediment bodies in fluvial deposits and their use in predicting the behavior of groundwater in aquifers in relatively young fluvial deposits, or of oil and gas in reservoirs in sand bodies formed as ancient fluvial deposits.

Fluid movement

SEDSIM can represent unsteady flow in irregular channels that merge as tributaries or diverge as distributaries, a capability based on adaptation of the Navier-Stokes equations. The simplified Navier-Stokes equations contain parameters that depend partly on flow conditions which are related to factors that include channel shape, fluid depth, surface slope, and flow velocity. These equations as adapted in SEDSIM involve the simplifying assumption that the velocities of individual fluid elements do not vary with depth at a given instant and geographic location. Although this may seem to be an oversimplification, flow velocities thus represented are more or less realistic because the velocity profile sets limits on the velocities of individual fluid elements.

The Navier-Stokes equations combine the continuity equation and the momentum equation to provide a complete mathematical description of flow for an isotropic Newtonian fluid whose properties are uniform in all directions and which follow Newton's laws of motion. SEDSIM involves the assumption that the fluid is homogeneous and incompressible at constant temperature, that the fluid density and viscosity are constant, that the Coriolis acceleration is ignored, and that pressure distribution is hydrostatic. SEDSIM thus cannot simulate large-scale circulation patterns where Coriolis acceleration is important, although SEDSIM could be modified to include it.

Numerical procedures for the Navier-Stokes equations employed in SEDSIM use a marker-in-cell method in two horizontal dimensions. The flow velocity and position of each fluid element is represented at a specific point during each time increment. A two-dimensional grid with square cells represents the depth of flow, the average velocity of fluid elements in each grid cell, the second derivative of velocity in the cell, and topographic elevation in the cell. SEDSIM is thus pseudo-three dimensional in its representation of flow, because the flow is "depth-averaged", with no variations of flow velocity with depth at a given time and location. The resulting formulation is more or less suitable for simulating free-surface flow under a variety of different flow conditions. Time is, of course, a fundamental aspect in the dynamic system represented by SEDSIM, and the flow calculations, using both grid parameters and marker parameters, are updated in increments of a few seconds.

The utilization of finite-difference equations in SEDSIM involves the following sequence of events by which the flow configuration is advanced from one time step to the next:

1) The complete field of velocities must be known at the beginning of the span of time over which an experiment is to run. The beginning velocities are known either from the results of an immediately preceding run, or are supplied as part of the initial conditions at the outset of an experiment.
2) The corresponding topographic elevation and the slope of the fluid upper surface are represented with hyperbolic paraboloid functions fitted an a cell-by-cell basis, a hyperbolic paraboloid being the simplest continuous function that fits the four edges of a rectangular cell.
3) The fluid depth at a specific cell is calculated from the individual fluid element volumes and the fractions thereof that are assigned at that moment to the four neighboring grid points at each cell.
4) Two components of acceleration and bottom friction are calculated. The products of these, coupled with the length assigned to each time increment, give the changes in velocity to be incorporated with previous values.
5) The marker fluid elements move according to the velocity components in their vicinities.
6) Adjustments are made for passage of fluid elements across cell boundaries.

Sediment movement

SEDSIM incorporates multiple types of sediment that can be transported in open channel flows. Currently, SEDSIM can handle up to four types of clastic sediment, which are treated with two equations that simulate erosion, transport, and deposition of sediment, namely the sediment *continuity* equation and the sediment *transport* equation. The sediment *continuity* equation ensures that all sediment that is eroded, transported, or deposited is properly accounted for, and the sediment *transport* equation describes how much sediment of the types that have been supplied can be transported by a flow of given hydraulic characteristics, including the rate at which sediment is transferred between the flow and the underlying stream bed. As represented in SEDSIM, these two equations cause sediment to move at the same velocity as the flow, a situation that does not accord with actual flows, where the total sediment load includes both bed load and suspended load which move at different rates. Sediment particles in the bed load move slower because they are confined to a thin layer immediately above the stream bed, where they move by rolling, jumping(saltation), and dragging(traction). The suspended load, however, is maintained by turbulence and moves at approximately the same rate as the flow (Reading, 1986). SEDSIM compensates for the slower movement of the bed load by reducing the proportion assigned to bed load and by increasing the fall velocities of bed load sediment through the fluid. Thus, even though the sediment-transport equation only roughly approximates actual transport rates, it is more or less suitable for our present experiments.

SEDSIM thus represents the total sediment load as a single component, although the load actually consists of bed load and suspended load. The transport of sediment both in SEDSIM and the real world depends on the balance between the flow's transport capacity and the effective sediment concentration in the flow. If the effective sediment concentration is greater than the transport capacity, deposition occurs at a rate proportional to the excess effective concentration. If the sediment concentration is less than the transport capacity, erosion occurs if the critical shear stress is exceeded. To contain deposited sediment, SEDSIM utilize cells in a three-dimensional "sediment grid" that contain specific volumes and types of sediments that are assigned to them as a consequence of successive intervals of deposition and erosion. Detailed materials-accounting procedures are required to deal with topographic changes caused by erosion or deposition as determined by changes in sediment load.

SEDSIM Experiment Involving a Bent Channel

As an example application of SEDSIM in creating fluvial deposits, we created a hypothetical "bent" channel, whose inital configuration is shown by a contour map and a fishnet diagram Figure 1, and then progressively altered the channel by allowing water to flow through it. The experiment is an extension of an earlier experiment described by Tetzlaff and Harbaugh (1989). The grid in the experiment is 41 columns wide and 41 rows long, defining 1681 square cells. Each cell is 50 by 50 m in extent, so that the total area of the grid is 2 by 2 km, which was sufficient for observing the responses to the flow. The initial topographic slope was 0.8 percent and there were four "sources" or locations from which the flows emanated, providing both the water and the contained sediment load. These sources were arranged laterally at the upper end of the channel, with the discharge rate at each source remaining constant at 10 cubic meters of water per second, and containing a sediment load of 2 kg of clastic material per cubic meter of water. The four sediment types were represented in the following proportions:

	Percent
medium sand	8
fine sand	15
silt	30
clay	47

The initial substrate over which the water flowed at the outset of the experiment, including the channel and the adjacent banks, consisted of a mixture of the four sediment types in the above proportions.

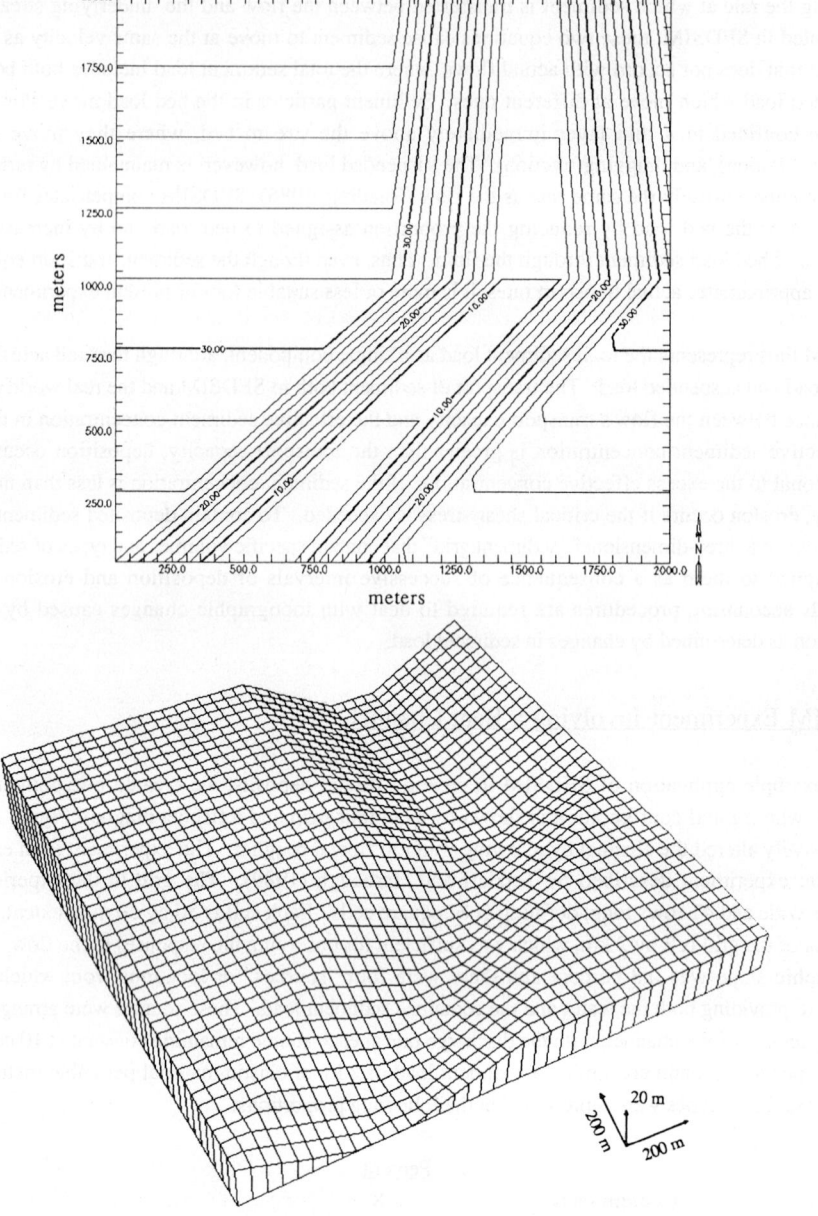

Fig. 1. Contour map and perspective fishnet diagram of bent channel showing initial conditions in simulation experiment. Grid spans 2 by 2 km, representing 41 by 41 cells, and topographic slope of 0.8 percent. Contour interval is 2 meters. Channel is 20 meters deep, and 1 km wide.

The experiment involved a simulated time span of ten years, and results are shown in two-year intervals in Figures 2 to 6. The average magnitude and the direction of flow velocities are shown in Figures 2a and 6a on a cell-by-cell basis by means of vectors superimposed on the topographic contours. The vector lengths are proportional to the average velocity of fluid elements in the cell surrounding each grid node, with flow velocities ranging from 0.05 to 0.7 meters per second. The various mixtures of the four sediment sizes are represented by patterns. After two years of simulated time (Figure 2b), there has been some erosion and deposition occurred in the channel, and underwater dunes or transverse bars formed, with downstream slip faces and gently sloping tops. After initial deposition of fine and medium-grained sand, the sandy deposits of the channel continued to migrate downstream as submerged dunes, with later deposition of silt or clay from suspension. The bedforms in the open channel resulted from the complex interactions of sediment supply and discharge at each grid point. High rates of supply of sediment led to local overloading of the stream and to progressive steepening of slopes, until a balance between transport and supply was achieved. The bedforms and the pattern of channels affected the frictional resistance to flow as an aspect of this complex balance (Coleman, 1969). In the experiment, medium sand and fine sand were transported mainly as bedload and dominate the deposits, whereas the still finer-grained sediments were transported mostly in suspension and did not contribute greatly to sediment accumulation, even though fine material was abundant in the overall load.

CONCLUSIONS

Simulation experiments using SEDSIM reproduced patterns of fluid flow and sedimentation typical of fluvial environments. In the bent-channel experiment, SEDSIM reproduced flow velocities , fluid depths, bottom shear stresses , and rates of sediment transport comparable to those in natural channels. The results also show that fluid velocities were high near the cut bank of the simulated channel, and that channel deposits occur as fining-upwards sequences. Coarse-grained sediments were deposited in the upper channel as submerged dunes or transverse bars that subsequently migrated downstream. After six simulated years, fine-grained sediments covered the coarse-grained sediment, and the sediment load and fluid flow reached equilibrium. Thus, SEDSIM appears to be reasonably effective in representing flow and sedimentation in meandering channels, and should provide insight in understanding the spatial distribution of sediment bodies in fluvial deposits and the internal structures within these bodies.

REFERENCES

Coleman J.M., 1969, Brahmaputra River: Channel processes and sedimentation, Sedimentary Geol., Vol. 3 (special issue), p. 129-239

Harlow, F.H., 1965, Numerical calculation of time-dependent viscous incompressible flow of fluid with free surface, Physics of Fluids, Vol. 8, No. 12, p. 2182-2189

Laudati, R.P., 1988, Computer Simulation of Secondary Hydrocarbon Migration and Entrapment: M.S. thesis , 158 p., Dept. of Applied Earth Sciences, Stanford University.

Lee, Y.H., Martinez P.A., & Harbaugh J.W., 1990, Dynamic 3-D graphics are critically important in Stanford's SEDSIM project, Geobyte, Vol. 5, No. 1, p. 37-38.

Martinez, P.A., 1987, Simulation of Sediment Transport and Deposition by Waves for Simulation of Wave vs. Fluvial-dominated Beach Environments: M.S. thesis , 406 p., Dept. of Applied Earth Sciences, Stanford University.

Martinez, P.A., 1989, Computer simulation of wave and fluvial-dominated nearshore environments, in: Applications. In: Coastal Modelling (Lakhan C.V. & Trenhaile A.S. eds.), Elsevier Oceanography Series 49, Elsevier Publishing Co., New York, p. 297-337.

Reading H.G., 1986, Sedimentary Environments and Facies, 2nd ed., Blackwell Scientific Pub., Oxford, 615 P.

Scott, N., 1986, Modern vs. Ancient Braided Stream Deposits: A Comparison between Simulated Sedimentary Deposits and the Ivishak Formation of the Prudhoe Bay Field, Alaska: M.S. thesis , 103 p., Dept. of Applied Earth Sciences, Stanford University.

Tetzlaff, D.M. & Harbaugh, J.W., 1989, Simulating Clastic Sedimentation: Van Nostrand Reinhold, New York, 202 p.

Wendebourg J. & Ulmer J.W.D., 1990, Modeling of compaction and isostatic compensation used in SEDSIM for basin analysis and subsurface fluid flow: this volume

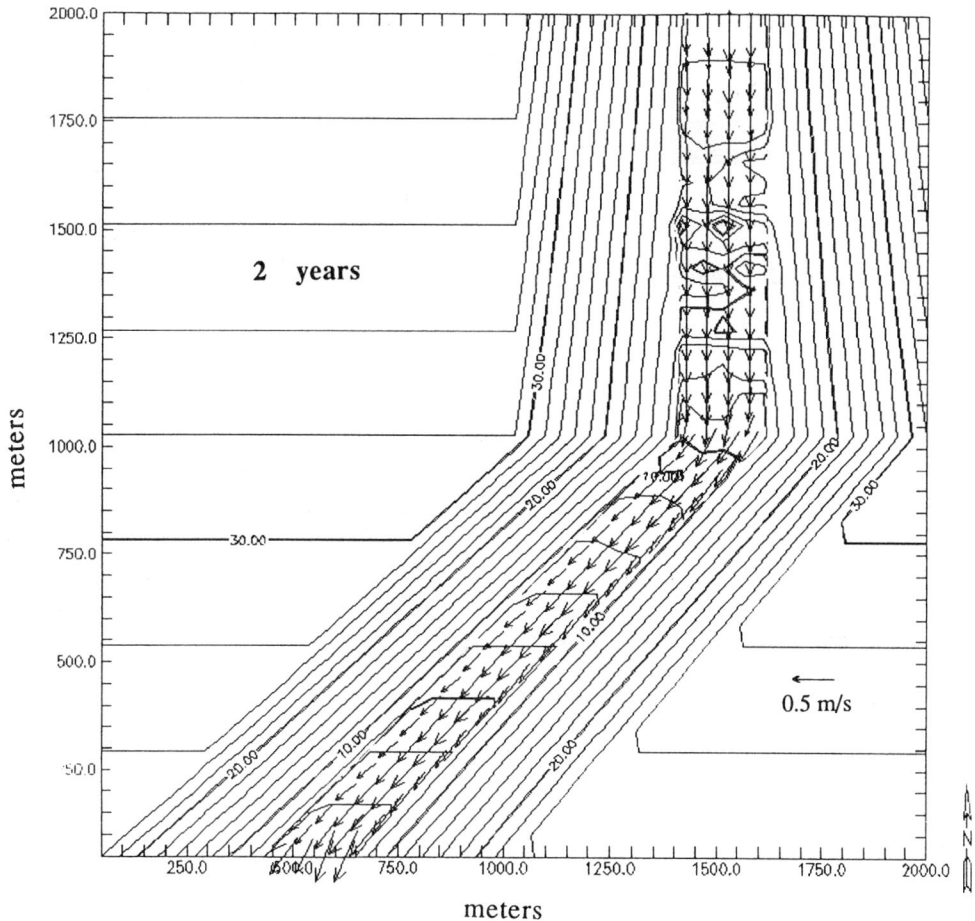

Fig. 2a. Map of bent channel showing topographic contours and flow velocities at two simulated years. Contour interval is 0.5 m within channel and 2 m outside channel. Flow velocites denoted by vector arrows vary in magnitude across channel width, especially at bend, thus mimicking flow variations observed in natural channels. Average velocity of currents represented by vectors is 0.5 m/sec.

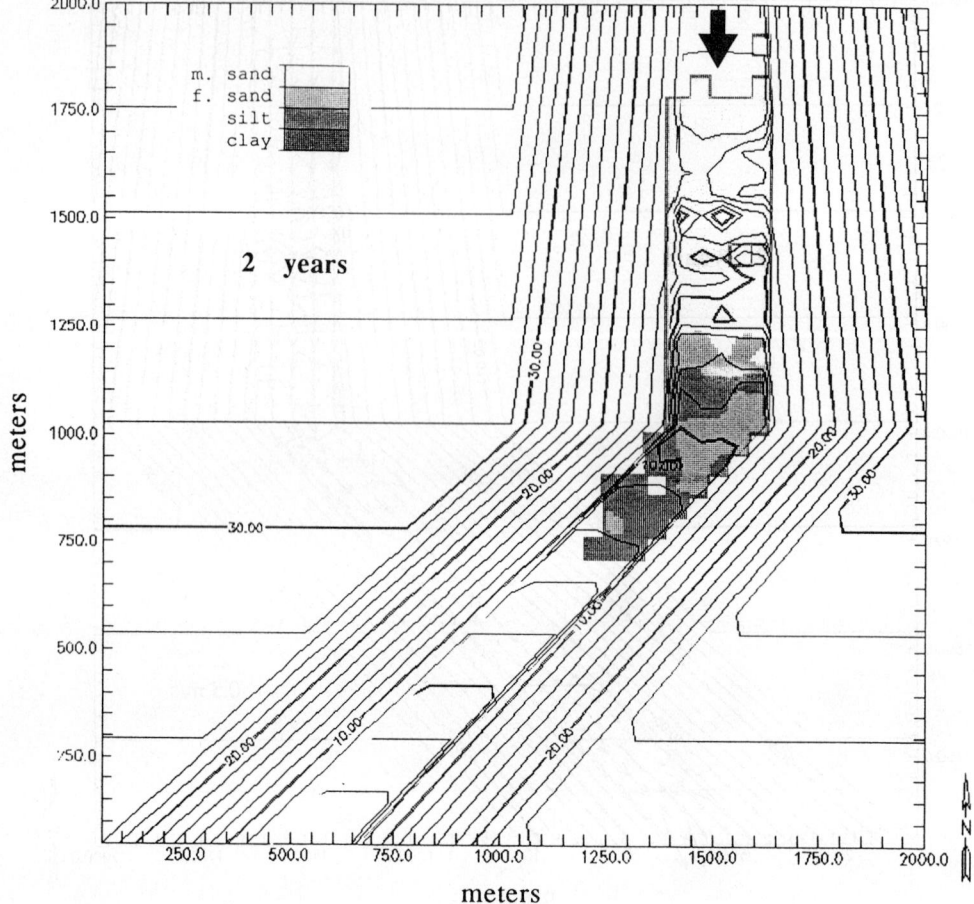

Fig. 2b. Facies map of channel deposits superimposed on topographic contours at two simulated years. Large arrow denotes location of fluid "sources". Composition of sediment is denoted by grayscale patterns. Ordinarily patterns are viewed in color. Coarse sediments have been deposited and submerged dunes have formed in upper part of channel.

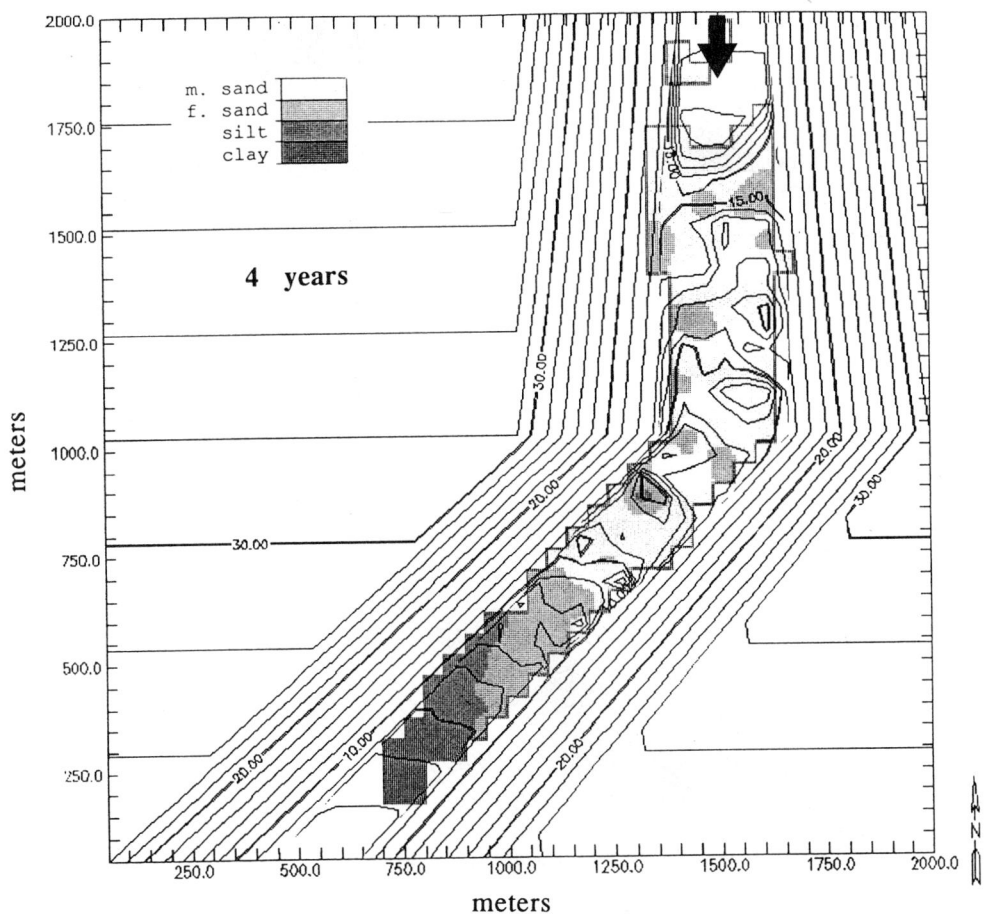

Fig. 3. Facies and topography at four simulated years. Erosion has occurred near fluid sources (arrow), coarse sediments have moved downstream, and submerged dunes occur in center and downstream parts of channel. Succession of facies maps representing sequences of time intervals can be viewed in rapid succession, showing simulation results in video-like form, allowing changes to be observed through successive time increments as if features were in motion.

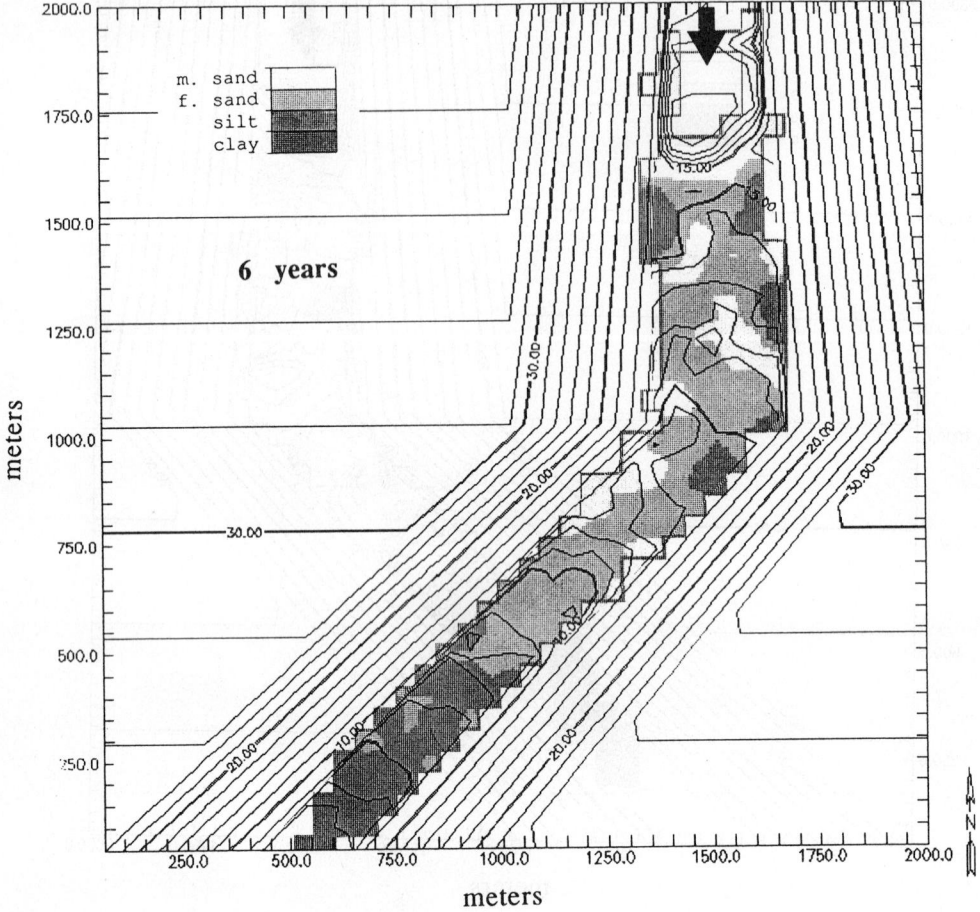

Fig. 4. Facies and topography at six simulated years. More erosion has occurred near fluid sources, and fine sediments (darker shades) have begun to cover coarse sediment.

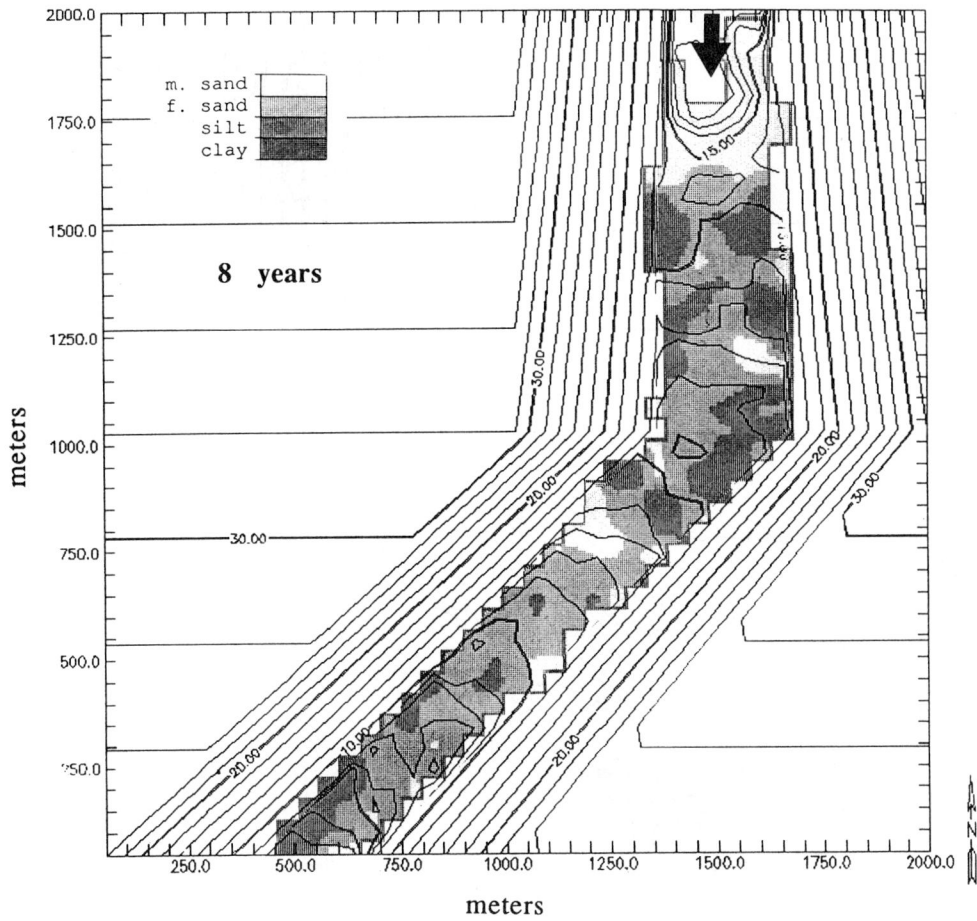

Fig. 5. Facies and topography at eight simulated years. Fine sediments cover coarse sediments and sediment load and fluid flow have reached equilibrium, as indicated by similarity with Figure 4.

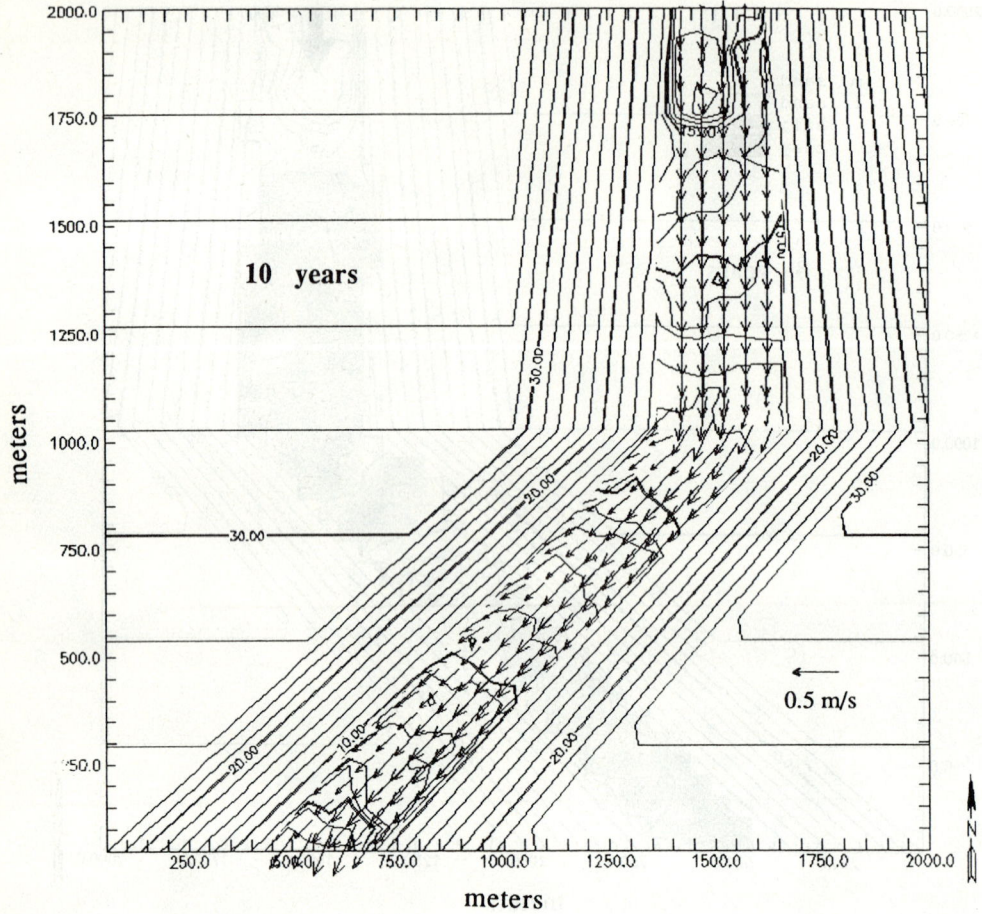

Fig. 6a. Flow map and topography at ten simulated years. Only small changes in flow velocites denoted by arrows are visible compared to initial starting condition (Fig. 2a), indicating that flow was in equilibrium. Discharge rates of fluid and sediment have remained constant throughout experiment.

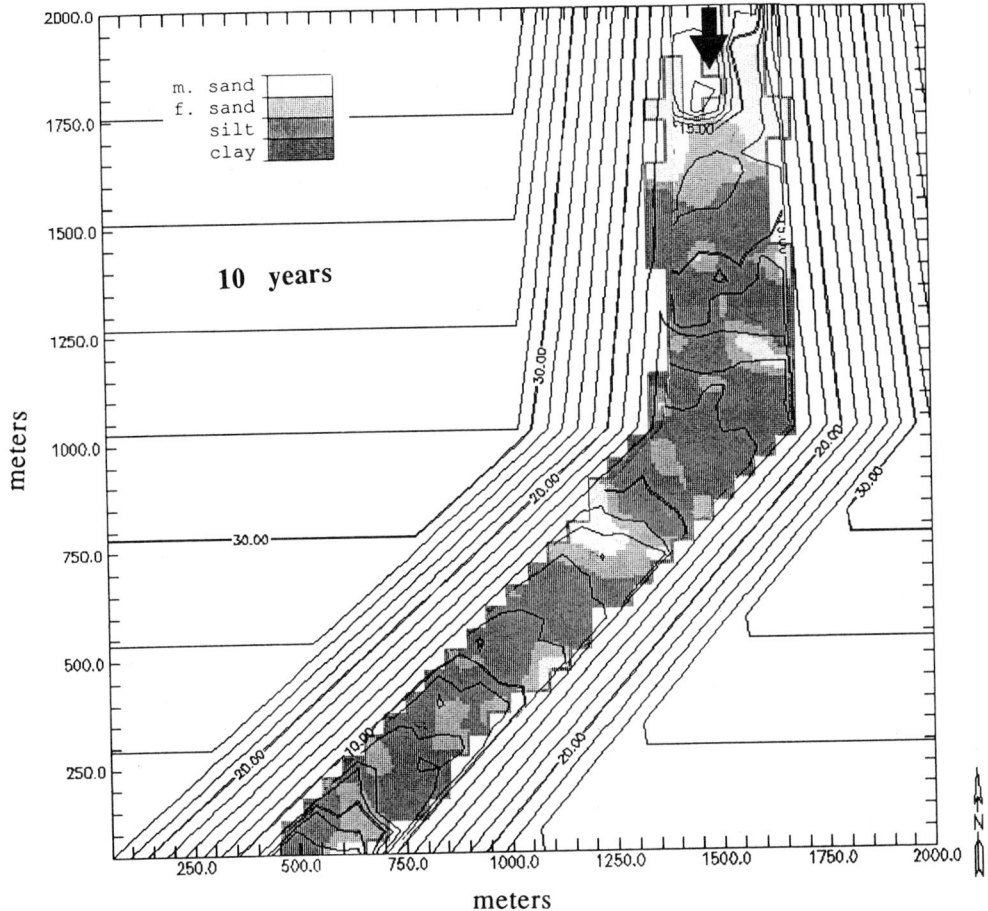

Fig. 6b. Facies and topography at ten simulated years. Geometric forms of channel deposits have remained stable, with additional deposition of silt and clay.

THREE-DIMENSIONAL SIMULATION OF LITTORAL TRANSPORT

PAUL A. MARTINEZ

Department of Applied Earth Sciences, Stanford University
Stanford, CA 94306

ABSTRACT

A computer model named "WAVE" has been developed to simulate the development of coastal environments affected by wave-induced erosion, transport, and deposition. The model simulates wave refraction, wave-current interaction, wave set-up, wind, oscillatory wave motion, longshore currents, rip currents, and nearshore sediment transport. Combining these processes into a unified model provides a process-response model that is dynamic in that the effects of wave energy on an evolving coastline are simulated through time. A series of equations based upon the principles of fluid dynamics as applied to coastal hydraulics, are used to calculate nearshore currents within a grid network representing a coastal area. The equations assume that mass and momentum are conserved as waves shoal towards shore, and are solved by integrating over the total water depth, and time-averaging over a single wave period. A finite-difference scheme provides solutions for the wave equations at each node in the grid. Once nearshore currents are calculated, sediment transport is initiated using empirical relationships that are effective in predicting rates of littoral transport.

KEYWORDS

Littoral transport, Sand transport by waves, Ocean waves, Computer modeling, Simulation.

INTRODUCTION

At present, many computers models exist for modeling fluid dynamics of nearshore current systems, but few attempt to simulate sediment transport by nearshore currents. WAVE simulates wave-induced currents in the nearshore area and is linked to algorithms that redistribute clastic sediment in response to these currents, thus creating a three-dimensional model that simulates depositional processes in nearshore areas. The computing procedures also incorporate procedures devised by other authors, producing a unified simulation model that simulates various aspects of clastic sedimentation, including sediment transport by currents of various sources, including those induced by waves.

WAVE can be used to simulate littoral transport at various geographic scales. Presented here is one application of WAVE involving simulation of sediment transport past a groin. Detailed simulations spanning small areas can focus on coastal engineering problems involving the erosion of beaches, or the effects of man-made structures on littoral transport. Research of littoral transport has often involved tracking the movement of tracer materials (such as sand dyed with distinctive colors) in the surf zone, providing an opportunity to compare simulated wave-induced littoral transport with actual littoral transport. In this simulation experiment, a marker-in-cell method can be used in conjunction with a finite-difference solution of wave parameters. Once wave hydrodynamics have been computed for each grid node, simulated tracers representing the dyed sediment grains can be introduced into the grid and are tracked as they move within the wave-induced current system, providing an accurate means of calibrating littoral transport rates. Such experiments can be particularly useful where engineering remediation steps are proposed, such as in the construction of groins to stabilize beaches.

WAVES AS A SOURCE OF NEARSHORE CURRENTS

Sand generally moves in the direction of wave-induced currents. Bedload will tend to move with bottom currents, whereas suspended load may be more affected by currents higher in the water column. In simulating sediment transport by waves, the properties of wave-induced currents are critical. Waves of all types, whether water waves, light waves, or radio waves, embody the propagation, transfer, and dissipation of energy. In simple terms, we can think of water waves as sources of energy, stemming from thermal energy transformed to wind, which in turn is partly transferred to and propagated through the water. It is this form of kinetic energy, supplied through waves, that does much of the work in a nearshore area. Some of this kinetic energy is dissipated as heat through friction, some is dissipated within the breaker zone by the turbulent, breaking motion of waves, and the remainder may create the "excess momentum fluxes" that are capable of producing longshore and rip currents and mean changes in sea-level within the surf zone. These nearshore currents, thus provide the power to move sediment, and it is therefore essential that the equations for simulating sediment transport by waves incorporate the conservation laws, including conservation of energy. Magnitudes of currents and rates of nearshore sediment transport are all a function of the wave energy supplied by the local wave climate.

Figure 1 describes some important terms and concepts pertaining to beach environments. Waves slow, shorten, and steepen as they travel from deep water into shallow water. Also, waves often approach the shore at an angle, causing paths of waves to bend or refract, creating complex nearshore currents that must be described in three dimensions. Thus, simulation models should be three-dimensional if they are to adequately represent this complex system. WAVE, being quasi three-dimensional, simulates nearshore circulation, and also represents wave refraction, orbital wave motions, wave-current interaction, longshore currents, rip currents, wave setup, convective accelerations, lateral mixing, and wind effects.

SAND TRANSPORT BY WAVES

The processes by which sand is transported changes across the shorezone, as do the nearshore currents generated there (Fig. 1). Onshore-offshore transport resulting from oscillatory motion of waves dominates seaward of the breaker zone, and involves pulsating motions generally parallel to the direction of wave propagation. By contrast, longshore transport dominates shoreward of the breaker zone, and generally involves continuous movement of water and sand parallel to shore. Rip currents produced by nearshore circulation also occur shoreward of the breaker zone, and provide a mechanism for moving sediment offshore.

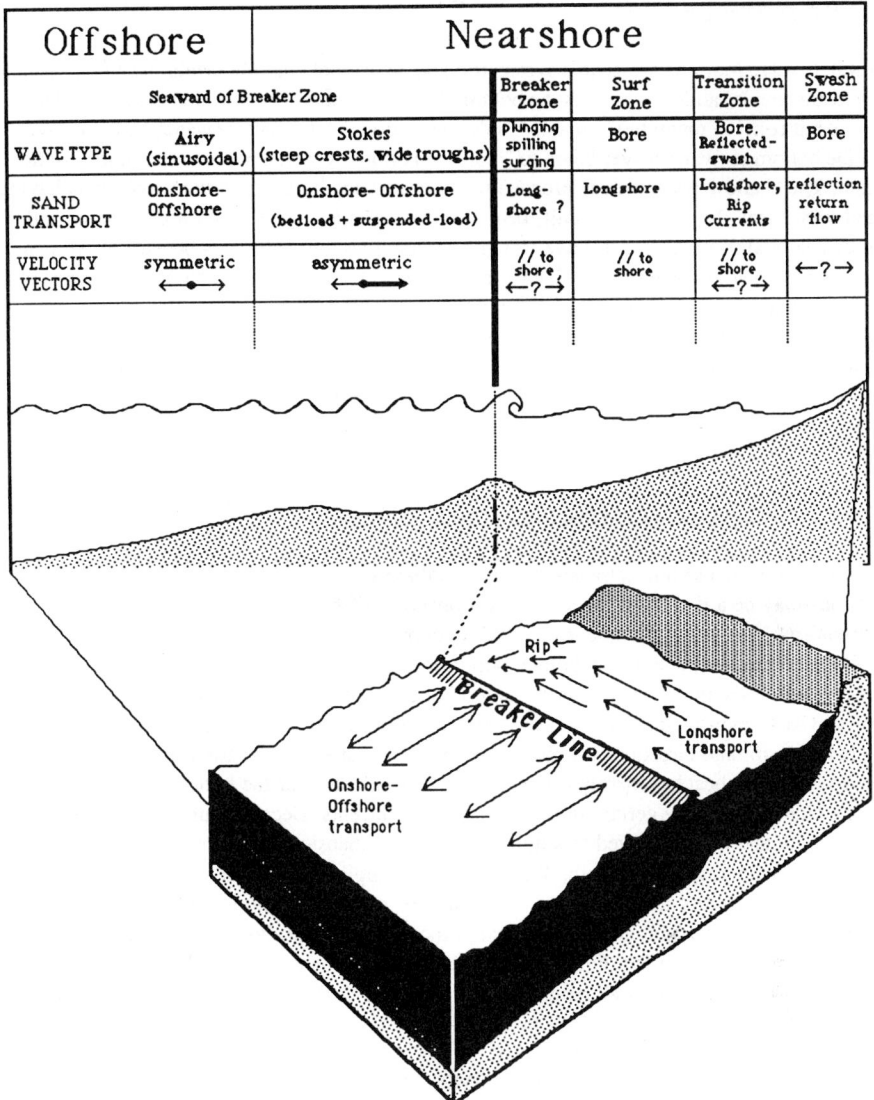

	Offshore		Nearshore			
	Seaward of Breaker Zone		Breaker Zone	Surf Zone	Transition Zone	Swash Zone
WAVE TYPE	Airy (sinusoidal)	Stokes (steep crests, wide troughs)	plunging spilling surging	Bore	Bore. Reflected-swash	Bore
SAND TRANSPORT	Onshore-Offshore	Onshore- Offshore (bedload + suspended-load)	Long-shore ?	Longshore	Longshore, Rip Currents	reflection return flow
VELOCITY VECTORS	symmetric ←•→	asymmetric ←•→	// to shore, ←?→	// to shore	// to shore, ←?→	←?→

Rip←

Breaker Line

Longshore transport

Onshore-Offshore transport

Fig. 1.   Dominant nearshore currents and mechanisms of sand transport in littoral zone. Sand movement generally reflects movement of wave-induced bottom currents. Compiled from ideas of Ingle (1966), Komar (1971), Longuet-Higgins (1970) and Stive and Battjes (1984).

Onshore-Offshore Sand Transport

In deep water, wave motions are nearly symmetric and sinusoidal, and the orbital motions associated with waves there are circular and do not affect the bottom. Bottom velocities created under wave peaks are opposite in direction, but equal in magnitude, to those under wave troughs. In shallow water, however, wave motions do affect the bottom as water waves begin to "feel" the bottom as orbital motions encounter friction from the sea bottom. The shallow water waves proceed to shore developing asymmetric waveforms having steeper crests and wider troughs, and as a result, orbital motions under wave crests are stronger and of shorter duration than orbital motions under wave troughs, creating bottom velocities beneath wave peaks that are greater than those below wave troughs. Thus, in shallow water, wave action alone can transport sediment shoreward because of the asymmetry of bottom velocities (Fig. 1). Larger grains may be moved shoreward by strong velocities, but return velocities may be too weak to move them back again. Lesser particles are also moved back and forth, and there may also be a slow net movement of particles in the direction of wave propagation toward shore. Fine particles, however, tend to be moved back and forth in suspension within wave orbitals, and can be moved seaward under the effects of gravity.

Sand Movement by Longshore Currents, Rip Currents, and Bores

Generalizations about net sediment transport due to oscillatory motion are greatly simplified. While oscillatory wave motions may be a dominant mechanism for onshore-offshore transport seaward of the breaker zone, other mechanisms such as longshore currents and rip currents are the dominant mechanisms for moving sand shoreward of the breaker zone. For example, within the breaker zone undertows or return flows associated with breaking waves are important in transporting sediment offshore. Within the shore zone, the change in the nature of waves as they reach shallow water, break, reform into bores, and swash onto shore, provides different mechanisms and directions for moving sediment. Figure 1 demonstrates that the changing nature of sand movement along beaches merely reflects the changing nature of the bottom currents. Different shore regimes are characterized by certain dominant types of currents. Because current velocities predicted by WAVE are depth and time-averaged to simplify the three-dimensional representation by dealing only with mean quantities with respect to depth, WAVE is most applicable for simulating sand transport in the shorezone, where currents and sediment transport are nearly continuous. Offshore, the time-varying oscillatory movements of waves are more critical in moving sediment shoreward and seaward, and WAVE's algorithms may be oversimplified. Furthermore, the extreme turbulence in the breaker zone and the run-up motion in the swash zone are not well represented by a depth and time-averaged solution.

FORMULATING WAVE'S COMPUTER MODEL

The principal governing equations in WAVE are described by Martinez (1987), Martinez and Harbaugh (1989), Tetzlaff and Harbaugh (1989), Longuet-Higgins (1970), Noda and others (1974), and Ebersole and Dalrymple (1979). These equations are formulated initially in differential form using finite-difference approximations and numerical solutions to solve the continuity equation and other sets of linear equations describing wave motion. These equations are time-averaged over a single wave period and are integrated over total depth at each grid location.

WAVE has been linked with SEDSIM, a computer model described by Tetzlaff and Harbaugh (1989) that simulates erosion, transport, and deposition of clastic sediments in natural environments. WAVE provides wave-induced current vectors, whereas SEDSIM simulates open-channel flow and erosion, transport, and deposition of clastic sediments. Changes in fluid flow and changes in deposits of clastic grains are simulated over intervals of geologic time, or as a progression of time increments. SEDSIM includes accounting procedures that allow it to represent simulated lithologies in three dimensions. Flow equations used in SEDSIM are solved using a "marker-in-cell" method involving a Langrangian method of representation, in which flow parameters are represented with respect to the moving "fluid elements" as a frame of reference (Fig. 2). A body of fluid is thus defined by an aggregate of individual "fluid elements" whose flow parameters are determined by assessing the positions and velocities of the individual fluid elements at each time step. The "marker-in-cell" method is advantageous because it allows representation of "tracers" in the form of fluid elements, to be followed as they move through the grid, and their motions observed as they are affected by wave-induced currents. These tracers, or fluid elements, can move either in steady or unsteady flow, and they can react to local currents and to changes in submerged topography caused either by erosion or by deposition in preceeding time increments. Thus, while WAVE simulates mean nearshore currents, SEDSIM provides the marker-in-cell system which tracks tracers as they move within the nearshore zone.

Program Execution

Conventions defining a typical experimental grid are shown in Figures 3 and 4. The grid is rectangular, with x-axis generally perpendicular to the shoreline, and y-axis parallel to the shore. Conventions for algebraic signs for the two axes are shown. Each grid row is of length Δx perpendicular to shore, and each column is of width Δy parallel to shore direction. Subscripts i and j are used to index the rows and columns in the grid or array and also serve as x and y cartesian coordinates, permitting values stored in arrays to be accessed by subscript numbers.

The characteristics of a propagating wave front are predicted by a "forward-finite-difference scheme" that evaluates characteristics in grid cell n+1 for the next time step, based wave characteristics during the previous time step n-1. Values of wave parameters for each time step are based on those of two immediately preceeding time steps. The first iteration provides a rough estimate of wave parameters throughout the grid, yielding initial values that the finite- difference solution scheme then utilizes.

The solution begins with a shoaling wave front that passes through the grid system, and assumes initially that the beach is smoothly sloping or planar in form (Fig. 5). Initially the model is at rest and all velocity components are zero. Snell's law provides an estimate of wave heights and refraction angles during the first iteration, yielding initial values over the entire grid. Subsequent iterations use a finite-difference scheme to update the initial values. Relaxation techniques then provide new velocities, wave numbers, wave setup values, and wave heights, based on those in previous time increments. Relaxation then causes the numerical solutions to converge. In the relaxation procedure, wave height is built up from zero to its full deep-water value over a specified number of iterations. Gradual buildup prevents "shock loading" and is analogous to gradual generation of waves in an experimental wave tank instead of suddenly propagating a large wave front through the tank.

Simulating wave processes at varying geographic scales

Experiments must be designed to accommodate the amount of resolution and accuracy that is desired by the experimentalist. Simulating small areas of coastline, with small time steps yields the most accurate results.

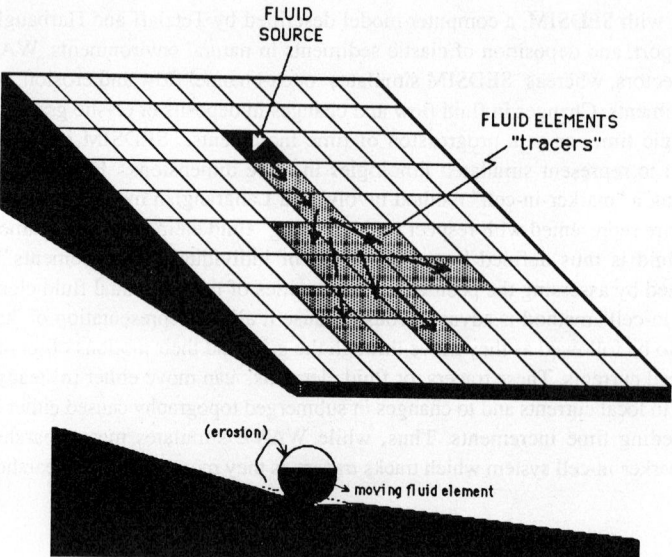

Fig. 2. Schematic representation of "marker-in-cell" method. Fluid elements flow through grid, representing flow of dyed sand or other tracers that can react to wave-induced currents. Fluid elements can also erode, transport, or deposit sediment.

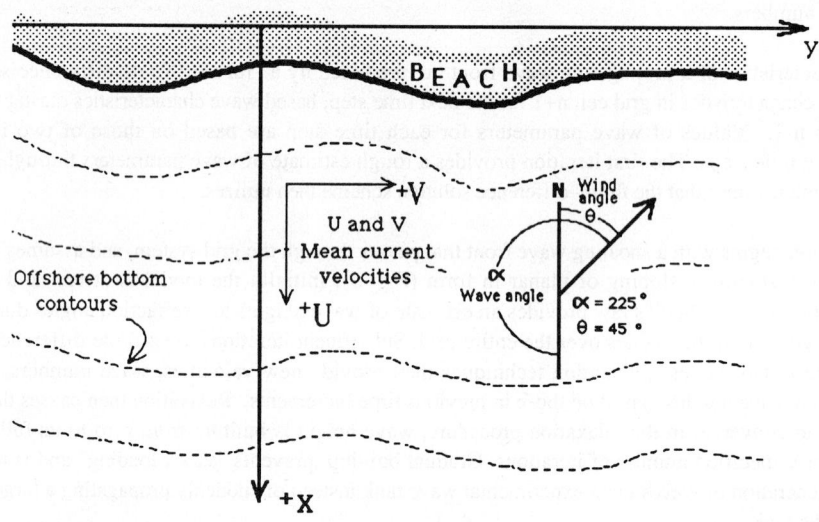

Fig. 3. Plan view of nearshore beach, illustrating terminology. Adapted from Noda and others (1974).

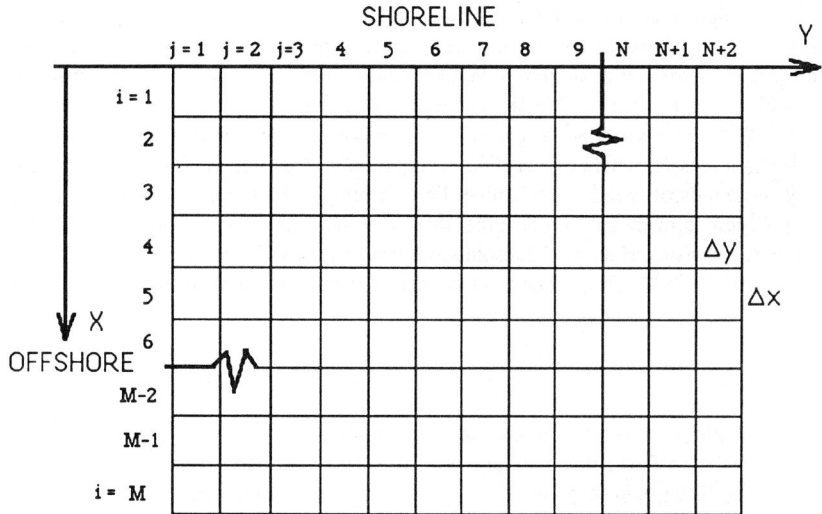

Fig. 4.   Grid used in WAVE in which M and N designate rows and column augmented to original grid. Rows and columns are indexed as i and j and serve as x and y coordinates for cells in grid. Grid represents three-dimensional coastline where each cell is assigned a specific depth and lateral dimensions, Δx and Δy.

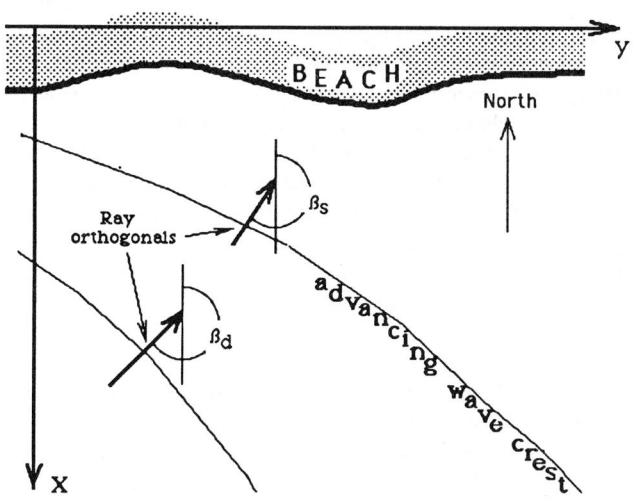

Fig. 5.   Terminology used to define wave refraction. β_d is wave angle in deep water, β_s is wave angle in shallow water. WAVE begins by propagating waves across grid representing beach topography, using Snell's Law to determine a first-approximation of wave parameters.

Loss of resolution might seem to inhibit WAVE's application to coastal environments such as large deltas whose longshore and offshore dimensions are commonly expressed in kilometers or tens of kilometers. However, the specific degree of resolution in WAVE's experiments merely reflects the degree of resolution reasonable at such scales. Loss of resolution may be acceptable where experiments may have goals that differ with geographic scale. An observer might be interested in onshore-offshore current fluctuations when observing a few hundreds of meters of coastline, whereas only gross trends of longshore transport might be important over thousands of meters of coastline. For example, if the focus is on gross trends of longshore transport along hundreds of meters of coastline, there may be little concern for minor effects of onshore-offshore velocity fluctuations. If increased resolution is critical, arrays in WAVE can be increased to provide a denser grid for simulations, at the expense, of course, of increased computer time required for the greater number of calculations.

Time Extrapolation

Simulation of "geologic time" over thousands of years requires that flow calculations determined for short time intervals be reused, or extrapolated, to represent larger intervals. This is accomplished by allowing flow and transport in a velocity field to be unchanged for a number of time steps, an extrapolation procedure informally called "compute and drift".

Equations for Sand Movement

Many equations have been proposed to describe movement of sediment (King and Seymour (1989). A major problem in adapting these to sediment transport in nearshore environments is that no single equation is relevant to the transport processes that occur there. There is general agreement that waves are primarily responsible for mobilizing sediment, but currents of different origin may affect the sediment once it is mobilized. For example, Longuet-Higgins (1970), Komar (1971) and Komar and Miller (1973) describe how longshore currents transport sand by littoral drift. Stive (1984) describes offshore sediment transport produced by return-flow and rip currents. Bagnold (1940,1963,1966) mathematically describes downslope movement by gravity, and Eagleson and Dean (1961), and Bowen (1980) describe the onshore and offshore transport of sediment. Each computational procedure yields a sediment transport rate, but none of the procedures have been successfully applied throughout the shorezone. Because of the interdependencies between the different transport processes, it is necessary to utilize generalized formulas for sediment transport that are not totally "process-specific", and instead depend upon only a few physical parameters. Thus, a generalized transport model is required by which sediment transport is predicted from basic physical parameters, regardless of variations in the processes at work.

WAVE and SEDSIM employ different methods for moving sediment within the simulated grid. WAVE, because it employs a finite-difference method of solution, moves sand from grid node to grid node, where the transport rate out of a node is a function of the thickness of the mobile bedload, average grain size, and current velocity. Other methods for determining sediment transport, based more upon a first-principles approach proposed by Komar (1971) and Komar and Miller (1973) are being tested.

SEDSIM, because it employs a marker-in-cell method, tracks the movement of sediment carried by individual fluid elements. Each fluid element has a "transport capacity", or ability to move sediment, based upon its velocity, fall velocities of sediments, and the concentration of sediment already being carried. Criteria for erosion are based on Shields and Reynolds numbers for incipient motion, and local bottom shear stresses. The distance that sediment particles are transported is a function of their fall velocities. In this way, SEDSIM simulates transport of suspended-load and bed-load by assuming that less dense grains are held in suspension

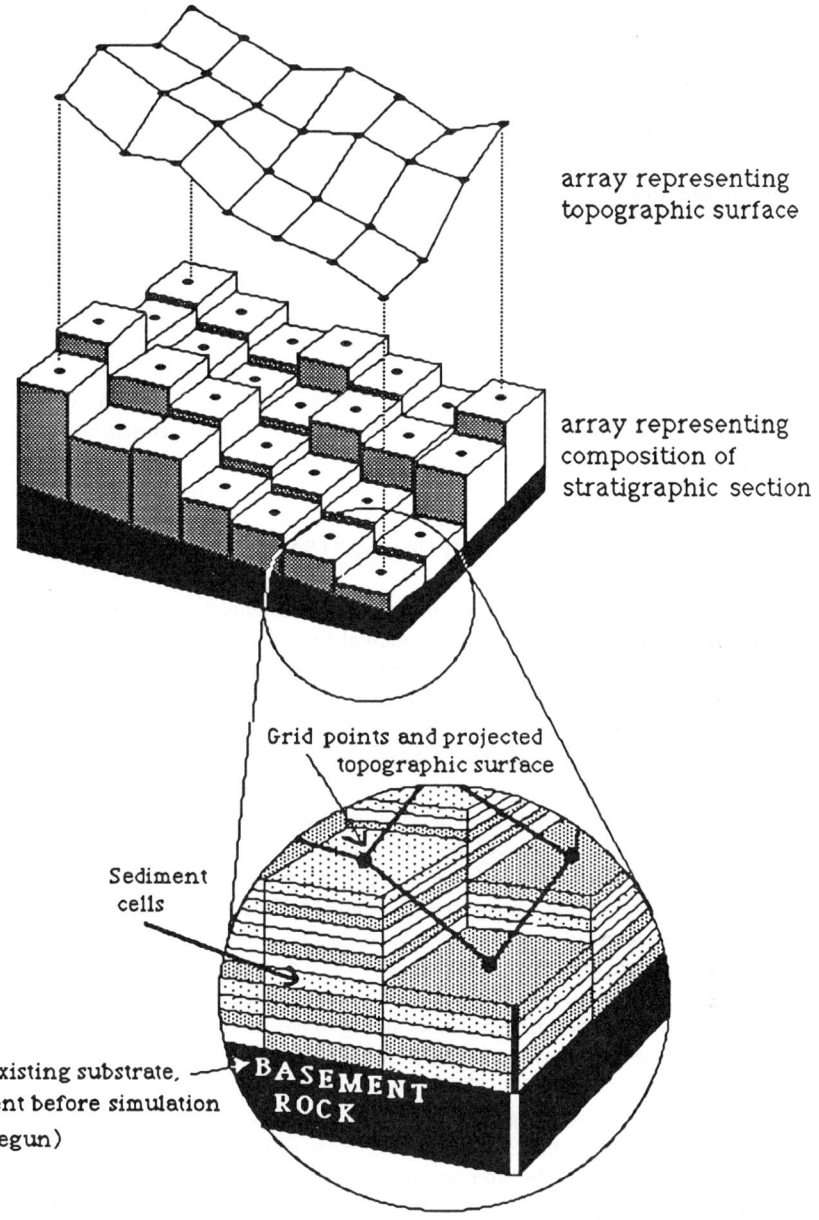

array representing
topographic surface

array representing
composition of
stratigraphic section

Grid points and projected
topographic surface

Sediment
cells

Pre-existing substrate,
(present before simulation
was begun)

BASEMENT
ROCK

Fig. 6.   Schematic representation of arrays used by WAVE and SEDSIM. Topographic surface is defined by array of grid cells. Flow characteristics are calculated at each grid point, sediment type is stored in separate array. Grid points correspond to centers of sediment cells. Thickness of sediment 'layers' or cells can vary, where each layer represents a variable mixture of four sediment types specified as input.

longer and transported farther than denser grains. Thus, the transport model created by combining SEDSIM and WAVE can simulate sorting of sediment by waves and other nearshore currents.

Three Dimensional Representation of Sand Transport by Waves

SEDSIM can simulate erosion, transport, and deposition of a combination of four sediment types whose grain diameters and grain densities are specified as input. Deposits formed by SEDSIM are represented in a three-dimensional array that represents only those deposits that are preserved (Fig. 6). A separate array defines topographic elevation as a fitted surface. Erosion is represented by removal of 'stacked' cells from the sediment array, whereas deposition is represented by 'stacking' additional cells containing sediment onto those containing previous deposits. Sediment, in the form of 'sediment cells', can be carried into or out of the area being simulated by moving fluid elements that define the flow. Characteristics of the flow (fluid elements) are determined at grid points that correspond to the centers of grid cells. WAVE and SEDSIM use "accounting" techniques that conserve mass, where sediment is neither created or destroyed (although sediment may pass into or out of the simulated area).

SIMULATING MOVEMENT OF BEACH SAND AROUND A GROIN

Groins represent one of the oldest and most commonly used structures for controlling littoral sand transport. Major engineering problems have resulted when natural littoral transport has been interrupted by groins because they can significantly change local wave-induced circulation and sand transport, creating excessive erosion along some portions of the coastline, and excessive sediment accumulation in others. Experiments can be designed to simulate the impact of groins of various designs and configurations on the local wave climate and nearshore circulation system, providing insight for better groin design and placement.

A large body of literature and engineering studies concerning groins provides a plentiful source of data for computer simulation experiments of beach environments. In many of these studies researchers use fluorescent- and radioactive- dyed sands as "tracers" to provide insight into the movement of beach sand in the vicinity of coastal structures. Figure 7 shows a groin studied by Ingle (1966), who used dyed sand grains to study the movement of beach sand around a groin. Sand accumulated on the up-current side of the groin, and waves, reacting to the change in submerged topography created by the groin, broke at different locations on either side of the groin. Magnitudes and directions of currents were also affected by the groin, as shown in Fig.7. Results of Ingle's tracer experiment shown in Fig. 8 can be compared to Figs. 9 and 10, produced from a simulation experiment using WAVE. The simulated wave climate is similar to that observed by Ingle, where waves approached at 15 degrees to shore (Fig. 7). Patterns of sediment dispersal and wave-induced circulation are similar, causing littoral transport of sediment against the groin, but the computer simulation can provide more versatility in studying the interaction between the wave climate and groin design. For example, Ingle noted that sand transport past the groin was most prevalent when the breaker line occurred beyond the seaward reach of the groin, whereas sand transport past the groin was inhibited or stopped when waves broke shoreward of the groin's extent. Simulation of different wave climates using WAVE could provide insight into selecting an optimum length for the groin. Such a determination cannot simply involve study of the local wave climate in the area where a groin is to be built because the building of the groin will immediately change the local submerged topography and alter wave-induced currents and rates of sand transport in ways which are difficult to predict.

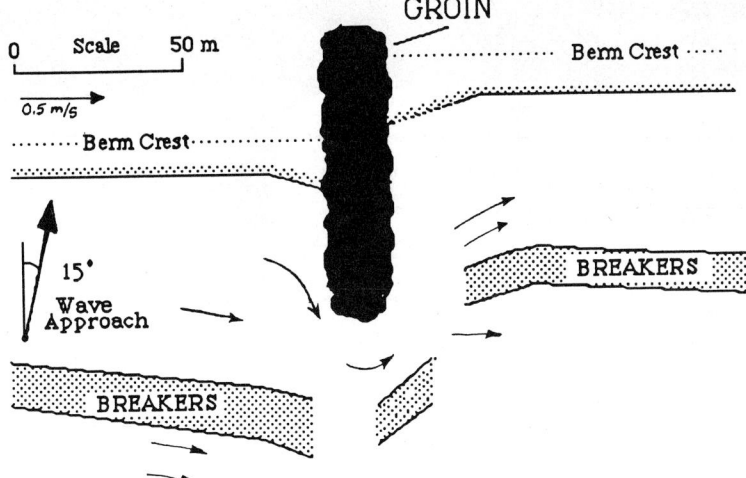

Fig. 7. Current directions, velocities, and nature of breaking waves around groin studied by Ingle (1966). Location of breakwater shifted with changing wave climate and tides.

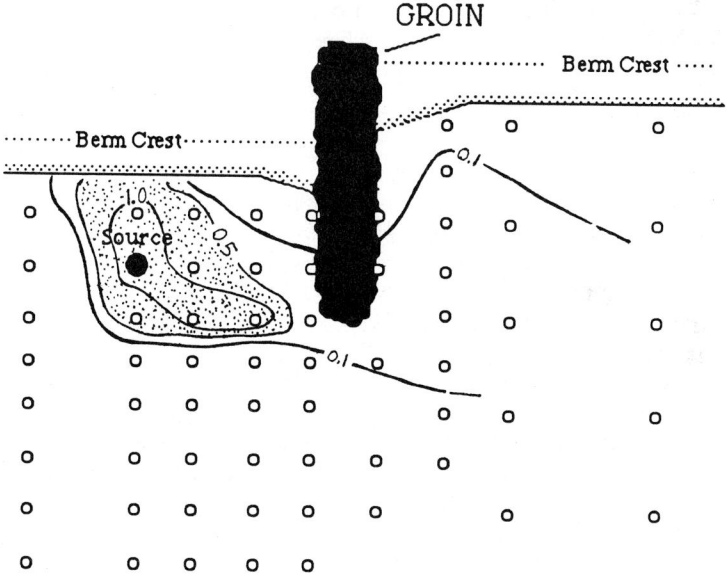

Fig. 8. Movement of fluorescent-dyed sand around groin. Sand was released at position of large filled circle, sample stations are denoted by open circles. Contours represent equal numbers of dyed grains per square inch, 90 minutes after release. Sand moved towards breaker zone and past groin. After Ingle (1966).

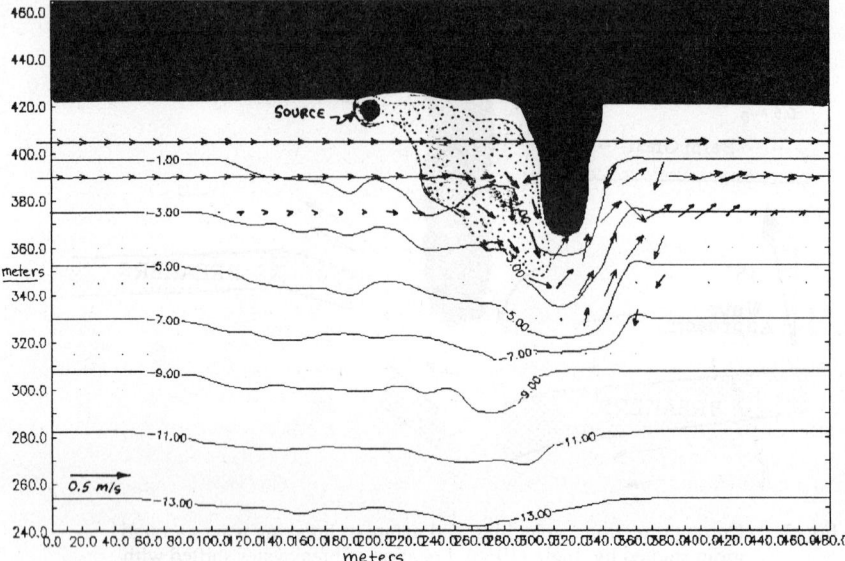

Fig. 9.   Plan view of simulated beach and groin showing results of WAVE compared to those of Ingle (1966), for a similar wave climate. Simulated sand "tracers" were released at position of large filled circle, stippled area represents path of moving tracers after approximately 7 hours of simulated time Contours representing submerged topography show evidence of littoral transport and accumulation of sand against groin, as groin changes local beach topography and wave climate.

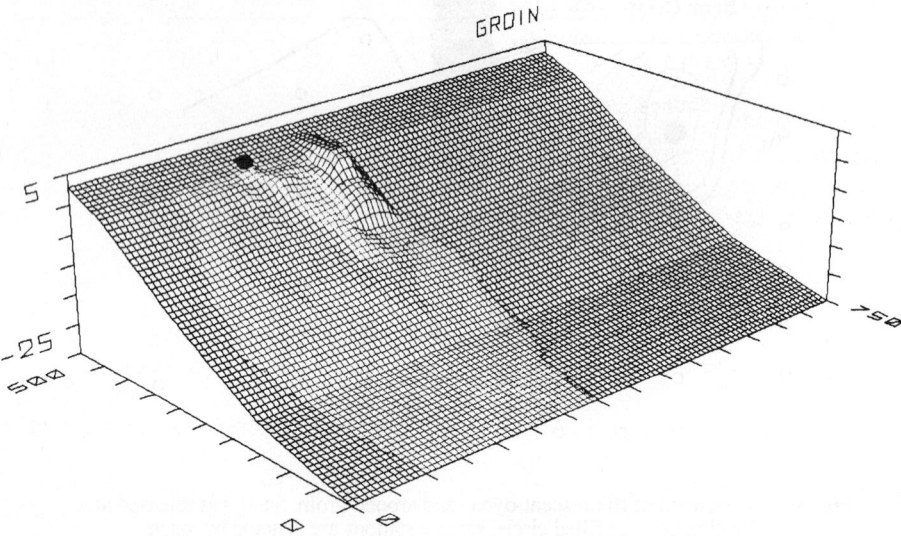

Fig. 10.   Three-dimensional diagram of groin depicted in Fig. 9., after 24 hours of simulated time Release point of tracers shown by filled circle. Patterns of sediment dispersal accumulating against groin are indicated by shades of grey.

CONCLUSION

The computer model described here simulates nearshore environments affected by wave-induced currents capable of eroding, transporting, and depositing sediments. These simulations incorporate theories that describe propagating water waves, open-channel flow, and sediment transport, as expressed mathematically and incorporated into computer subprograms. By joining many subprograms into a single integrated program, we have produced a computer model that simulates complex, dynamic processes, which when operated in concert produce experimental results helpful in understanding interdependencies between physical processes that are almost impossible to deal with in purely conceptual models.

WAVE allows a relatively unlimited framework for experimentation, where a variety of variables can be selectively changed or held constant, under readily controlled conditions. Physical models such as wave tanks or small replicas of larger systems do not allow such versatility. For example, it is much easier to simulate and isolate the effects of a single parameter such as wave height, if we can hold all other variables constant. Computer models provide this ability; whereas on a real beach, wave periods, wind speeds, and offshore wave height constantly change, making it difficult to isolate and study the effects of any one variable. Similarly, simulation of complex beach environments allows the experimentalist to change a single variable, such beach slope or water depth, without having to consider possible variations in a host of other parameters. Simulations of nearshore environments also provide insight in understanding complex, three-dimensional geologic features. For example, the experiment presented here reproduces many morphologic features and facies relationships characteristic of man-made structures in the shorezone. Computer graphics procedures allow these morphologic features and facies relationships to be observed with the aid of cross sections, topographic maps, and facies maps (Figs. 9, 10). By contrast, studies of modern beaches may only provide a few sections or maps because data are often limited and difficult to obtain. Dynamic simulation models such as WAVE and SEDSIM provide alternative procedures to the use of static, idealized, and largely conceptual models that geologists currently use to study dynamic nearshore environments.

REFERENCES

Bagnold, R. A. (1940). Beach formation by waves; some model experiments in a wave tank. In: *Beach Processes and Coastal Hydrodynamics* (J.S. Fisher and R. Dolan, eds.), Dowden, Hutchinson & Ross, Inc., Stroudsburg, Pennsylvania, Benchmark Papers in Geology , v. 39, p. 281 - 303.

Bagnold, R. A. (1963). Mechanics of marine sedimentation. In: *The Sea* (M. N. Hill, ed.), John Wiley & Sons Inc. - Interscience Pub. Division , New York, v. 3, p. 507- 528.

Bagnold, R. A. (1966). An approach to the sediment transport problem from general physics. *U. S. Geological Survey Professional Paper* , no. 422-I, 37 p.

Birkemeier, W. A. and R. A. Dalrymple (1975). Nearshore water circulation induced by wind and waves. In: *Proc. of the Symposium on Modelling Techniques* : Am. Soc. Civil Engineers, San Francisco, CA., p. 1062-1081.

Bowen, A. J. (1980) Simple models of nearshore sedimentation; beach profiles and longshore bars. In: *The Coastline of Canada - Littoral Processes and Shore Morphology* (J. S. McCann, ed.), Canadian Govt. Pub. Centre, Hull, Quebec , Geological Survey of Canada Paper, no. 80 - 10, p. 1- 11.

Eagleson, P. S. , and R. G. Dean (1961). Wave induced motion of bottom sediment particles. *Am. Soc. of Civil Engineers* , Transactions, v. 126, p. 1162 - 1189.

Ebersole, B. A., and R. A. Dalrymple (1979). A numerical model for nearshore circulation including convective accelerations and lateral mixing. Ocean Engineering Report no. 21, Dept. Civil Engineering,

Univ. of Deleware, Newark, Del., Tech. Report no. 4, Office of Naval Research Geography Programs, 87 p.

Ingle, J. C. (1966). *The Movement of Beach Sand* . Elsevier Pub., New York, 221 p.

King, David B. and R. J. Seymour (1989). State of the Art in Oscillatory Sediment Transport Models. In: *Nearshore Sediment Transport* (R.J. Seymour, ed.), Chap. 16, pp. 371-385. Plenum Press, New York.

Komar, P. D. (1971). The mechanics of sand transport on beaches. *J. Geoph. Res.*, v. 76, no. 3, p. 713 - 721.

Komar, P. D., and M. C. Miller (1973). The threshold of sediment movement under oscillatory water waves. *J. Sediment. Petrol.*, v. 43, no.4, p. 1101 - 1110.

Longuet-Higgins, M. S. (1970). Longshore currents generated by obliquely incident sea waves, 1 and 2. *J. Geoph. Res.*, v. 75, no. 33, p. 6778 - 6801.

Martinez, P.A. (1987). *Simulation of Sediment Transport and Deposition by Waves, for Simulation of Wave Versus Fluvial-Dominated Deltas*. Masters thesis, Stanford University, Stanford, California.

Martinez, P.A. and J. W. Harbaugh (1989). Computer simulation of wave and fluvial-dominated nearshore environments. In: *Applications in Coastal Modelling*. (C.V. Lakhan and A.S. Trenhaile, eds.), Elsevier Oceanography Series 49, Elsevier Publishing Co., New York, p. 297-337.

Noda, E. K., J. C. Sonu, V. C. Rupert, and J. I. Collins (1964). Nearshore circulations under sea breeze conditions and wave-current interactions in the surf zone. Tetra Tech, Inc., Pasadena, Ca., Technical Report No. TC-149-4, 213 p.

Stive, M. J. F., and J. A. Battjes (1984). A model for offshore sediment transport. In: *Proceedings 19th International Coastal Engineering Conference* : Am. Soc. Civil Engineers, New York, v. 2, p. 1420 - 1436.

Tetzlaff, D. M. and J. W. Harbaugh (1989). *Simulating Clastic Sedimentation*: Van Nostrand Reinhold, New York, 202 p.

MODELING COMPACTION AND ISOSTATIC COMPENSATION IN SEDSIM FOR BASIN ANALYSIS AND SUBSURFACE FLUID FLOW

Johannes WENDEBOURG and J.W. Dominik ULMER
Department of Applied Earth Sciences, Stanford University
Stanford, CA 94305, U.S.A.

ABSTRACT

The clastic sedimentation simulation program SEDSIM has been extended for basin-scale applications to model compaction and isostatic compensation. A "microgeometrical" model after Marion (1990) was used to assign initial porosity to simulated synthetic sediment layers consisting of continous mixtures of different grain sizes. This model has been confirmed under different stress conditions by experimental data (Yin et al. 1988) and can be used to model compaction as the consequence of changes in sediment porosity due to overburden and pore fluid pressure. Permeability, which is an important parameter for subsurface fluid flow, is determined by the Kozeny-Charman equation (Palciauscas 1986). Isostatic compensation is modeled as flexure of an inhomogeneous elastic plate, involving the buoyancy force of the underlying mantle. Burial history of sedimentary sequences simulated by SEDSIM in turn can be analysed using conventional basin analysis techniques. As an example application, the development of a delta was modeled over one million years and the resulting stratigraphic sequence was then buried beneath 3,000 meters of sand-rich sediments.

KEYWORDS

Computer simulation, basin analysis, compaction, isostasy, porosity, permeability.

INTRODUCTION

Compaction and isostatic response of the lithosphere to changes in sediment load are important, and therefore in simulation models of basins, they should be coupled with sedimentation processes. These processes affect topography and in turn they also affect sedimentation. In addition, compaction is an important aspect of fluid flow modeling, modifying the magnitude and spatial distributions of rock properties, such as porosity and permeability. Compaction and isostatic compensation are scale dependent. SEDSIM, Stanford's SEDimentary Basin SIMulation project, can model a variety of depositional environments at varying geographic scales, depending on grid size, time-step resolution, and flow velocities (Lee & Harbaugh 1990). This paper describes how rock properties, compaction and isostatic compensation are implemented in SEDSIM, and also describes how differences in scale that affect localized features formed by sedimentation processes can be linked with large-area basin-forming processes that include compaction and isostatic compensation.

ROCK PROPERTIES

Calculating petrophysical properties that affect flow in heterogeneous media is critical in modeling subsurface fluid flow. Movements of subsurface fluids have large influence on geological and sedimentological phenomena and therefore need to be quantified despite lack of data and poorly understood sedimentary record of past fluid activities. Permeability is the most important parameter that affects fluid flow, and its variation in the 3-D distribution of permeability in rocks is the result of primary sedimentary structures and subsequent diagenetic changes (Weber 1982). Frequency distributions of permeability are often multimodal and non-continuous (Deutsch 1987). Effective permeability is sensitive to spatial distribution, and as an example, permeabilities within ancient distributary channels formed in deltaic environments may change over several orders of magnitude within short distances (Dreyer et al 1990). Effective permeability, as well as other rock properties, can be calculated by averaging or "upscaling" techniques applied to observational data (Deutsch 1987). Alternatively, it is also possible to use quantitative computer models of geologic processes and thereby simulate specific depositional environments that yield simulated lithologic distributions, which in turn provide three-dimensional spatial distributions of rock properties. SEDSIM can provide synthetic stratigraphic records to which petrophysical properties may be assigned (Tetzlaff & Harbaugh 1989). SEDSIM assumes that appropriate petrophysical relationships are available by which rock properties can be related to the deposits formed by specific processes operating in specific sedimentary environments.

Porosity

Within a SEDSIM experiment, a certain amount of sediment may be deposited or eroded during each timestep within a grid cell. Empirical relationships between flow velocity, sediment load, shear stress, and grain size frequency distributions govern the sedimentation response (Lee & Harbaugh 1990). When deposition occurs, an initial porosity can be assigned to the deposit that has been formed representing the water content by volume of newly deposited, unconsolidated material. Assignment of an initial porosity for either pure clays or pure sands is more or less straightforward (Bethke 1989) but initial porosities of mixtures of clastic sediments are more difficult to specify.

SEDSIM, however, can produce synthetic deposits in which continuously ranging sediment mixtures containing up to four distinct grain sizes are formed. Deposits consisting of mixtures cannot be described simply as either clay or sand, but a "microgeometrical" model can be used for intermixtures which can be described as sands, clay-rich sands, sandy clays, and clay (Marion 1990, Fig. 1).

The relationship of initial porosity to mixtures of grain sizes is influenced by the manner in which the grains are packed together. If the clay content of a deposit is less than the volume of porosity defined by the interstices between the sand grains, the clay is dispersed randomly within the framework of the sand grains. As the clay content increases, however, porosity decreases rapidly and is least where the clay content equals the volume of space between sand grains, which can be defined as the "sand porosity". If the clay content is greater than the sand porosity, the clay becomes the matrix, and the sand grains "float" within the clay. With progressively increasing shale content, clay replaces the volume that might otherwise be occupied by sand grains, increasing porosity. If we assume that the packing of sand grains remains constant, which is justified if the dispersed clay particles are small enough, a function relating porosity to clay content can be derived from "effective medium theory" (Palciauskas 1986) and can be confirmed with laboratory data (Yin et al 1988). In applying such relationships to SEDSIM, the four grain sizes used in SEDSIM must be grouped into grain sizes which represent the sand framework, and those which represent the clay matrix, employing a linear weighting procedure.

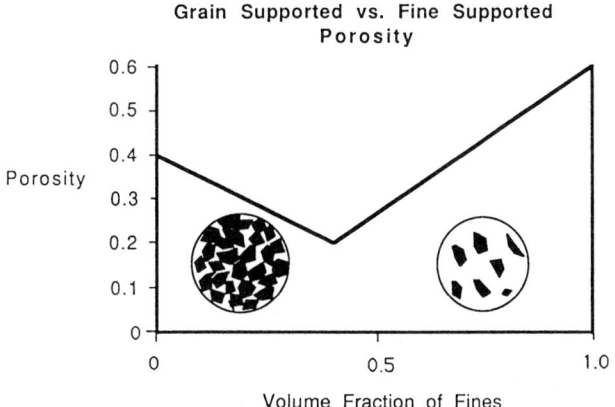

Fig. 1. Initial porosity as function of volume fraction of fine grained matrix (after Palciauskas 1986). Porosity minimum occurs when pores within framework formed by coarse grains are filled to maximum extent with fine grains.

Permeability

Intrinsic permeability depends only on rock properties and not on fluid properties, and is an important petrophysical parameter that is supplied to synthetic deposited material produced in SEDSIM experiments. In conventional sedimentary basin analysis with actual data, permeability is commonly derived from an empirical log-linear relationship that relates porosity and permeability (Bethke 1989), although observed permeabilities and porosities are often poorly correlated, with much scatter when data plots are made (Wendt et al 1986). Only unconsolidated packed sand yields experimental data in which permeability is a well-defined function of porosity, grain size, and sorting (Beard and Weyl 1973). Alternatively, permeability can also be estimated using the Kozeny-Carman equation (Palciauskas 1986), where permeability is a function of porosity, clay content, tortuosity, and specific surface area. Tortuosity is related to the geometry of the grain network, while specific surface area is related to the sizes of grains, being measured per unit volume of grains and expressed such that it is roughly inversely proportional to mean grain radii. The Kozeny-Carman equation is appropriate for use in SEDSIM because it is valid for non-spherical and non-ideally packed, statistically uniformly distributed particles, and when extended it is valid for bimodal frequency distributions of sediment grain sizes (Palciauskas 1986, Marion 1990). The Kozeny-Carman relationship also compares well with sandstone permeabilities with varying clay content measured in Gulf Coast wells (Fig. 2). The Kozeny-Carman model yields a permeability range of about five orders of magnitude from 5 to 30 percent clay content. A permeability minimum is accounted at about 30 percent clay content by weight. With clay content higher than 30 percent, permeability is almost constant.

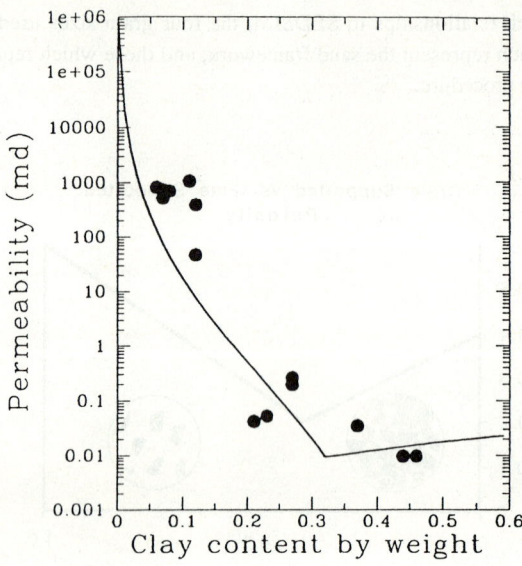

Fig. 2. Permeability versus clay content in Gulf Coast sandstones measured as air permeabilities (from Marion 1990).

COMPACTION

Compaction results in irreversible loss of porosity by changing the packing configuration, reducing primary porosity as overburden increases. The proportion of porosity loss is a function of total stress, which is the difference between deformational stresses in the rock matrix and the stress due to pore fluid pressure (Grün et al. 1989). In SEDSIM, grain compressibility is considered negligible and grain-size changes due to diagenetic processes are not considered. Furthermore, we assume in a SEDSIM simulation that the three-dimensional deformational stress field can be approximated by considering only its vertical component , because lateral stress variations stemming from differences in topography are small enough to be ignored (Bayer 1989). Thus, in SEDSIM, compaction is a function only of lithostatic pressure, which is increased as each new deposited layer of sediment is added to the overburden.

In conventional basin modeling, porosity decreases as an exponential function of either depth (Bethke 1989) or effective stress (Doligez et al 1986), and can be readily calculated for either pure sands or clay. In SEDSIM, however, mixtures of grain sizes are produced, and relationships to describe initial porosity as a function of clay content can be extended also as functions of effective stress. Functions relating porosity and effective stress have been determined experimentally (Fig. 3). Two observations can be made, namely that increasing lithostatic pressure shifts the porosity minimum to correspond with a higher clay content, and that porosity loss decreases with increasing pressure, most conspicuously for high clay content. Porosity data from Gulf Coast sandstones for varying depths indicate this porosity model for continous grain size mixtures (Marion 1990), and these relationships are implemented in SEDSIM by means of a bilinear lookup table.

In SEDSIM, compaction takes place under equilibrium conditions with hydrostatic pore pressures, and is simulated as a function of lithostatic pressure. This relationship works well at shallow depths for sediments undergoing progressive burial of normal sedimentation rates. If sedimentation rates and subsidence are very high, however, pore fluids cannot dissipate fast enough, so that fluid pressure remains above hydrostatic, with resulting abnormally low porosity loss. Therefore, compaction must be coupled with transient pore pressure calculations (Bethke 1989), so that it is linked with the degree to which pore-fluids are lossed. These relationships will be incorporated in SEDSIM.

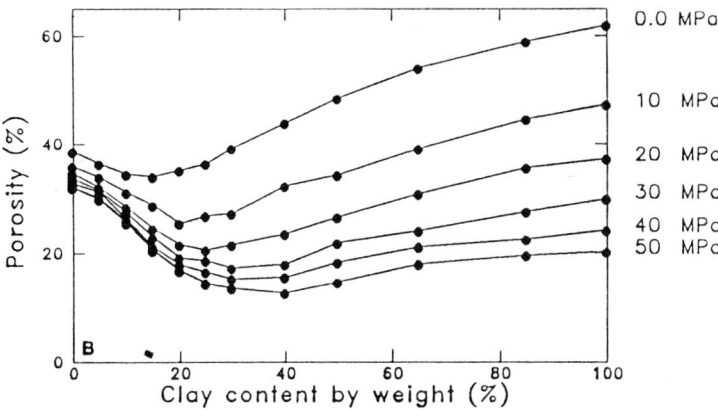

Fig. 3. Porosity versus clay weight fraction at various lithostatic pressures (data after Yin et al 1988 and from Marion 1990).

ISOSTATIC COMPENSATION

In our context, isostatic compensation involves subsidence of the crust when sediment load increases, or uplift if sediment load decreases. Subsidence modifies basin geometry, which in turn affects sedimentation. A modified Airy-model of isostasy is often used in sedimentary basin modeling, and assumes that the weight of water, sediment, and lithosphere are in Archimedes equilibrium with the asthenosphere (Turcotte & Schubert, 1982). In some basin simulation models, individual cells are employed that behave like blocks of ice in seawater that do not interact with each other, each adjusting instantaneously to achieve equilibrium when a change occurs (Fig. 4).

Such an assumption involves a crust without strength, thereby neglecting lateral effects of changes in load as subsidence or uplift occurs. In SEDSIM, however, isostatic compensation assumes that the crust has strength, which can be represented by flexure of an elastic plate overlying an inviscid fluid, employing the following governing differential equation (Ranalli, 1987)

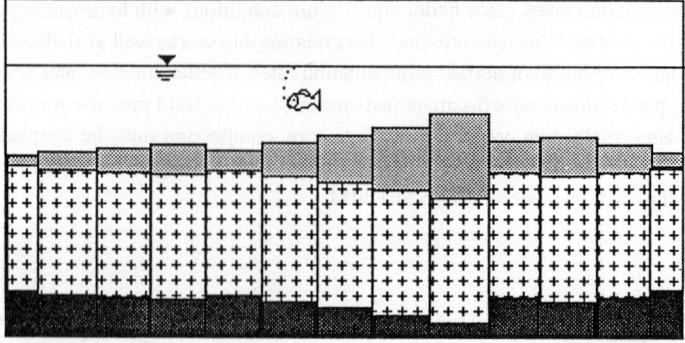

Fig. 4. Schematic diagram of Airy model. Columns containing water, sediment, and other lithosphere float independently on asthenosphere.

$$\nabla^2 \left(D \, \nabla^2 w \right) \;=\; P \;-\; (\rho_m - \rho) \, g w$$

(1)

where

w	= deflection (m)
D	= flexural rigidity of plate in Newton times square meter (Nm^2)
P	= load per unit area (Pa)
ρ_m	= density of asthenosphere (kg/m^3)
ρ	= density of material above lithosphere (kg/m^3)
g	= acceleration of gravity (m/s^2)

Flexural rigidity D (Walcott, 1970) describes the mechanical strength of the plate

$$D \;=\; \frac{E \, (EET)^3}{12 \left(1 - v^2 \right)}$$

(2)

where

E	= Young's modulus
EET	= Effective plate thickness in meter
v	= Poisson's ratio

In basin modeling, Equation (1) is often applied to the basement topography in a cross section which is one dimensional, either by superposition of an analytical solution for a line load with homogeneous mechanical properties after Turcotte & Schubert (1982) (Jordan & Flemings 1990), or by Fourier transform methods (Zoetmeijer et. al, 1990). In SEDSIM, Equation (1) is solved numerically in two dimensions with inhomogeneous spatial distribution of rigidity. A finite-difference scheme is employed and solved with a direct elimination technique. Alternative boundary conditions consist of either (a) a plate fixed between two walls, or (b) a plate lying upon two walls (Fig. 5), the latter procedure being incorporated in SEDSIM. To reduce boundary effects, the area involved in isostasy calculations covers up to four times the area of the grid used for the sedimentation simulations.

Although Equation (1) represents steady state conditions, quasi-transient behavior of isostatic subsidence can be simulated. In SEDSIM experiments, with sediment and water column loads continuously changing over small time spans, isostasy calculations are made in timesteps of 5,000 to 10,000 years to allow for a sufficient lag time for the crust's response.

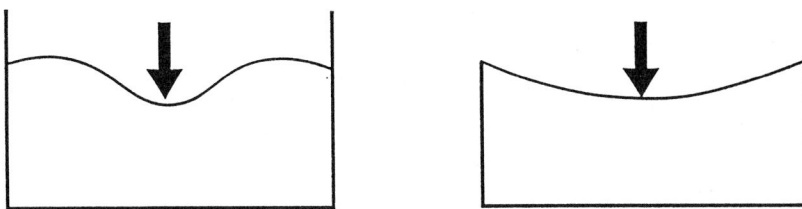

Fig. 5. Two alternative means of treating boundary conditions for bent elastic plates.

SCALE DEPENDENCE

In describing rock properties of porous media, three or sometimes four different scales are commonly distinguished (Bayer 1989). The "microscopic" scale pertains to pore size properties, the "macroscopic" scale pertains to core and laboratory measurements, the "megascopic" scale pertains to field size properties, and the "gigascopic" scale to overall basin properties. SEDSIM can simulate a wide range of scales of sedimentary environments and petrophysical properties with resolution depending on grid node spacing and on "time averaging" procedures for sediment deposition (Lee & Harbaugh 1990). In an example experiment (Martinez 1989) with node spacing ranging from 10 to 100 m, detailed sedimentary features including channel deposits and overbank deposits were created, including local variations in porosity and permeability. Such details require timesteps ranging from 10 to 100 years, which generally limits the aggregate time in an experiment to a range from 1,000 to 10,000 years, and to a maximum sediment thickness measured in tens of meter. However, if node spacing is measured in kms, permitting sedimentation patterns to change more gradually over such larger areas, deposition then can be averaged over timesteps ranging from 1,000 to 10,000 years, with aggregate simulation times of 100,000 to 1,000,000 years, and with maximum sediment thickness of hundreds of meter. However, as the time resolution decreases, the spatial resolution of rock properties likewise decreases. At present however, 3-D simulations of basin histories spanning tens of million of years is barely feasible, even with supercomputers, but procedures will be developed later utilizing SEDSIM's algorithms for basin experiments that should span much greater geologic time and permit simulations over large basin areas involving substantial thicknesses of deposits.

In conventional basin analysis, burial history is often analyzed in two main steps. The first step often involves "backstripping" of the sediment cover, which is performed with a "decompaction" algorithm which operates in reverse of actual processes. Sedimentation rates and the "lost geologic time" at unconformities, as well as water depths, must be specified in advance of the backstripping calculation, since only the amount of total subsidence may be known for a sequence of deposits in a basin which must be separated into individual compactional, isostatic, and tectonic components (Van Hinte & Deighton 1987). As a second step, the basin's history is then run in a forward mode, where thermal history and flow of pore fluids are represented (Bethke 1989). Backstripping, of course, is not an actual process, but it permits the analyst to represent large thicknesses of sediment over tens of million of years. SEDSIM, by contrast, is entirely a process-oriented model, and SEDSIM can run only in a forward mode because all actual processes are irreversible and must be simulated in a forward mode. In a computer time saving procedure to simulate thick deposits of sediments, we can assume that a sedimentary deposit once formed in a specific depositional environment, is not reworked later and undergoes only compaction and diagenetic changes. Then, conventional basin analysis techniques can be applied to the subsequent burial history of simulated deposits (Fig. 6), freeing computer time for other time-intensive modules that deal mainly with subsurface processes, such as pore water flow and hydrocarbon migration.

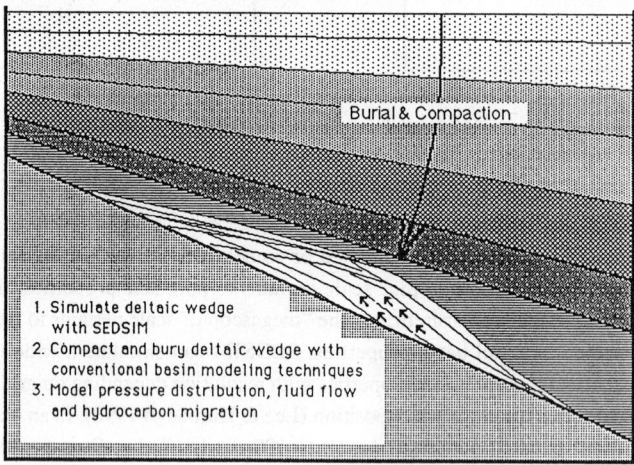

Fig. 6. Schematic diagram of procedure to simulate compaction and pore-fluid pressure with SEDSIM.

EXAMPLE APPLICATION

To verify implementation of procedures described here, a SEDSIM simulation involving the Upper Cretaceous Buda shelf in East Texas was run that spanned 1,000,000 years in which the resulting sediment body was buried and compacted beneath several kms of subsequently deposited sediment. The basement topography in the experiment is based on interpretation of the paleogeography as reconstructed from seismic sections and well data (Vail et al 1977). The simulated area extends over a rectangular grid, 70 by 86 km in extend, with a grid node spacing of 2 km (Fig. 7).

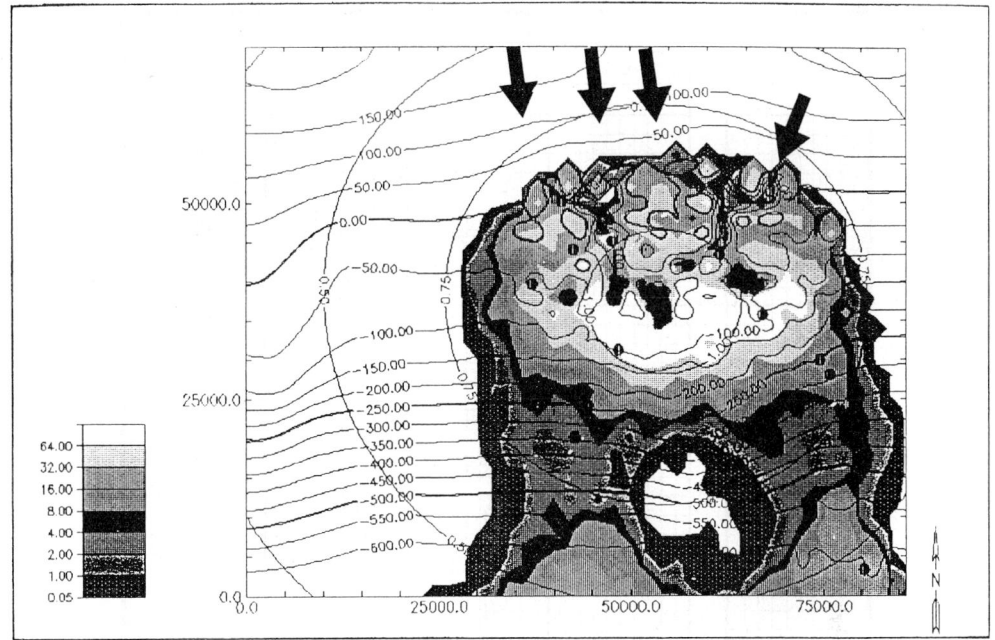

Fig. 7. Delta experiment involving Upper Cretacious Buda Shelf extending over one million years. All units are in meters. Geographic coordinates are given for simulation area with origin in lower left. Topographic contours are relative to shore line at 0.0 m, circular shaped contours represent vertical changes in elevation caused by isostatic compensation during span of experiment. Shaded contour intervals represent sediment thickness. Arrows denote locations where rivers enter the area. Dots represent locations of individual fluid elements.

In the experiment, four rivers carrying clay-rich sediment debouched into the marine basin. Hypothetical sea level changes are represented by an oscillatory or saw-tooth function with wave lengths of 100,000 years and amplitude of 100 m. Flexural rigidity of the lithosphere is $5 \cdot 10^{22}$ Newtton on square meter (Nm^2).

The experiment simulated delta-plain progradational wedges with thicknesses up to about 120 m (Fig. 7). Isostatic subsidence is only about 1 to 1.5 m, reflecting the small amount of deflection resulting from the geographic distribution of sediment and the fact that areas of thick sediment are small relative to the overall area. In an equivalent two-dimensional sedimentation model, in which only vertical sections can be represented, subsidence due to loading might be overestimated because lateral distribution of sediment is not possible in a two-dimensional model permitting excessively thick sediment to accumulate locally. Furthermore, in a two-dimensional approach, we assume that high loads extend indefinitely perpendicular to the cross-section, yielding even greater crustal deflections.

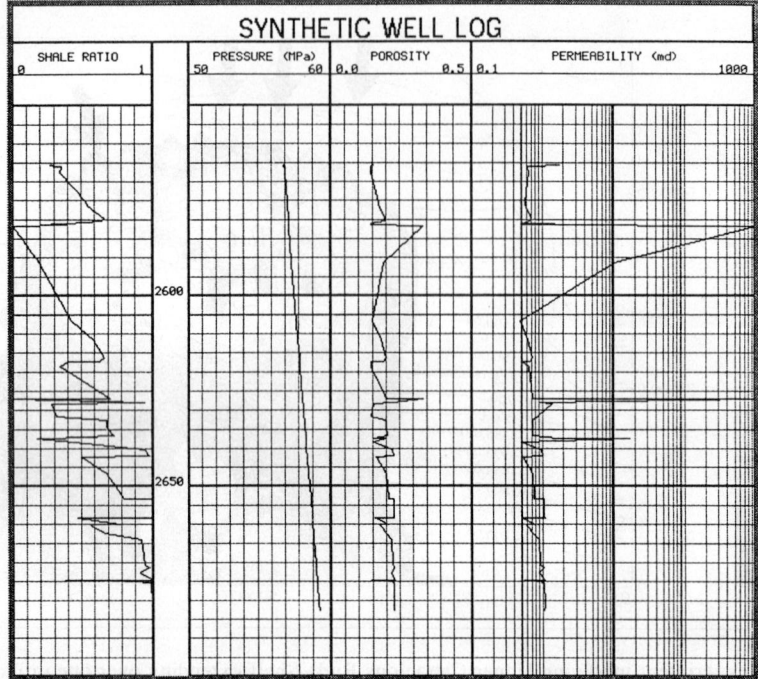

Fig.8. Synthetic well logs for a grid node located at 67000 m west and 35000 m north as shown in Fig. 7. Sediments were compacted with 3,000 m sand. Depth is in m below surface. First column displays sand/shale ratio as calculated from four grain sizes simulated by SEDSIM, second column represents lithostatic pressure within this interval, third column displays porosity, and fourth column displays permeability on logarithmic scale as derived from Kozeny-Carman equation extended to sand/clay mixtures.

In the Buda shelf experiment, burial was simulated by adding 3,000 m of sand-rich sediment on top of the earlier deposits. The simulated deposits' petrophysical properties can be displayed with three-dimensional colored fence diagrams, or alternatively with synthetic well logs at selected grid nodes as in Fig. 8. SEDSIM can create logs of shale-ratios, lithostatic pressure, porosity, and permeability. The grid node chosen for the log in Fig.8 is close to the delta front. The shale-ratio log reflects sea-level changes by sudden increases in sand/shale ratio, and progradation of the delta is reflected by a general coarsening-upward trend, and within this trend, by single coarsening-upward sequences. Porosity and permeability do not change significantly for the large range of shale percentages between 0.4 and 1.0, reflecting the petrophysical model used in SEDSIM. By contrast, a thick sandy unit at the top of the sequence, and several smaller ones below, indicate single prograding events during the build-out phase of the delta, when sea level dropped slowly.

REFERENCES

Bayer, U. (1989). Sediment compaction in larger scale systems. Geol. Rundsch., 78/1, 155-169.

Beard, D.C. & Weyl, P.K. (1973). Influence of texture on porosity and permeability of unconsilidated sand. Am. Ass. Petroleum Geol Bull., 54 (2), 339-369.

Bethke, C.M. (1989). Modeling subsurface flow in sedimentary basins. Geol. Rundsch., 78/1, 129-154.

Deutsch, C.V. (1987). A Probabilistic Approach to Estimate Effective Absolute Permeability. MS thesis, Stanford University.

Dreyer, T., Scheie, A. & Walderhaug, O. (1990). Minipermeameter-based study of permeability trends in channel sand bodies. Am. Ass. Petroleum Geol. Bull., 74, 359-374.

Doligez, B., Bessis, F., Burrus, J., Ungerer, P. & Chenet, P.Y. (1985). Integrated numerical simulation of the sedimentation heat transfer, hydrocarbon formation, and fluid migration in a dedimentary basin: the Themis Model. In: Thermal Modeling in Sedimentary Basins (J. Burrus, ed.). Collection Colloques et Seminaires, 44, pp. 173-195, 1st Institut Français du Pétrol Exploration Research Conference, Carcans.

Grün, G.U., Wallner, H. & Neugebauer, H.J. (1989). Porous rock deformation and fluid flow - numerical FE-simulation of the coupled system. Geol. Rundsch. 78/1, 171-182.

Jordan, T.E. & Flemings, P.B (1990). From geodynamic models to basin fill - a stratigraphic perspective. In: Quantitative Stratigraphy (T.A.Cross, ed.), Chap. 9, pp. 149-163. Prentice Hall, Englewood Cliffs.

Lee, Y.H. & Harbaugh, J.W. (1990). Stanford's SEDSIM project: dynamic three-dimensional simulation of geologic processes that effect clastic sedimentation. This convention.

Marion, D. (1990). Acoustical, Mechanical, and Transport Properties of Sediments and Granular Materials. PhD thesis, Stanford University.

Martinez, P.A. (1989). Computer simulation of wave and fluvial-dominated nearshore environments, in:Applications. In: Coastal Modelling (C.V. Lakhan and A.S. Trenhaile eds.), Elsevier Oceanography Series 49, pp. 297-337. Elsevier Publishing Co., New York,

Palciauscas, V. V. (1985). Models for thermal conductivity and permeability in normally compacting basins. In: Thermal Modeling in Sedimentary Basins (J. Burrus, ed.). Collection Colloques et Seminaires, 44, 1st Institut Français du Pétrol Exploration Research Conference, Carcans.

Ranalli, G. (1987). Rheology of the Earth. Allan & Unwin, Winchester.

Tetzlaff, D.M. & Harbaugh, J.W. (1989). Simulating Clastic Sedimentation. Van Nostrand Reinhold, New York.

Turcotte, D.L. & Schubert, G. (1982). Geodynamics, Application of Continuum Physics to Geological Problems. Wiley, New York.

Vail, P.R., Todd, R.G. & Sangree, J.B. (1977). Seismic stratigraphy and global changes in sea level, part 5: Chronostratigraphic significance of seismic reflections. In: Seismic Stratigraphy-Applications to Hydrocarbon Exploration (C.E. Payton, ed.). Am. Ass. Petroleum Geol. Mem., 26, pp. 99-116.

Van Hinte, J.E. & Deighton, I.C. (1987). Burial and Thermal Geohistory Modeling of Sedimentary Basins. Notes for JAPEC Course No. 57, Imperial College, London.

Walcott, R.I. (1970). Flexural rigidity, thickness, and viscosity of the lithosphere. J. Geoph. Res., 75, 3941-3954.

Weber, K.J. (1982). Influence of common sedimentary structures on fluid flow in reservoir models. J. Pet. Technol., 34, 665-672.

Yin, H., Han, D.H. & Nur, A. (1988). Study of Velocity and Compaction on Sand-Clay Mixtures. Stanford Rock and Borehole Project, Vol. 23.

Zoetmeijer R., Desegaulx, P., Cloetingh, S., Roure, F. & Moretti, I. (1990). Lithospheric dynamics and tectonic-stratigraphic evolution of the Ebro basin. J. Geoph. Res., 95, B3, 2701-2711.

LANDFORMS DEVELOPING AND BASINS FILLING: THREE-DIMENSIONAL SIMULATION OF EROSION, SEDIMENT TRANSPORT, AND DEPOSITION.

JOHN C . TIPPER

Geology Department, Australian National University, GPO Box 4,
Canberra ACT 2601, Australia

ABSTRACT

It is possible, in principle, to model landform development and basin filling at the microscopic level, by using fundamental physical principles to predict the exact patterns of erosion, sediment transport, and deposition that will take place at any time. In practice, however, for the spatial and temporal scales with which the geologist is concerned, macroscopic-level modelling seems more promising. Here such a model is developed, based on the existence of what is termed the 'degradation potential' field. (The degradation potential at any point is a reciprocal function of the expected survival time of any sediment deposited there.) The model predicts the evolution in time of this degradation potential field, and so enables the evolving topography of a landform or basin floor to be calculated. The model is implemented for simulation, using a finite-difference approximation over an irregular triangular mesh.

KEYWORDS

Sedimentary basins; Surface processes; Stochastic processes; Finite-difference methods; Modelling; Triangular meshes.

INTRODUCTION

In looking at how the Earth's surface evolves, it is easy to focus on how complex are most of the processes that work there, and to overlook how fundamentally simple are some of their results. Surface processes certainly are complex when analysed in detail (the mechanisms of denudation that develop landforms, the sedimentation mechanisms that fill sedimentary basins), and many of their results are certainly complex in detail too (the small-scale geometries of landforms, the depositional fabrics of sedimentary sequences). Yet it is only because these processes <u>do</u> work at the Earth's surface that analysis in such detail can even be contemplated. Were the processes to be less accessible, is it not likely that other, far simpler results might stand out more?

To illustrate what is meant by this, consider the contrasting viewpoints that the sedimentologist and the stratigrapher have when looking at the one sedimentation system. The sedimentologist, on the one hand, is concerned with features that relate to the environment of deposition and to the nature of the sediments that are produced (the physical and chemical conditions, the actual depositional mechanisms, the textures and internal structures, etc). The stratigrapher, on the other, is concerned only with the thickness of the sediments that are put down, and its exact relationship to time. Whereas the sedimentologist concentrates on different features for different sedimentation systems, the stratigrapher's concern never changes, and it is this constant concern for thickness and time that points through the complexity of the surface processes themselves to the fundamental simplicity of their results — all surface processes operate in time to alter the elevation of the depositional/erosional surface.

In this paper, I use a stochastic representation of sedimentation systems to develop a simple framework within which the development of landforms and the filling of basins can be studied in the same way. Using this framework, the changing elevation of the depositional/erosional surface can be modelled quantitatively, and simulated.

MODELLING SURFACE SYSTEMS

Landform development and the filling of sedimentary basins are two sides of the same story. Each follows tectonism — landform development the uplift of the land surface, the filling of basins their formation by basement subsidence. Each involves the same three basic processes — erosion, sediment transport, and deposition. Finally, each can be seen as the response of a system disturbed from its equilibrium, a system seeking a steady state. The mountains are worn down, the sediment carried to where it rests, at last, in the seas.

Any such system will undoubtedly be a complex one, and the problems that arise in analysing it will be those faced in analysing other complex systems: (1) it will be large in size, even when limited in time and restricted to some manageable geographic area; (2) it will involve very many variables, many of which cannot be observed or measured, either properly or at all; (3) its structure (which may not be entirely self-evident) will evolve in time. The standard approaches used in analysing complex systems (Vemuri, 1978) are applicable: of these, model building and simulation would appear to be of particular value. (The term 'model', as used here, describes a precisely defined system, usually a simplification and generalisation of some other, incompletely understood, parent system, which can be manipulated in such a way that predictions about its behaviour can be tested. From these, inferences can be made about the behaviour of the parent system. A simulation is just a dynamic working model.)

Modelling of surface systems (landform development or basin filling) can be carried out at several levels (Huggett, 1985; Syvitski, 1989). At the lowest, fundamental physical principles are used to predict from known physical conditions the exact patterns of erosion, sediment transport, and deposition that will take place at any time. At the highest level, those patterns are treated directly. The two extremes may be termed the 'microscopic' and the 'macroscopic' levels respectively. Naturally, there are intermediates between them.

In selecting the level at which to model any particular system, a variety of factors must be considered. The first is the degree to which the basic structure of the system is understood, i.e. the degree to which the behaviour of the system can be predicted from fundamental physical principles. The second factor is the amount and quality of the data that is available to constrain the model, either in setting its parameters or in testing predictions that result from it. The third factor is the physical size of the system, both in space and in time. For the same spatial and temporal resolution, a large system will naturally require more computation to model than a small one, and the computational complexity may be much worse than linear. The physical size of the system may thus often be the limiting factor in any modelling exercise, because of its implications for computation.

In principle, it is certainly possible to model most surface systems at the microscopic level, but in practice it may be very difficult. Consider, for example, the problem of modelling sediment transport systems on the continental shelves. The basic structure of such systems is certainly understood, and the fundamental physical theory behind the processes that are involved (erosion, sediment transport, and deposition) has been known for some considerable time (Smith, 1977). Numerical models can readily be set up to implement this theory (Tetzlaff and Harbaugh, 1989). But data is the limiting factor — the data-base available to constrain any microscopic-level model of shelf sediment transport will undoubtedly be neither large enough nor precise enough. Smith has stressed the need to know the near-bottom velocity field accurately before working in this way, '[without which] there is no point in proceeding further' (Smith, 1977, p540). Even for modern shelves, adequate data on the near-bottom velocity field is hardly available: for ancient shelves (and, by inference, for any ancient basin), it would be surprising if it could ever be obtained.

For landform development, modelling at the microscopic level is, if anything, even more unpromising than it is for sediment transportation and for the filling of basins. Culling, in 1960, concluded that it was then 'too ambitious an undertaking ... to construct a theory [of erosion] from fundamental physical principles' (Culling, 1960, p336). This is still very much the case (Rose and Hairsine, 1988), and there is little prospect of progress — for one fundamental reason. Erosion, by its very nature, removes information on the prior states of a system, as that system evolves in time. The initial conditions needed to set up any model that incorporates erosion can thus never be determined from the final state of a system, other than by guesswork.

Microscopic-level modelling of surface systems certainly does have its place (in addition to its obvious attraction as a long-term goal). Short-term forecasting in small, well-known, especially artificial basins is a clear instance, and Lepetit and Hauguel (1979) have provided an elegant example which shows how the evolution of the sea-bottom topography in the vicinity of a new coastal structure can be predicted. In general, however, microscopic-level modelling of surface systems holds relatively little promise at the spatial and temporal scales with which the geologist is concerned. Where lateral distances of tens or hundreds of kilometres must be considered, where thicknesses (of deposition or erosion) of hundreds of metres are usual, where time-spans of millions of years must be covered, and where real data sets must be used as constraints, it seems more promising to model surface systems at the macroscopic level.

Modelling a system at the macroscopic level demands the assumption that the behaviour of the system can be described adequately by state variables that relate to that behaviour, but are not necessarily the obvious controls of it. These state variables usually refer to bulk properties of the system; sometimes they are entirely synthetic variables; sometimes they are space- or time-averages. The relationships between the state variables and the system's behaviour are commonly also stochastic in nature (in marked contrast to the determinism that characterises modelling at the microscopic level). Swift et al, in 1972, described one of the earliest macroscopic-level models of shelf sediment transport, and two sentences that they wrote then can advantageously be quoted to give the flavour of this approach to modelling: 'At this scale of observation [the whole shelf, over substantial periods of time], the deterministic laws of sediment transport, as elucidated by hydraulic engineers and experimental sedimentologists, are not directly useful. Modeling of sediment-transport systems calls rather for a sort of statistical mechanics in which the element processed is seen as a population of particles whose behaviour is best predicted by a probabilistic model' (Swift et al., 1972, p197). Note that the laws of sediment transport that apply to individual sediment particles are not described as being in any way wrong, but are simply 'not directly useful'. Thus in this model, as it is populations of particles that are used as the elements of the system, rather than the individual particles themselves, the basic variables that control how the individual particles are entrained, moved, and deposited (the near-bottom fluid velocity field, for instance) are just not relevant. The requirement that these variables be known accurately everywhere before modelling can begin (e.g. Smith, 1977) is thus side-stepped.

There must, of course, be another side to this story, for if macroscopic-level modelling could produce the same results as microscopic-level modelling, with just a fraction of the trouble, why would we ever bother with fundamental physical principles? Naturally, the results are not the same. What is lost in modelling at the macroscopic level are the implicit spatial and temporal linkages that modelling at the microscopic level automatically provides.

Consider, for instance, the flow of sediment-carrying fluid in a sedimentary basin. Everywhere, and at all times, four governing equations must be satisfied (momentum, energy, sediment continuity, fluid continuity). There is thus both spatial linkage in this system (between what goes on at any one point and what goes on at the same time at points adjacent to it), and temporal linkage (between what goes on at any point now and what will occur there immediately afterwards), and these linkages are automatically built into

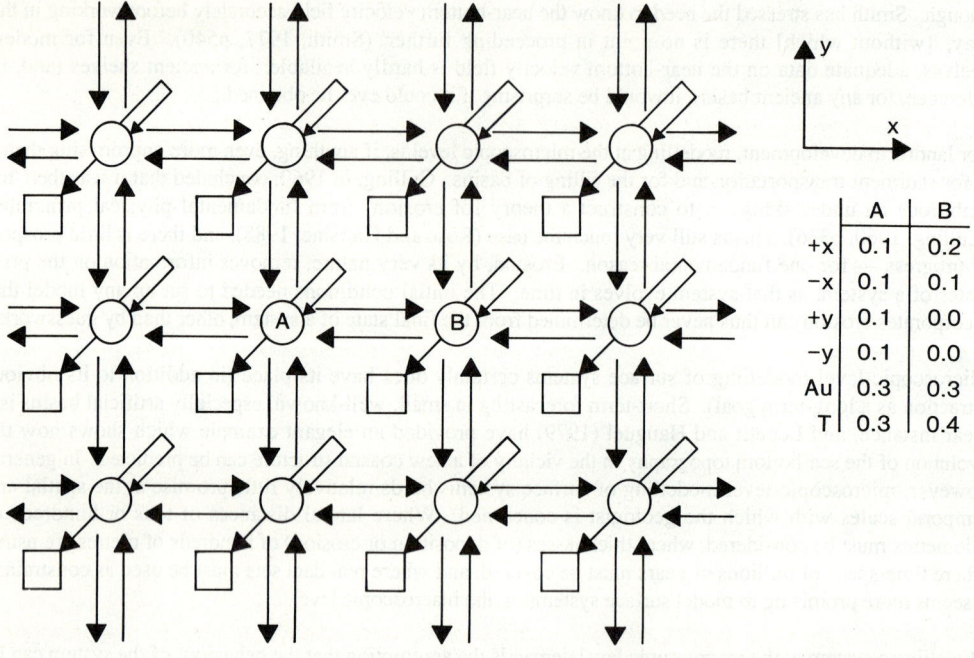

	A	B
+x	0.1	0.2
−x	0.1	0.1
+y	0.1	0.0
−y	0.1	0.0
Au	0.3	0.3
T	0.3	0.4

Fig. 1. Probabilistic shelf sediment transport model of Swift et al (1972). At each time-step, the sediment in transport at a site (circle) may (1) move to an adjacent site in the +x, -x, +y, or -y directions, (2) remain in transport at the site (auto-transition: Au), or (3) be trapped at the site (square: T). Table at right shows sample transition probability matrices for sites A and B.

sediment transport models that are based on solutions of the governing equations. In contrast, macroscopic-level models do not implicitly recognise these linkages. For instance, for the probabilistic shelf sediment transport model described by Swift et al (1972; see also Price, 1976), in which the spatial relationships between adjacent points are represented by a two-dimensional Markov process (Fig. 1), there is no necessary correlation either between the transition probability matrices at adjacent points at the same time, or between successive transition probability matrices at the same point. Arbitrarily assigned transition probability matrices can easily be used. It is hardly surprising that this model can give almost any required result, provided that suitable transition probability matrices are chosen.

This conclusion is true of macroscopic-level models in general. They are sufficiently flexible that almost anything can be achieved by tuning their parameters in an appropriate way. Thus if there is one problem to be overcome in modelling surface systems at the macroscopic level, it is this: to find ways in which models can be used which have spatial and temporal linkages implicit in their structure, but which do not appeal (at least, not directly) to fundamental physical principles. The theories of erosion and slope development due to Ahnert (1976), Culling (1960, 1963, 1965), Kirkby (1976), Scheidegger and Langbein (1966), and Scheidegger (1970, in part) give rise to models of this type. What is needed, however, is similar theory that can be applied across the whole range of surface environments, depositional as well as erosional.

STOCHASTIC MODELS FOR SEDIMENTATION SYSTEMS

As mentioned earlier, one characteristic of models that are built at the macroscopic level is that the relationships within them tend to be stochastic in nature rather than deterministic. This renders them very appropriate for studying sedimentation systems, for Scheidegger (1970) and Schwarzacher (1972, 1975, 1976) have both shown how valuable the stochastic approach can be in analysing erosion and deposition. Tipper (1989) has even demonstrated that a stochastic analysis of single grain entrainment is necessary in order to interpret correctly the form of the conventional entrainment threshold function — an application of stochastic modelling in the very heartland of determinism. It will now be shown, using a simple stochastic process, how spatial and temporal linkages can be defined within a macroscopic-level model of a general sedimentation system.

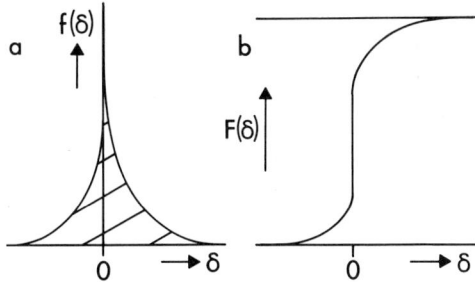

Fig. 2. Frequency distribution of thickness change, δ, for a sedimentation system of depositional regime ($E[\delta]>0$). Prob($\delta<0$) = 0.22. Prob($\delta=0$) = 0.44. Prob ($\delta>0$) = 0.34. (a) Frequency function, $f(\delta)$. Note spike of infinite height at $\delta = 0$, corresponding to non-deposition. (b) Distribution function, $F(\delta)$.

The stochastic process is the one-dimensional random walk in discrete time and continuous space (Cox and Miller, 1965). This has been widely used before to describe sedimentation systems (Kolmogorov, 1951; Rivlina, 1970; Mizutani and Hattori, 1972; Schwarzacher, 1975; Dacey, 1976; Tipper, 1983). The space variable is sediment thickness, and the thickness change, δ_t, during time interval t is taken to be distributed according to some frequency distribution that characterises the particular sedimentation system concerned (Tipper, 1983; Fig. 2). In doing this, an assumption is tacitly made that sedimentation systems can be characterised in this way, and on first sight this seems unreasonable. Every sedimentation system, after all, is specific both to its time and to its location, and so can have only one, unique realisation. Any frequency distribution for thickness change can thus refer at best only to a family of like systems (Tipper, 1983; McCrae, 1990). Despite this, characterising individual sedimentation systems in this way has been found to be a useful approach, and estimating the form of the frequency distributions for real sedimentation systems is a long-term research goal (Friend et al, 1989). Needless to say, much more work is needed, both on ancient sedimentation systems (cf. Mizutani and Hattori, 1972) and on modern ones.

The length of the basic time interval in the random walk process is taken to be that of the shortest distinct episode in the particular sedimentation system concerned. For clastic depositional systems, this is controlled by the turbulence of the fluid boundary layer, and may range from a few seconds to a few hours (Jackson, 1975). In contrast, for carbonate and evaporite depositional systems, and for systems that are predominantly erosional, the basic interval length may be very much longer. Irrespective of the nature of the particular system, however, one point must be stressed — that although in many sedimentation systems it is non-deposition (or at least the absence of readily discernible deposition or erosion) that is normal, with states of significant deposition or erosion being relatively brief and relatively rare, it is, nevertheless, these rare events that do produce the significant deposition and erosion. It must thus be the typical duration of these events that determines the basic interval length for the random walk process. Prolonged periods of deposition, erosion, or non-deposition should then be treated just as multiples of the basic time interval.

For time interval t, the expected thickness change, $E[\delta_t]$, describes how, on average, the state of the system will change, and defines the two regimes of sedimentation. If $E[\delta_t]$ is positive (the depositional regime), a bed of that thickness (on average) will be deposited: if $E[\delta_t]$ is negative (the erosional regime), that thickness of sediment (on average) will be eroded. If $E[\delta_t]$ is zero, the system is at base-level. In any sedimentation system, the value of $E[\delta_t]$ will generally vary, both laterally (across a land surface or basin floor) and in time. Thus as a basin fills or a land surface evolves, it should be expected that most points will change their regime, perhaps many times, and that the instantaneous pattern of depositional and erosional areas will continually evolve.

Denoting by $H_{t,p}$ the elevation of the depositional/erosional surface at point p at the start of time interval t, then $H_{t+1,p} = H_{t,p} + E[\delta_{t,p}]$. The set $H = \{H_{1,p} ... H_{n,p}\}$, for all p, describes the way that, on average, the whole depositional/erosional surface changes through n successive time intervals. If the set of initial elevations, $\{H_{1,p}\}$, is given, for all p, then H can be determined, provided only that the values of $E[\delta_{t,p}]$ can be calculated for all t and p. The problem is to calculate these values, ensuring both that the values at adjacent points are linked together properly at each time interval, and that there is a proper linkage between sucessive values at each point. As stressed earlier, arbitrarily imposed spatial and temporal linkages are not adequate.

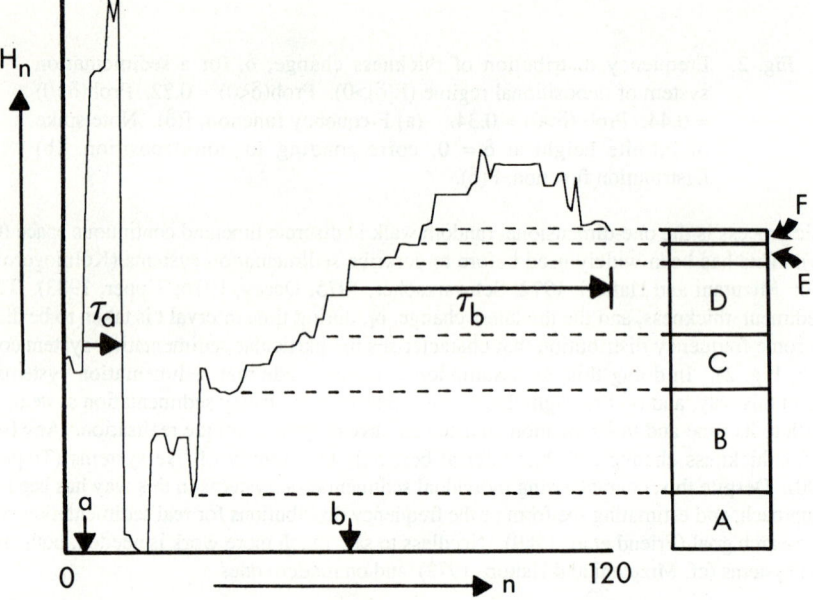

Fig. 3. Barrell diagram for a sedimentation system with prob($\delta<0$) = 0.3 and prob($\delta>0$) = 0.5. H_n is the elevation of the depositional/erosional surface at the end of (discrete) time interval n. Beds A-F are the stratigraphic record (vertical column). Horizontal arrows show the survival times, τ_a and τ_b, for time horizons a and b, the former eroded, the latter still preserved.

One solution to this problem is to introduce the concept of what is termed here the 'degradation potential' at a point. This can be developed in the following way. For any point, p, the set H can be plotted as a Barrell diagram (Fig. 3), and this diagram can also be used to show what can be termed the 'survival time', τ, for any time horizon. This is the length of time that that horizon will survive in the stratigraphic record at p before being removed by erosion. (Note that there is no need for that horizon to necessarily be represented by sediment in order to have a non-zero survival time. Horizons corresponding to intervals of non-deposition can also survive.) The expected survival time, $E[\tau_p]$ (which can be obtained from the frequency distribution of δ at p), is an important indicator of the sedimentation conditions at p, showing basically the degree to which the sedimentation system there is not degradational. The reciprocal of $E[\tau_p]$ is thus a variable, D_p, which will be high in value if the sedimentation system at p is highly degradational, and will be low in value if p is a sediment sink. D_p measures what can be thought of as the 'degradational energy level' of a sedimentation system: D_p is termed the 'degradation potential' at p.

The existence of the degradation potential — and it can, in principle, be measured — suggests that the evolution in time of the depositional/erosional surface might advantageously be modelled as a process of potential-driven flow, analogous for instance to the flow of fluid through a porous medium. The form of the degradation potential field at any time would then define how the values of $E[\delta]$ are linked spatially, and the time-evolution of the field would define how the successive values of $E[\delta]$ at each and every point are linked in time. What would then be provided would be a macroscopic-level model, applicable across the range of erosional and depositional environments, with spatial and temporal linkage implicit in its structure.

THE POTENTIAL-DRIVEN FLOW MODEL

By analogy with the time-dependent, two-dimensional flow of fluid through a porous medium (Wang and Anderson, 1982), the fundamental equation for the potential-driven flow model may be written:

$$\frac{\partial^2 D}{\partial x^2} + \frac{\partial^2 D}{\partial y^2} = \frac{1}{T[x,y,t]} \left(S[x,y,t] \frac{\partial D}{\partial t} - Q[x,y,t] \right) \tag{1}$$

where x and y are space coordinates, and t denotes time.

In this model, the transport of sediment laterally is controlled by the 'conductance' of the transport path and by the difference in degradation potential at the path ends. The transport parameter, $T[x,y,t]$, is thus directly analogous to the conductance of a porous medium, and expresses the ease with which any transport path can move sediment, either through a basin or across a land surface. An obvious first approximation for T is to take it to be a linear function of the slope of the depositional/erosional surface along the particular transport path, as was done by Culling (1960) in developing his analytical theory of erosion. Other functions, however, must certainly be used where sediment transport is channelled, for instance, or where flow is other than as bed load.

The 'storage' parameter, $S[x,y,t]$, the change produced in the elevation of the depositional/erosional surface by a unit change in the degradation potential, seems analogous to the storage coefficient of a porous medium. It must be stressed, however, that this is not necessarily the case. $S[x,y,t]$ is certainly zero for any sedimentation system that is at base-level, i.e. for which $E[\delta]$ is zero, but for systems away from base-level it need not always be positive (as is the storage coefficient of a porous medium). If S, in fact, is positive, the degradation potential is implied to increase with increasing elevation, and this is not unreasonable for sedimentation systems in the depositional regime and for some erosional situations (Scheidegger, 1970; see also Lambeck and Stephenson, 1986). If, however, the erosion of a land surface is slope-dependent (which is also not unreasonable in many situations), then the degradation potential may decrease with increasing elevation, and S may be negative. Generalising, the form of S must be dependent on the (relatively) local geometry of the depositional/erosional surface. The nature of this dependence requires more and detailed investigation, and it may well be profitable to do this in the context of the fractal geometry of the depositional/erosional surface (e.g. Culling and Datko, 1987).

The third parameter, Q[x,y,t], represents fluxes of sediment into and out of the sedimentation system. These may be, for instance, the supply of clastic sediment to a basin by rivers at its margins, or the precipitation of carbonate muds on a basin floor. Dissolution of material in surface waters is also covered by this parameter.

The relationship of this model to the established diffusion-degradation model for landscape evolution (Culling, 1965) is both important and straightforward. If the parameter S is taken to be identically equal to unity, then the degradation potential field is everywhere identical to the elevation of the depositional/erosional surface. Equation (1) then becomes the diffusion equation (e.g. Culling, 1965, eqn 10), with the parameter T as the diffusion coefficient. What is done in the potential-driven flow model, in keeping the parameters S and T independent and distinct (although, of course, the ratio S/T is effectively a single parameter, and Q/T another), is to provide a single framework within which can be analysed the separate effects of the 'storage' and transport factors that each contribute to the evolution of the surface elevation. On a landform, for instance, dominance of S corresponds to weathering-controlled denudation, whereas dominance of T corresponds to denudation controlled by the rate of material transport. The diffusion-degradation model, which applies when the transport rate is the only control, is thus a special case of the more general potential-driven flow model.

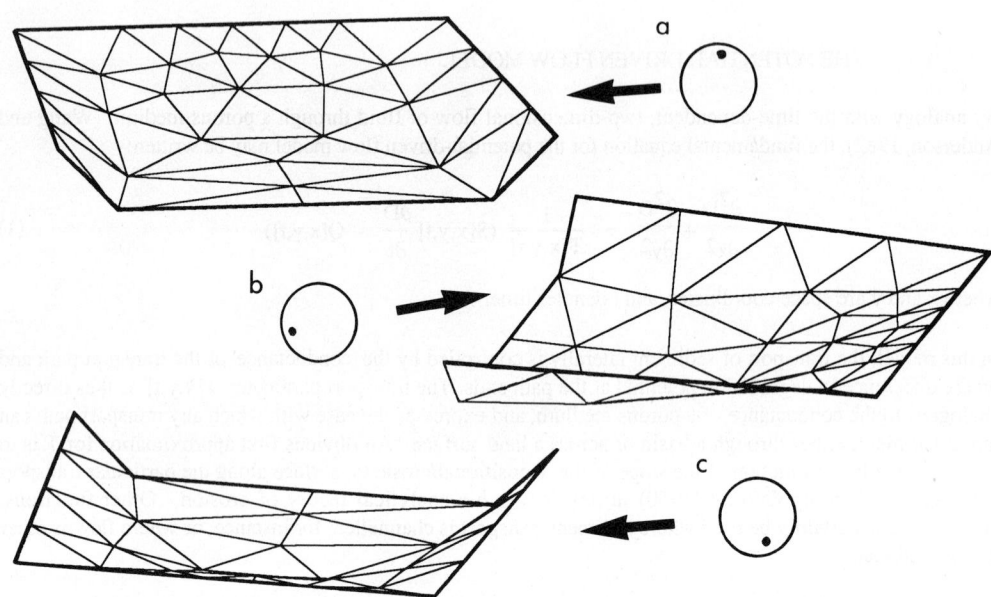

Fig. 4.   Oblique views of an irregular triangular mesh used to represent the geometry of a depositional/erosional surface (cf. Fig. 5). Circles are hemispherical projections to show viewing angle. (a) Azimuth = 15°; Inclination = 70° (b) Azimuth = 250°; Inclination = 75° (c) Azimuth = 165°; Inclination = 70°.

APPLYING THE MODEL

Ultimately, for this as for any other model, it is not the formulation of the model that is of interest, but its application to real data, either natural or synthetic. In this case, the application will be to study the time-evolution of an actual landform or basin, and so the model must be implemented in such as a way that this time-evolution can be simulated. This implementation is remarkably straightforward. The geometry of the depositional/erosional surface is represented by an irregular triangular mesh (Fig. 4), and an explicit finite-difference approximation used to solve a version of equation (1) at each of the triangle vertices (MacNeal, 1953; Tyson and Weber, 1964). Tipper (submitted) discusses at some length the rationale for selecting the irregular triangular mesh as the spatial framework for implementing this model. Briefly, its advantages are: (1) that it allows a time-explicit finite-difference scheme to be used to solve the set of equations (as opposed to a computationally more demanding time-implicit finite-element scheme); (2) that it readily permits the parameters T and S to be varied both spatially and in time; and (3) that it is compatible with standard surface-fitting techniques, which can thus readily be used to specify the initial form of the depositional/erosional surface and for graphic output.

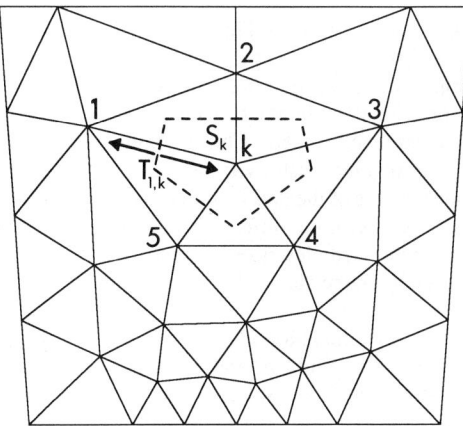

Fig. 5.   Configuration of triangular mesh used to solve Equation (2) at vertex k. Vertex k is connected to 5 (Voronoi) neighbours, i.e. $n_k = 5$. These neighbours are vertices 1-5. $T_{i,k}$ (see text), as shown, applies to the transport path $1 \leftrightarrow k$. S_k (see text), applies to the Voronoi polygon about vertex k (dashed line).

The finite-difference approximation to equation (1), for vertex k at the end of the $(j + 1)$'th time-step is:

$$\sum_{i=1}^{n_k} \left(D_i^j - D_k^j \right) T_{i,k}^j = S_k^j \left(D_k^{j+1} - D_k^j \right) / \Delta t - Q_k^j \tag{2}$$

where n_k is the number of other vertices in the mesh to which vertex k is connected;

$T_{i,k}^j$ is the conductance of the transport path between vertex k and its i'th neighbour (Fig. 5) at the end of the j'th time-step;

S_k^j is the storage parameter applying to the area immediately surrounding vertex k
(Fig. 5) at the end of the j'th time-step;

Q_k^j is the flux parameter applying to vertex k at the end of the j'th time-step;

Δt is the length of the time-step.

At each time-step, new values are calculated for S and T, for each vertex and transport path respectively: the value of Q for each vertex is either recalculated at each time-step, or supplied as a boundary condition. Equation (2) is then solved to obtain D_k^{j+1}, for all k. The new elevation of the depositional/erosional surface is then calculated:

$$H'^{j+1}_k = H_k^j + S_k^j \left(D'^{j+1}_k - D_k^j \right). \tag{3}$$

An example of the use of this model is shown in Figure 6, a simulation of coastal margin sedimentation. This particular simulation also illustrates some of the problems that are met in using this approach. The physical situation in this example is a coastal gulf, fed with sediment from rivers at two of its margins. The topography of the initial depositional/erosional surface is relatively well known (Fig. 6a), although considerably more so in some areas than in others. This is common in most real-world situations. The first problem is then to generate a triangular mesh over the basin floor to represent the initial topography effectively (allowing for the unevenness of the data control), and which will serve as the basis for the finite-difference solution of equation (2). (The finite-difference solution, in fact, requires that the triangular mesh be strictly acute-angled: Tipper (submitted) describes how such a mesh can be constructed.) The mesh used for this present simulation is given in Figure 6b.

The second problem in setting up the simulation is to specify the initial configuration of the degradation potential field. About this, little can be said. As indicated earlier, characterising actual sedimentation systems by frequency distributions for δ is now little more than a long-term goal. Until the form and parametrisation of those distributions is known, the values of $E[\tau]$ and D can only be conjectured. (Some initial simulation results are summarised in Table 1.) For the present simulation, two alternatives for the initial potential field seem reasonable. One is to take it to be zero across the entire basin: the field is then found to rather rapidly assume a configuration with peaks around the river mouths (Fig. 6c). The second alternative is to use one of these initial peaked configurations, perhaps the final potential field from some earlier base run. This was the approach used by Syvitski (1989) in his simulations of prograding deltas.

Specifying the boundary conditions is the third problem in setting up the simulation. In principle, either fixed potential (Dirichlet), fixed flow (Neumann), or mixed boundary conditions may be used. Given, however, that there is little likelihood that actual values of the degradation potential will be known (at least at present), fixed flow conditions are the most sensible in practice. This means that values of Q must be specified for all vertices at the mesh boundary. For the present simulation, values of Q are assumed to be known for the vertices at the river mouths: elsewhere on the boundary, Q is set to zero. (Setting the two ocean boundaries in the present simulation to be no-flow boundaries inevitably distorts the results, especially late in the simulation as sediment from the rivers finally crosses to those boundaries and is trapped against them. The solution to this is to set the no-flow boundaries even further oceanward.)

The fourth problem in setting up the simulation, selecting forms for the parameters T and S, has been referred to briefly earlier. As mentioned then, the form for T will almost certainly depend on the way that the sediment is transported. Areas in which different transport mechanisms operate will probably have different values for T, and where several mechanisms operate together, the value for T will reflect the balance of these transport mechanisms. T is, however, a parameter that applies to any transport mechanism, and so the value

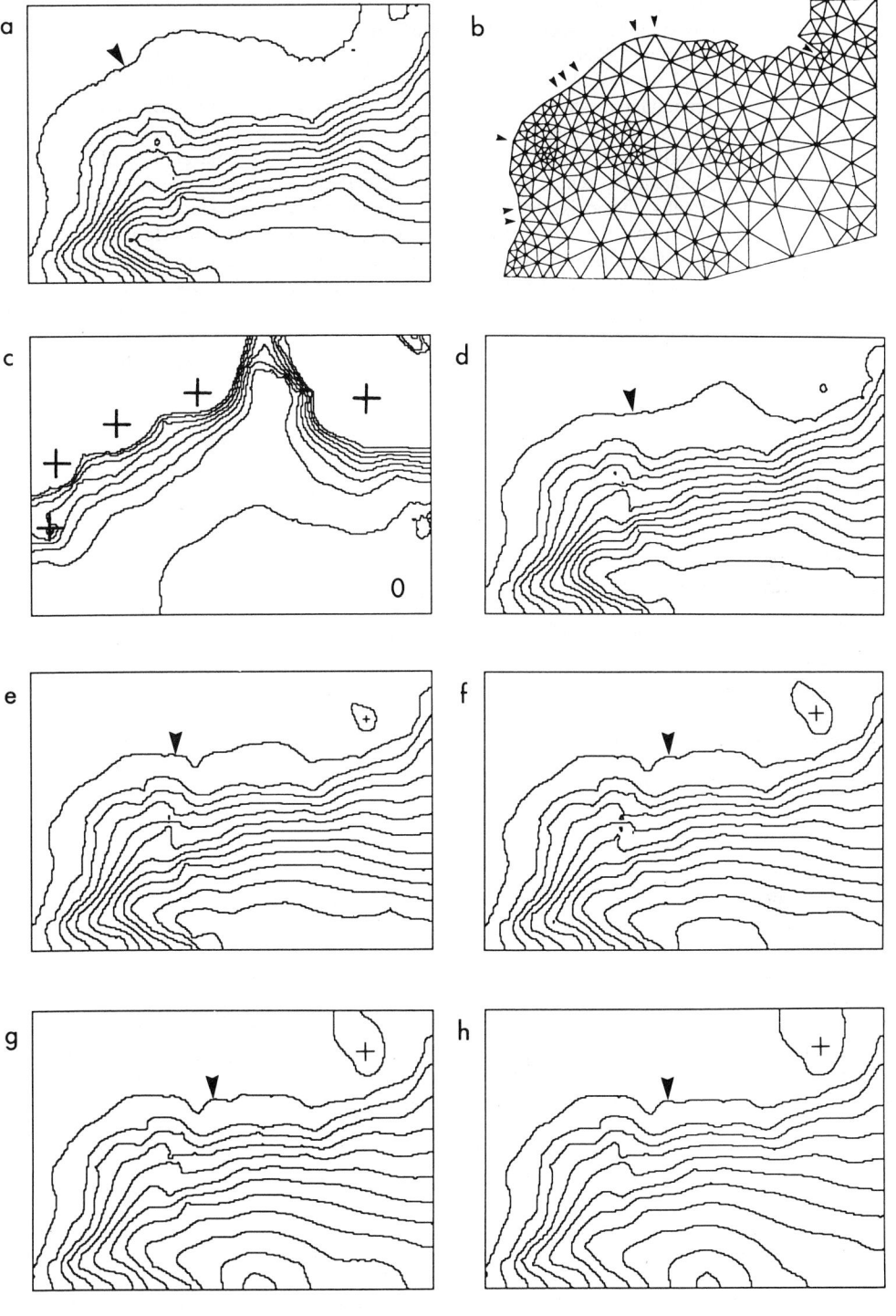

Fig. 6.

Fig. 6. A simulation of coastal margin sedimentation. (a) Initial topography of depositional/erosional surface. Sea-level contour marked by arrow. Other contours (unlabelled) show depth of surface below sea-level increasing systematically towards lower-right of figure. (b) Triangular mesh used for simulation. Note uneven density of data points. All boundaries are no-flow except at arrowed points on coastline, at which rivers supply sediment. (c) Topography of degradation potential field at time t = 500. Positive areas are around river mouths. Otherwise the field slopes smoothly down to zero at lower-right of figure. (d) - (h) Topographic maps of developing depositional/erosional surface. In each case the sea-level contour is arrowed. (d) t = 1000. Build-out of delta platforms by major rivers. (e) t = 3000. Build-up of exposed area (marked +) on main delta. (f) t = 5000. (g) t = 7000. Fill concentrated on right-hand side of figure, due to no-flow boundary there. (h) t = 9000. Ponding of sediment against all no-flow boundaries.

of T can be expected to be a continuous function of position, even across areas in which the predominant transport mechanism changes abruptly. There is no need to deal with each mechanism independently (cf. Syvitski, 1989, p. 507). (In the present simulation, the simple slope-dependence of T is used throughout.)

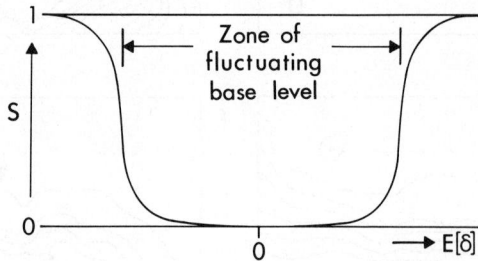

Fig. 7.   A base-level-dependent function for S (see text). For convenience, S is defined in the range 0 → 1.

Estimating the form of S, and its value for each vertex, is the final problem in settting up the simulation, and one that is far greater than that of estimating T. On-going studies of modern sedimentation systems as stochastic processes hold the key to how this can best be done. For the present simulation, the function illustrated in Figure 7 is used. This corresponds to a simple base-level-dependency for erosion and deposition, but recognises that there is a pronounced elevation range on either side of the mean base-level surface, within which the actual base-level commonly fluctuates.

SOME ADVANTAGES OF USING THIS APPROACH

Perhaps the greatest advantage of this approach is that it focusses attention on the basic stratigraphic result that is produced by surface processes — the change in time of the elevation of the depositional/erosional surface. Attention is shifted from the details of the processes themselves. A second advantage is that the theory applies across the whole range of environments, from areas of extreme denudation far above the highest excursions of base-level, through areas across which sediment is by-passed, to sediment sinks in

Table 1. Simulation results relating $E(\tau)$ to probabilities of erosion ($\delta<0$), non-deposition ($\delta=0$), and deposition ($\delta>0$). n is number of time-steps for the random walk process (Fig. 3). Distribution function, $F(\delta)$, is:

$$\delta = 0.1\,(\tan(x)-x), \text{ where } x = \frac{\Pi}{2}\left(\frac{F(\delta)}{\text{Prob}(\delta<0)}-1\right), \text{for } F(\delta) < \text{Prob}(\delta<0); \quad \delta = -0.1(\tan(x)-x),$$

$$\text{where } x = \frac{\Pi}{2}\left(\frac{1-F(\delta)}{\text{Prob}(\delta>0)}-1\right), \text{ for } F(\delta) > 1-\text{Prob}(\delta>0), \text{ For Prob } (\delta<0) = 0,\ 2*E[\tau]/n = 1.$$

			2*E[τ] /n			
Prob($\delta<0$)	Prob($\delta=0$)	Prob ($\delta>0$)	n = 99	n = 399	n = 699	n = 999
0.1	0.0	0.9	0.723	0.726	0.693	0.730
0.1	0.1	0.8	0.709	0.660	0.699	0.676
0.1	0.2	0.7	0.672	0.645	0.611	0.626
0.1	0.3	0.6	0.651	0.540	0.563	0.558
0.1	0.4	0.5	0.458	0.434	0.482	0.444
0.1	0.5	0.4	0.439	0.336	0.354	0.363
0.1	0.6	0.3	0.291	0.248	0.227	0.238
0.1	0.7	0.2	0.176	0.124	0.121	0.109
0.1	0.8	0.1	0.142	0.086	0.038	0.033
0.1	0.9	0.0	0.017	0.004	0.003	0.002
0.2	0.0	0.8	0.540	0.410	0.420	0.479
0.2	0.1	0.7	0.486	0.386	0.437	0.387
0.2	0.2	0.6	0.456	0.377	0.351	0.327
0.2	0.3	0.5	0.375	0.284	0.276	0.240
0.2	0.4	0.4	0.324	0.218	0.191	0.211
0.2	0.5	0.3	0.198	0.128	0.140	0.100
0.2	0.6	0.2	0.125	0.063	0.078	0.043
0.2	0.7	0.1	0.045	0.018	0.011	0.012
0.2	0.8	0.0	0.015	0.004	0.002	0.001
0.3	0.0	0.7	0.311	0.295	0.235	0.248
0.3	0.1	0.6	0.327	0.211	0.218	0.172
0.3	0.2	0.5	0.167	0.163	0.157	0.140
0.3	0.3	0.4	0.129	0.103	0.081	0.090
0.3	0.4	0.3	0.156	0.051	0.063	0.049
0.3	0.5	0.2	0.081	0.042	0.031	0.013
0.3	0.6	0.1	0.045	0.010	0.009	0.005
0.3	0.7	0.0	0.012	0.003	0.002	0.001
0.4	0.0	0.6	0.215	0.147	0.128	0.109
0.4	0.1	0.5	0.162	0.103	0.085	0.083
0.4	0.2	0.4	0.116	0.077	0.048	0.047
0.4	0.3	0.3	0.095	0.047	0.023	0.021
0.4	0.4	0.2	0.062	0.020	0.013	0.009
0.4	0.5	0.1	0.033	0.006	0.006	0.003
0.4	0.6	0.0	0.010	0.002	0.001	0.001
0.5	0.0	0.5	0.133	0.061	0.032	0.037
0.5	0.1	0.4	0.077	0.038	0.021	0.022
0.5	0.2	0.3	0.070	0.019	0.011	0.009
0.5	0.3	0.2	0.036	0.011	0.006	0.006
0.5	0.4	0.1	0.025	0.005	0.004	0.003
0.5	0.5	0.0	0.008	0.002	0.001	0.001
0.6	0.0	0.4	0.061	0.027	0.023	0.014
0.6	0.1	0.3	0.034	0.017	0.009	0.012
0.6	0.2	0.2	0.037	0.007	0.005	0.004
0.6	0.3	0.1	0.019	0.004	0.003	0.002
0.6	0.4	0.0	0.006	0.001	0.001	0.001
0.7	0.0	0.3	0.039	0.010	0.007	0.004
0.7	0.1	0.2	0.015	0.005	0.003	0.002
0.7	0.2	0.1	0.015	0.003	0.002	0.001
0.7	0.3	0.0	0.004	0.001	0.001	0.000
0.8	0.0	0.2	0.019	0.005	0.003	0.001
0.8	0.1	0.1	0.009	0.002	0.001	0.001
0.8	0.2	0.0	0.002	0.001	0.000	0.000
0.9	0.0	0.1	0.005	0.001	0.001	0.001
0.9	0.1	0.0	0.001	0.000	0.000	0.000
1.0	0.0	0.0	0.000	0.000	0.000	0.000

which whatever sediment is supplied is deposited. A third advantage is practical, for the potential-driven flow model is computationally very economical. It allows real-time, three-dimensional simulation to be carried out. If the computer implementation is sufficiently flexible, it is a simple matter to experiment with different types of sedimentation system, different parameter values, and different initial and boundary conditions etc, to achieve a best-fit to real data.

Finally, this approach has the advantage that the predictions that it makes of landform development or basin fill patterns are expectations, and that at each time-step there is thus an associated confidence interval for the estimate of δ at each point, in principle at least. This means that predictions of basin fill history can at last be put on a probabilistic footing. Having said that, however, it must be stressed that this confidence interval can be calculated from the present theory only if the frequency distribution of δ at a point is uniquely determined by the value of $E[\tau]$ there (and hence by the value of D). This certainly does not seem reasonable, unless there turns out to be some unexpected simplicity in those distributions, irrespective of environment. If that is not found, another independent relationship must be sought for the dispersion of δ. Once that is obtained, a full, macroscopic-level model of landform development and basin filling will exist.

REFERENCES

Ahnert, F. (1976). Brief description of a comprehensive three-dimensional process-response model of landform development. *Zeit. Geomorph. Neue Folge, Suppl,* 25, 29-49.

Cox, D.R. and Miller, H.D. (1965). *The Theory of Stochastic Processes.* Chapman & Hall, London.

Culling, W.E.H. (1960). Analytical theory of erosion. *J. Geol.,* 68, 336-344.

Culling, W.E.H. (1963). Soil creep and the development of hillside slopes. *J. Geol.,* 71, 127-161.

Culling, W.E.H. (1965). Theory of erosion on soil-covered slopes. *J. Geol.,* 73, 230-254.

Culling, W.E.H. and Datko, M. (1987). The fractal geometry of the soil-covered landscape. *Earth Surface Processes and Landforms,* 12, 369-385.

Dacey, M.F. (1979). Models of bed formation. *Math. Geol.,* 11, 655-668.

Friend, P.F., Johnson, N.M. and McRae, L.E . (1989). Time-level plots and accumulation patterns of sediment sequences. *Geol. Mag.,* 126, 491-498.

Huggett, R.J. (1985). *Earth Surface Systems.* Springer-Verlag, Berlin.

Jackson, R.G., II. (1975). Hierarchical attributes and a unifying model of bed forms composed of cohesionless material and produced by shearing flow. *Bull. geol. Soc. Am.,* 86, 1523-1533.

Kirkby, M.J. (1976). Deterministic continuous slope models. *Zeit. Geomorph. Neue Folge, Suppl,* 25, 1-19.

Kolmogorov, A.N. (1951). Solution of a problem in probability theory connected with the problem of the mechanism of stratification. *Am. math. Soc. Transl.,* No. 53, 1-8.

Lambeck, K. and Stephenson, R. (1986). The post-Palaeozoic uplift history of south-eastern Australia. *Aust. J. Earth Sciences,* 33, 253-270.

Lepetit, J.P. and Hauguel, A. (1979). A numerical model for sediment transport. In: *Marine Forecasting* (J.C.J. Nihoul, ed.), pp. 453-463. Elsevier, Amsterdam.

MacNeal, R.H. (1953). An asymmetrical finite difference network. *Q. Appl. Math.,* 11, 295-310.

McCrae, L.E. (1990). Paleomagnetic isochrons, unsteadiness, and non-uniformity of sedimentation in Miocene fluvial strata of the Siwalik Group, northern Pakistan. *J. Geol.,* 98, 433-456.

Mizutani, S. and Hattori, I. (1972). Stochastic analysis of bed-thickness distribution of sediments. *Math. Geol.,* 4, 123-146.

Price, W.E., Jr., (1976). A random-walk simulation model of alluvial-fan deposition. In: *Random Processes in Geology* (D.F. Merriam, ed.), pp. 55-62. Springer-Verlag, New York.

Rivlina, T.S. (1970). A stochastic model of stratification. (The case of unlimited interstratal erosion.) In: *Topics in Mathematical Geology* (M.A. Romanova and O.V. Sarmanov, eds), pp. 142-150. Consultants Bureau, New York.

Rose, C.W. and Hairsine, P.B. (1988). Processes of water erosion. In: *Flow and Transport in the Natural Environment* (W.L. Steffen and O.T. Denmead, eds), pp. 312-326. Springer-Verlag, Berlin.

Scheidegger, A.E. (1970). *Theoretical Geomorphology.* 2nd. ed. Springer-Verlag, Berlin.

Scheidegger, A.E. and Langbein, W.B. (1966). Probability concepts in geomorphology. *US Geol. Surv. Prof. Paper,* No. 500, C1-C14.

Schwarzacher, W. (1972). The semi-Markov process as a general sedimentation model. In: *Mathematical Models of Sedimentary Processes* (D.F. Merriam, ed.), pp. 247-268. Plenum, New York.

Schwarzacher, W. (1975). *Sedimentation Models and Quantitative Stratigraphy.* Elsevier, Amsterdam.

Schwarzacher, W. (1976). Stratigraphic implications of random sedimentation. In: *Random Processes in Geology* (D.F. Merriam, ed.), pp. 96-111. Springer-Verlag, New York.

Smith, J.D. (1977). Modeling of sediment transport on continental shelves. In: *The Sea* (E.D. Goldberg, I.N. McCave, J.J. O'Brien and J.H. Steele, eds), Vol. 6, Chap. 13, pp. 539-577. Wiley, New York.

Swift, D.J.P., Ludwick, J.C. and Boehmer, W.R. (1972). Shelf sediment transport: a probability model. In: *Shelf Sediment Transport: Process and Pattern* (D.J.P. Swift, D.B. Duane and O.H. Pilkey, eds), Chap. 9, pp. 195-223. Dowden, Hutchinson & Ross, Stroudsburg, PA.

Syvitski, J.P.M. (1989). Modelling the sedimentary fill of basins. *Geol. Surv. Canada, Paper,* No. 89-9, 505-515.

Tetzlaff, D.M. and Harbaugh, J.W. (1989). *Simulating Clastic Sedimentation.* Van Nostrand Reinhold, New York.

Tipper, J.C. (1983). Rates of sedimentation, and stratigraphical completeness. *Nature,* 302, 696-698.

Tipper, J.C. (1989). The equilibrium and entrainment of a sediment grain. *Sedim. Geol.,* 64, 167-174.

Tipper, J.C. (submitted). Surface modelling for sedimentary basin simulation. *Am.Assoc. Petrol. Geol.*

Tyson, H.N., Jr. and Weber, E.M. (1964). Ground-water management for the nation's future — computer simulation of ground-water basins. *J. Hydraul. Div. Am. Soc. civ. Engrs.*, 90, 59-77.

Vemuri, V. (1978). *Modeling of Complex Systems*. Academic Press, New York.

Wang, H.F. and Anderson, M.P. (1982). *Introduction to Groundwater Modeling*. Freeman, San Francisco.

3-D COMPUTER GRAPHICS IN MODELING PLUTON EMPLACEMENT

GIOVANNI GUGLIELMO JR.

Earth Sciences Board, University of California*
Santa Cruz, CA 95064

ABSTRACT

Structures which occur around plutons provide crucial data by which to analyze the deformational history in orogenic belts, and computer modeling is a powerful method by which to study the genesis of these structures. The complex interplay of variables during pluton emplacement dramatically increases the complexity of the simulation. To overcome these difficulties, a combination of computer techniques such as 3-D viewing, rendering, and an interactive graphical user interface was used. These techniques allow simulation of pluton expansion in a crust undergoing tectonic deformation. 3-D viewing was critical for establishing the orientation, distribution and position of structures around the pluton. The interactive nature of the simulation allowed not only reading of complex input parameters and reliable evaluation of complex output, but also the real-time analysis of progressive deformation. Finally, rendering allowed both the realistic visualization of strain and structures around the pluton, and the creation of a new strain diagram that integrates the analysis of various strain parameters. As a result of the simulation, the strain patterns and structures which were modeled provided useful insights on the processes and structures related to pluton emplacement and tectonic deformation in orogenic belts. In addition this simulation demonstrated that computer graphics techniques can be not only convenient but may be required for the proper visualization and interpretation of complex geological simulations.

KEYWORDS

Pluton, rendering, orogenic belts, computer modeling, graphical user interface, deformation, strain, computer graphics, structural geology.

---

*Current address: Dept. of Geological Sciences, University of California
Santa Barbara, CA 93106

INTRODUCTION

Cross-cutting relationships between plutons of known age and tectonic structures are commonly used to determine the age of deformation in orogenic belts. However, there is no single reliable criterion to determine if a pluton was emplaced before, during or after deformation. This lack of criteria is largely due to the difficulty in determining whether the structures around a pluton were caused by the emplacement of the pluton, by regional deformation, or some combination. In an attempt to determine the origin of these structures I carried out field studies of strains in wall rock in the vicinity of a natural pluton and compared them with a three-dimensional computer model that simulates strains associated with pluton emplacement in a crust undergoing regional tectonic deformation. Comparison of measured strain fields with computer-generated strain fields provides new insights into the role of regional- versus intrusive-related deformation, specifically in establishing the timing of emplacement of plutons and the correct interpretation and timing of structures present in wall rocks in orogenic belts. In this paper I describe both the programming and the computer graphics techniques used in the computer model. These techniques were critical for the simulation, visualization, and interpretation of the strains and structures associated with pluton emplacement and tectonic deformation. Further details on strain analysis and mathematical proofs will be published in a forthcoming paper.

The goals of this computer simulation were to determine which structures in the country rock can be attributed to pluton emplacement in a crust undergoing regional deformation, and to determine the types, 3-D distribution, and orientation of structures produced by the interaction of local pluton-related strain fields with regional strain fields.

Clay and computer models of pluton emplacement have been described in the literature (Ramberg 1970; Ramberg and Sjostrom 1973; Dixon 1975; Schwerdtner and Troeng 1978; Morgan 1980; Soula 1982; Cruden 1988; Schmeling, *et al.* 1988; Cruden 1990). However, only one previous study considered the effect of tectonic deformation during pluton emplacement (Brun and Pons, 1981), and their 2-D program simulated only simple shear. The model I developed is a three-dimensional simulation of how regional shortening, extension, and simple shear can interact with pluton emplacement. The model assumes that the pluton expands during emplacement (Holder 1979; Brun and Pons 1981; Ramsay 1989) and that the regional tectonic regime was dominated by homogeneous deformation.

MATHEMATICAL BASIS, PROCESSES, AND RESULTS

The regional deformation displacement types I have modeled to date assume three kinematic schemes: simple shear, pure shear during uniaxial shortening and pure shear during uniaxial extension. The country rock around the pluton was divided into a cubic simulation grid (Fig. 1) and for each element of this grid the program calculates the resulting strain ellipsoid with increasing strain and plots its strain parameters (strain magnitude and shape). The calculated strain ellipsoid parameters are useful in predicting the type, position, intensity, and orientation of extensional and compressional structures around the pluton

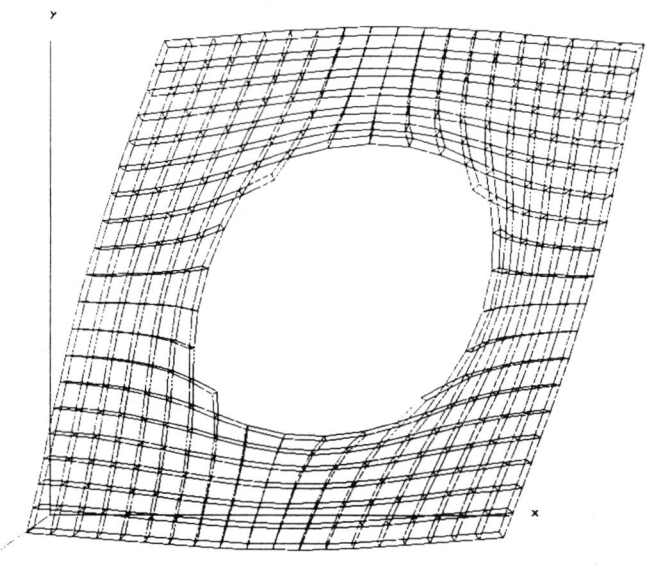

Fig. 1. Deformed simulation grid. Only a two-dimensional slice of the 3-D model is shown for clarity.

During each iteration of the program, each point of the simulation grid is displaced by an incremental strain tensor. The deformation can be described by

$$F = I - C \tag{1}$$

where F is the final displaced 3-D vector; I is the vector displaced by an incremental strain tensor; and C is an offset due to the changing position of the pluton's center during deformation. For the simulation of pluton emplacement under simple shear we have

$$I = V \times S \tag{2}$$

where V is the initial 3-D vector, S is the expansion incremental strain tensor coeval with simple shear, and is given by

$$S = \begin{vmatrix} sBsB & 0 \\ 0 & sB & 0 \\ 0 & 0 & sB \end{vmatrix}$$ (3)

where s is the simple shear rate, and B the expansion rate given by

$$B = 1 + \frac{G}{d^2}$$ (4)

where G is an expansion parameter, and d is the distance from the center of the pluton to the point being displaced. The offset C is given by

$$C = \frac{GM}{d^2}$$ (5)

For the simulation of pluton emplacement under pure shear we have

$$D = V \times P$$ (6)

where V is the initial 3-D vector P is the expansion incremental strain tensor coeval with pure shear, and is given by

$$P = \begin{vmatrix} pB & 0 & 0 \\ 0 & pB & 0 \\ 0 & 0 & pB \end{vmatrix}$$ (7)

where p is the pure shear rate. p values smaller than 1 represent uniaxial compression, whereas p values greater than 1 represent uniaxial extension.

For every element on the grid the program calculates: strain magnitude, symmetry, orientation of lineation, and orientation of foliation. All these parameters also can be calculated from field measurements, which allows good monitoring of the simulation's reliability.

Strain magnitude (Nadai 1963) is a measurement of the amount of deformation, and it is given by:

$$\overline{\varepsilon_s} = \frac{1}{\sqrt{3}} \left((\varepsilon_1 - \varepsilon_2)^2 + (\varepsilon_2 - \varepsilon_3)^2 + (\varepsilon_3 - \varepsilon_1)^2 \right)^{1/2}$$ (8)

where $\varepsilon_1, \varepsilon_2$, and ε_3 are principal logarithmic strains given by

$$\varepsilon_1 = \ln(1 + e_1); \ \varepsilon_2 = \ln(1 + e_2) \text{ and } \varepsilon_3 = \ln(1 + e_3)$$

where $(1 + e_1)$, $(1 + e_2)$, and $(1 + e_3)$ are the lengths of the strain ellipsoid semi-axes and e_1, e_2, and e_3 are the values of principal strains.

Strain symmetry is a measurement of the shape of the strain ellipsoid, and can be expressed by the Lode's parameter (Lode 1926):

$$v = \frac{2\varepsilon_2 - \varepsilon_1 - \varepsilon_3}{\varepsilon_1 - \varepsilon_3}$$

(9)

Lode's parameters can assume values between -1 and 1, where negative values represent constriction and positive values represent flattening. Lode's parameter equals to zero represents plane strain.

Although strain magnitude and Lode's parameter characterize the finite strain ellipsoid's intensity and shape, the rock's fabric can define the strain orientation. The orientation of lineation and foliation is calculated based on the orientation of the axis of the finite strain ellipsoid. Lineations are defined by the orientation of longest axis, whereas foliation is defined by the plane that contains the intermediate and longest axis. Once these strain parameters are calculated they can be displayed and viewed in three dimensions.

Preliminary results along 2-D slices of the 3-D model (Figs. 2 to 5) show that the distribution and shape of contour lines shown on maps of strain ellipsoid symmetry (Fig. 2) constrain the origin of regional strain fields formed during syn-tectonic pluton expansion. Regional uniaxial extensional produces Lode's parameter contour lines that run approximately parallel to the expanding pluton boundaries, while during uniaxial shortening and simple shear, these lines form high angles with the pluton boundaries. The distribution and orientation of lineations around the pluton help us distinguish between simple shear and uniaxial shortening. Whereas simple shear produces subhorizontal lineations on the sides of the pluton, uniaxial shortening produces vertical lineations virtually all around the pluton.

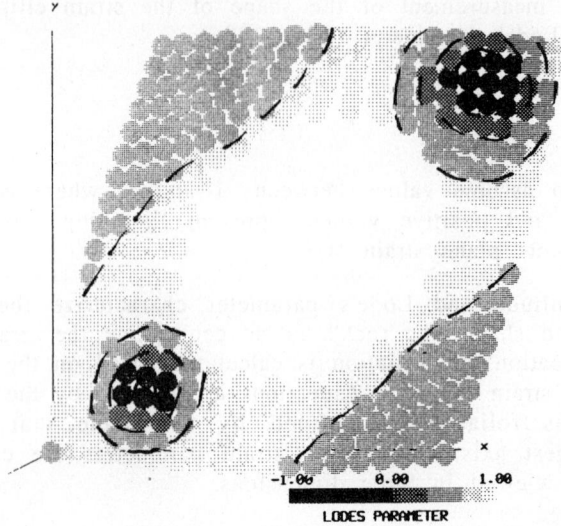

Fig. 2. Contour map of strain ellipsoid symmetry. Darker areas represent constriction whereas lighter areas represent flattening.

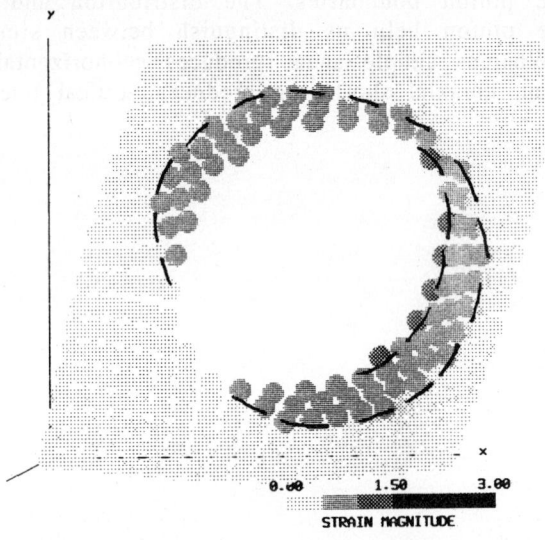

Fig. 3. Contour map of strain magnitude. Darker areas represent higher strain magnitudes.

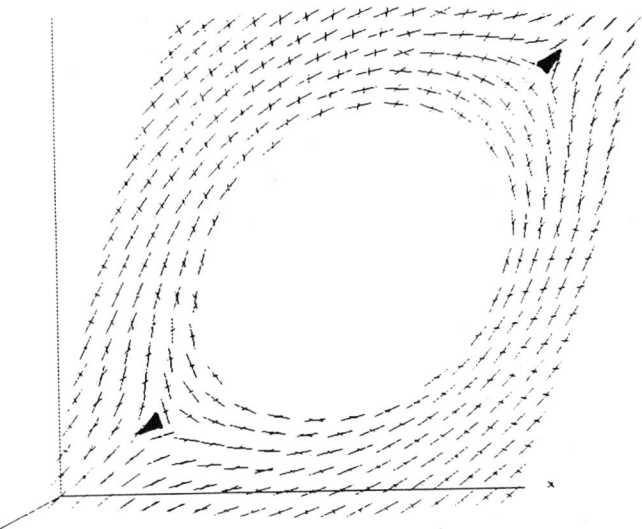

Fig. 4. Foliation patterns. Note foliation triple points at "ends" of pluton.

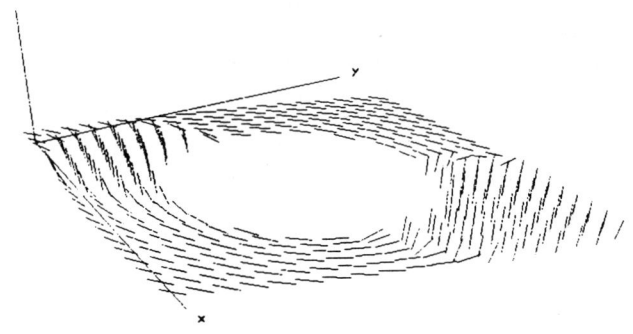

Fig. 5. Lineation patterns. View point is different from preceding figures to clarify the 3-D orientation and distribution of lineations.

The modeled strain patterns may differ slightly from natural patterns because of the influence of parameters such as structural anisotropies in the wall rocks, variable rock rheologies, variable regional strain fields etc. In addition, to date only the syn-tectonic situation has been addressed; modeling of pre- and post-tectonic situations are in progress. Although analysis of natural strain fields should be carried out in conjunction with radiometric ages' determinations, structural, microstructural and strain measurements, the geometry of strain patterns modeled to date are, by themselves, diagnostic of the regional strain fields that caused them. Therefore, the simulated strain patterns offer guidelines for field work and help us predict possible orientations and positions of structures and strains in the wall rock around a pluton. They can tell which structures are crucial to measure and where to increase the density of our strain and structural measurements. In addition, they can suggest which parameter of the strain ellipsoid (shape, orientation, or magnitude) is potentially most significant and where to measure it in order to better constrain possible deformation paths.

The simulation of pluton emplacement under tectonic deformation is complex. One needs to consider how both expansion and tectonic deformation simultaneously affect the resulting strain patterns. In addition, several distinct strain paths are possible for both of these variables, and changes in strain rates can occur. Furthermore, under the influence of several combinations of all parameters mentioned, the simulation of pluton emplacement before, during and after tectonic deformation, would result into an infinite number of redundant images. Besides, the output of the simulation should show in 3-D how the orientation and distribution of four variables (lineations, foliations, strain magnitude, and symmetry) changes as a function of all the parameters mentioned previously. This level of complexity required an algorithm design more complex than the ones used in traditional non-interactive simulations. The simulation used in this work needed to satisfy several criteria, namely a simple user interface, optimum visualization of results, and speed and reliability on the elimination of redundant images without missing critical information. The next sections describe the algorithm design and how I dealt with the implementation problems described above.

OVERVIEW OF THE PROGRAM

To deal with the complexity of this simulation, I used a combination of techniques such as 3-D viewing, rendering, multiple window viewing, and an interactive graphical user interface. In addition, I used the graphics capabilities of a Silicon Graphics IRIS workstation. The user interacts with the program in the following way:

The program loops forever waiting for one of the following user selections to: (1) do a deformation experiment; (2) change the viewing parameters, or (3) save the results.

If you decide to perform a deformation experiment the program prompts you with a selection of tectonic displacement type (simple shear, pure shear, expansion, etc.) Once this mode is selected, the input parameters are read using the mouse and a "2d-slider". This 2d-slider was critical to optimize interactively the reading of a complex combination of input variables. The importance of this slider and how it operates is described next (subheading 1). Once the input parameters are read in they are used to deform the simulation grid (subheading 2) and to calculate the output strain parameters.

Conversely, if you decide to change the viewing parameters, the program prompts you with viewing operation choices such as "rotate", "scale", "translate", etc. The ability of real-time 3-D viewing (subheading 3) is critical in order to describe the orientation and distribution of structures around the pluton as is illustrated by figure 5. If you select to save the results the program creates binary files that store the deformation grid and the output parameters in a format that can be read by rendering programs. The importance of rendering for pluton emplacement simulation and strain analysis is described in subheading 4.

1- Reading complex multiple parameters as an input

A thorough kinematic simulation of pluton emplacement requires a multiple set of input parameters. For example: several combinations of pluton related and tectonic related strains are possible. Furthermore, deformation can happen before, during or after expansion, and several kinematic schemes also are possible. I wanted to be able to change any of these input parameters at any time during deformation. In addition I wanted to take in consideration changes in strain rates as the deformation proceeded, and this posed the exciting question of "How to change the simulation equations at the same time as the simulation is happening?"

To handle such a complex input, and to interactively control the simulation I created what I named a "2d-slider". This user interface tool consists of a xy axis and an active cursor (Fig. 6). The xy axis of the slider represents 2 kinematic schemes: The y axis represents the pluton-related deformation whereas the x axis represents the tectonic-related deformation. These axes are set by a pull-down menu, whereas the cursor is dragged interactively with a mouse. The cursor's position, speed, and the sequence by which this position and speed are changed, are transmitted to the program and interpreted to interactively change four of the program's simulation variables: tectonic strain rate, pluton expansion rate; rate of changes in strain rate; and timing of pluton-related versus tectonic-related deformation.

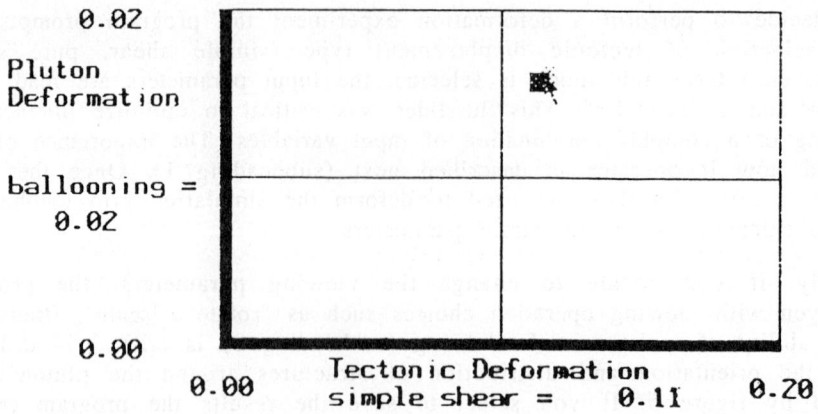

Fig. 6. 2d-slider used to modify interactively a complex set of input parameters by changing the cursor's position; speed; and sequence by which this position and speed are changed.

The strain rate is represented by the position of the cursor with respect to the xy axis of the 2d-slider. The x coordinate of the cursor records the current tectonic strain rate, whereas the y coordinate records the current pluton expansion rate. Therefore, positioning the cursor with the mouse changes both local and regional strain rates simultaneously.

The timing of pluton-related versus tectonic-related deformation is established by the sequence by which the cursor is moved (Fig. 6). For example: A cursor resting at position (a) simulates tectonic deformation only, happening at the strain rate given by the cursor's x coordinate. Similarly, a cursor resting at position (b) simulates expansion only, at a rate given by the cursor's y coordinate. Finally, a cursor resting at position (c) simulates syn-tectonic expansion. More complex deformation paths can be simulated by increasing the number and order of deformation events. For example: a cursor resting at (a) for a period of time and then brought to rest at (b) simulates post tectonic expansion. Conversely, a cursor resting at (b) and then at (a) simulates pre-tectonic expansion. The final total deformation depends on how long, in real-time, the cursor stays at one particular position, and on the strain rates associated with that position.

It is during the simulation of changing strain rates that the 2d-slider really shines. To account for changes in strain rates during deformation, the equations that transform the simulation grid have to change at the same time as simulation proceeds. These changes in equations are controlled by the speed by which the mouse drags the 2d-slider's cursor. For example: a cursor dragged slowly from (d) to (a), left to rest at (a) for a while and then brought back to (d) simulates the following tectonic scenario: initial tectonic quiescence (initial position at d) disturbed by gradually increasing tectonic deformation (increasing strain rates while moving from d to a) which level off by reaching a plateau of constant deformation (constant strain rates during resting period at a) and, finally the deformation stops (rest at d position where the strain rate is equal to zero). This sequence could represent a simplified kinematic version of a tectonic collision event, which is to be expected in orogenic belts where the majority of plutons intrude. The simulation of an expanding pluton could start at any time during this simulated sequence just by letting the 2d-slider cursor rest for a period of time at position (b). In a similar fashion, other tectonic settings can be studied by combining different sequences of deformation and timing of pluton emplacement, and by changing the expansion and tectonic strain rates. Because the 2d-slider allows the interactive input of these variables simultaneously, the user interface is simple and gives flexibility and speed on the control of the simulation of a wide variety of possible geological scenarios.

2- Deforming the simulation grid

To simulate kinematic deformation I constructed a 3-D cubic simulation grid (Fig. 1) able to deform by various kinematic schemes. This grid is able to: (1) be deformed by heterogeneous strain; (2) be modified by the insertion of a pluton interactively at any time during deformation; and (3) store information about the state of strain at each of its points.

Deformation of the model is accomplished by the displacement of the points that make up the elements of the grid. These elements are initially cubes, which, after heterogeneous deformation, become irregular 6-sided polyhedra. The deformation of each element, and the calculation of strain parameters for each element is done at each iteration of the program. During each iteration the 8 corners of each element are displaced by a finite incremental strain tensor described by Equations (1) to (7). The finite strain ellipsoid for each element is calculated based on the x, y, and z coordinate components of each corner of the element. The running of several iterations, each one controlled by the current input in the 2d-slider, produces the effect of real-time interactive deformation.

To simulate pre-, syn-, and post-tectonic pluton expansion, we should be able to emplace a pluton at any time during the simulation of regional deformation. To do so, the position and initial radius of the pluton are input interactively. Then, all grid elements within this radius are defined as "pluton elements" and all elements outside the pluton's radius are defined as "wall rock". The center of the pluton is the origin of the radial displacement vector field that simulates expansion. However, because non-coaxial rotational deformation, such as simple-shear, changes the absolute position of the pluton's center at each incremental deformation, a new center for the pluton is defined at each iteration. The updated center is calculated based on the average position of all the elements placed at the contact pluton/wall rock.

Using the data types available in the C programming language, I created a data structure to keep track of the information about each element of the grid at each incremental deformation. This data structure is composed of a 3-D array of "structs". Each cell of the array corresponds to one grid element, and the structs for each cell of the array store information on the deformed element. This information includes the element's strain ellipsoid's shape, orientation, and position. Once a finite incremental strain is applied to the grid, the strain ellipsoid information stored in each element of the grid is used to calculate the output strain parameters.

3- Display and viewing of output

Visualization of large amounts of data generated during the simulation required the use of techniques such as multi-window viewing, 3-D viewing, and rendering.

Each of the four strain parameters (stain magnitude, Lode's parameter, foliation pattern, and lineation pattern) is displayed in a separate active window (Figs. 2 to 5). The analysis of 4 strain parameters displayed simultaneously spawn more scientific insights than the sum of the insights produced with the same four parameters analyzed individually. In addition, the four windows are updated in real-time during the simulation, which provides significant insights on how the strain parameters change with respect to each other during progressive deformation.

3-D viewing helps us describe and interpret the orientation and distribution of strain and structures around the pluton. Lode's parameter and strain magnitude are displayed as contour maps (Figs. 2 and 3), whereas lineations and foliation patterns are shown as 3-D projections of line segments that have length proportional to the strain magnitude (Figs. 4 and 5). These contour maps and fabric projections are plots that can be compared with strain and fabric plots obtained from field measurements. In turn, this comparison can be used to guide further field and laboratory work, and to test tectonic models.

In addition, the information obtained from 3-D viewing is essential for the interpretation of geological maps. Structural patterns in conventional geological maps are a product of the intersection between 3-D structures and the present erosional surface. The poor understanding of these 3-D structures, contributes to faulty interpretations of map patterns, which, in turn, produces faulty structural and tectonic interpretations.

4- Rendering

Rendering techniques allow the realistic representation of complex 3-D objects and provide the basis for the creation of a new diagram to plot strain data. At the time of this writing this diagram was in final stages of programming, therefore, complete samples' plots will be given in a forthcoming paper; however, a brief description is given here.

Traditional diagrams for strain description take in consideration only one aspect or, at most, three aspects of the strain ellipsoid at the time, however, a "rendering plot" can show four aspects of strain in a single three-dimensional plot. Flinn diagrams (Flinn 1978) and Hsu diagrams (Hsu, 1966) can represent shape and intensity of the strain ellipsoid. Strain maps usually represent the 2D distribution or strain. For example, foliation and lineation maps give information on the distribution and orientation of the strain ellipsoid, whereas contour maps of strain parameters show distribution and an additional parameter such as either strain magnitude or shape. Similarly, strain ellipses' maps (Ramsay and Hubber 1983) show either XY, XZ, or YZ ellipses, which give only 2D information on strain shape and magnitude.

The "rendering" plot allows the 3-D analysis of four strain parameters simultaneously in one diagram: shape, magnitude, distribution, and orientation. The strain shape is represented by an irregular 3-D surface that connects strain ellipsoids of the same shape (Fig. 7 and 8). We may select this shape, for example, to be plane strain. This plain strain surface would divide the 3-D space into two regions: a region of constriction and a region of flattening. Alternatively, we may select to use the 3-D surface to connect ellipsoids of any given shape, or even several translucent surfaces connecting several strain shapes. For example, one translucent 3-D surface would connect all ellipsoids with Lode's parameter equals to zero (plane strain), whereas another surface, of a different color and transparency, would connect all points with Lode's parameter equals to 0.9 (high constrictional strains).

The 3-D "rendering plot" represents strain ellipsoid orientation by 3-D line segments along the major axis of each strain ellipsoid. These segments represent the position and orientation of expected mineral stretching lineations. Alternatively, strain orientation also can be represented by rectangles drawn along two axes of each stain ellipsoid. The rectangle along the XY axis of the strain ellipsoid gives the 3-D position and orientation of the expected foliation plane. The length of each lineation segment and the size of each foliation rectangle is proportional to the strain magnitude, which gives an intuitive feeling for how developed the mineral stretching lineations or foliations are, which, in turn, help us better interpret the structural significance and tectonic implications of these structures. Furthermore, any other rock property may be represented by assigning colors to the lineation segments, foliation squares, or strain shape surfaces, which help us analyze how properties, such as lithology, chemical variations, etc., change with respect to strain parameters. This new ability to represent all four aspects of strain in 3-D, and to analyze how these aspects change with respect to any rock property allows structural geologists to take the integrated approach fundamental for thorough strain analysis.

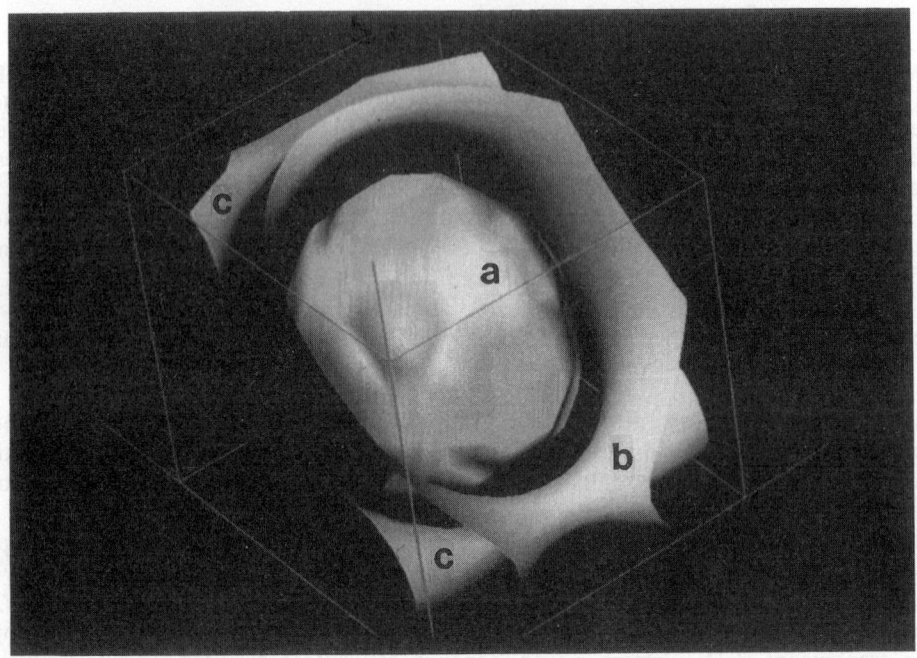

Fig. 7. Rendered plot showing pluton *"a"* surrounded by surfaces that connect points of same strain ellipsoid shape. Surface *"b"* connects points of plane strain (Lode's parameter equals to zero), and separates domain of flattening (region between surface "b" and pluton) from region of constriction. Two separate smaller surfaces *"c"* connect points of high constrictional strains (Lode's parameter equals to 0.5).

CONCLUSIONS

This simulation has provided insights not only on pluton emplacement but also on the use of computer graphics as a geological modeling tool. The simulation helped us constrain end-member tectonic regimes such as homogeneous syntectonic expansion. In addition, these patterns aid in establishing possible natural strain paths and in evaluating the relative contribution of local pluton-related deformation versus regional deformation. From a computer graphics perspective, the use of 3-D visualization, interactive user interface, and volume rendering techniques were critical for the proper visualization and interpretation of the model. 3-D visualization was essential not only for the description and analysis of the position, orientation, and distribution of geological structures around a pluton, but also for the correct interpretation of strain and structural patterns shown in geological maps. In addition, the interactive nature of the simulation was critical for the handling of complex geological data input and output, and for the real-time analysis of progressive deformation. Finally, The use of volume rendering made possible the creation of a new, more complete and intuitive, strain plot that will help structural geologists to perform thorough integrated strain analysis.

ACKNOWLEDGEMENTS

This work was supported by National Science Foundation grant number EAR-8904706. I would like to thank Othmar Tobisch for very helpful comments on the manuscript and enlightening discussions. I also thank Karen McNally and Jane Wilhelms for the use of their computers.

REFERENCES

Brun, J. P. and Pons, J. (1981). Strain patterns of pluton emplacement in a crust undergoing non-coaxial deformation, Sierra Morena, Southern Spain. *Journal of Structural Geology, 3*, 219-230.

Cruden, A. R. (1988). Deformation around a rising diapir modeled by creeping flow past a sphere. *Tectonics, 7*, 1091-1101.

Cruden, A. R. (1990). Flow and fabric development during the diapiric rise of magma. *Journal of Geology, 98*, 681-698.

Dixon, J. M. (1975). Finite strain and progressive deformation in models of diapiric structures. *Tectonophysics, 28*, 89-124.

Flinn, D. (1978). Construction and computation of three-dimensional progressive deformation. *Journal of the Geological Society of London, 135*, 291-305.

Holder, M. T. (1979). An emplacement mechanism for post-tectonic granites and its implications for their geochemical features. *Origin of granite batholiths Geochemical evidence*. Orpington U. K., Shiva Publishing Limited. 116-128.

Hsu, T. C. (1966). The characteristics of coaxial and non coaxial strain paths. *Journal of Strain Analysis, 1*, 216-222.

Lode, W. (1926). Versuche uber den Einfluss der mittleren Hauptspannung auf das fliessen des Matalle Eisen, Kupfer, and Nickel. *Zeitschr. Physik., 36*, 913-939.

Morgan, J. (1980). Deformation due to the distension of cylindrical igneous contacts: a kinematic model. *Tectonophysics, 66*, 167-178.

Nadai, A. (1963). *Theory of flow and fracture of solids*. New York, McGraw-Hill. 705 p.

Ramberg, H. (1970). Model studies in relation to intrusion of plutonic bodies. In: *Mechanism of igneous intrusion*. Liverpool, Gallery Press. 261-286 p.

Ramberg, H. and Sjostrom, H. (1973). Experimental Geodynamical models relating to continental drift and orogenesis. *Tectonophysics, 19*, 105-132.

Ramsay, J. G. (1989). Emplacement kinematics of a granite diapir: the batholith, Zimbabwe. *Journal of Structural Geology, 11*, 191-209.

Ramsay, J. G. and Hubber, M. I. (1983). *The techniques of modern structural geology*. London, Academic Press. 700 p.

Schmeling, H., Cruden, A. R. and Marquart, G. (1988). Finite deformation in and around a fluid sphere moving through a viscous medium: implications for diapiric ascent. *Tectonophysics, 149*, 17-34.

Schwerdtner, W. M. and Troeng, B. (1978). Strain distribution within arcuate diapiric ridges of silicone putty. *Tectonophysics, 50*, 13-28.

Soula, J. C. (1982). Characteristics and mode of emplacement of gneiss domes an Plutonic domes in central-eastern Pyrenees. *Journal of Structural Geology, 4*, 313-342.

Chapter 3

Economic Applications

AN INTERACTIVE 3-D MODELING SYSTEM
FOR INTEGRATED INTERPRETATION IN
HYDROCARBON RESERVOIR EXPLORATION AND PRODUCTION

THOMAS J. LASSETER

Tech·Logic, Inc.
15325 189th Avenue NE
Woodinville, Wa. 98072 U.S.A.

ABSTRACT

Integrated interpretation requires the representation of the reservoir as a single consistent model which can be displayed and interactively edited in three dimensions. To be practical, high-performance 3-D graphics systems and the requisite support software are essential. Integrated data display and editing tools allow interpreters from the various disciplines to view and edit the model from their unique perspectives.

The model representation must support more complex topologies than the conventional "layer-cake" model if reservoir models and their associated performance predictions are to be realistic. Surfaces represented by detailed grids derived from maps and seismic interpretation are used to form the "framework" for the geological interpretation. The use of parametric curves and surfaces provides a compact and editable representation of the geometry allowing the interpreter to quickly modify the model when new data becomes available or alternative "realizations" are required.

KEYWORDS

Reservoir modeling; integrated interpretation; 3-D graphics; seismic interpretation; petrophysical interpretation; geological modeling; reservoir simulation; parametric surfaces.

INTRODUCTION

Integration and interactivity are the keys to improved reservoir interpretation. Only through integration of the associated interpretation activities can the data be completely and consistently used. Only through interactivity can the interpretation activities be efficient enough to be practical.

The development of high-performance three-dimensional computer graphics systems coupled with high-performance computational systems has made integrated, interactive 3-D reservoir interpretation systems possible. This new technology has facilitated the development of innovative approaches to reservoir interpretation.

APPROACHES FOR INTEGRATED INTERPRETATION.

It is often stated that integrated interpretation can be achieved by simply having applications communicate with each other. As an example, if seismic interpretation and petrophysical interpretation applications both generate maps in a standard format which they can exchange, then under this definition they are integrated. The problem is that both applications build separate and generally inconsistent models. It remains for the interpreters from the separate (and contending) disciplines to negotiate a consistent model. This "negotiation process" is time-consuming and inefficient. Neither interpretation domain (geophysical or petrophysical) provides the framework for representing and interpreting data from the other domain, so the two interpretation "factions" are left with paper logs, paper sections, and the "blackboard" to work out a consistent model.

The essence of integrated interpretation is a single, consistent model. It is not sufficient for applications to simply communicate with each other or share a common database. Each discipline has a unique perspective on the model based upon the associated data with unique contributions to make to the model definition. These perspectives overlap significantly, however, so model definition and editing must be done in a way which allows each discipline to evaluate whether proposed changes remain consistent from their perspective.

In order to provide a well-integrated consistent interpretation system, the following guidelines have been used:

    1) all interpretation is done on a single self-consistent model;

    2) graphical displays are designed to integrate as many data sources as is practical;

    3) the use and juxtaposition of multiple windows provides additional integration where a single window would be impractical.

In subsequent sections, the details of this interpretation approach will be described.

INTEGRATED GEOPHYSICAL INTERPRETATION

Three-dimensional seismic interpretation has been conventionally done using vertical or time-slice sections. A major problem in seismic interpretation is making effective use of the well data. Wells are often highly-deviated making it difficult to tie the well data to planar seismic sections, particularly in complex geology. A better approach facilitated by high-performance 3-D graphics hardware is to construct ribbon sections between deviated wells and "cut" the 3-D seismic cube with these ribbon sections. This provides accurate ties between the well data and the seismic data as shown in Figure 1. Seismic data can also be displayed in the well-window allowing the interpreter to view the data both globally and in detail as shown in the figure. Synthetic seismograms could also be displayed to aid in the interpretation.

In order to generate these integrated 3-D seismic and well data displays, the well deviation survey must be converted from depth to time. Strictly speaking, a complete model is required to do this correctly. In practice, it seems reasonable to use sonic data, vertical seismic profiles, etc to achieve an approximate depth-to-time conversion. Once the well data is displayed in time in the well window together with the 3-D seismic data, the well data can be interactively adjusted so that geological markers match seismic markers. This effectively edits the velocity distribution which can be evaluated for consistency. Having accurately tied the 3-D seismic data and well data together at the wells, we can start from seismic markers at the wells and interpret events on the ribbon sections. This interpretation involves identifying both faults and horizons. When faults have been identified in wells, this can be used as important information for locating faults accurately on the connected ribbon sections.

Having defined and edited the time-to-depth relationship for each of the wells, we can convert each ribbon section from time to depth in several ways. The time-to-depth relationship implies a velocity which can be used to convert the complete section. Where the section is defined in terms of correlatable layers (as we will discuss), the velocity distribution can be extracted from the interpolated distribution for each layer. Each of the sections used in the seismic interpretation could then be mapped to depth with significantly less effort than inverting the entire seismic cube. Interactive editing of the velocity distribution would allow the interpreter to modify and presumably improve the time-to-depth conversion.

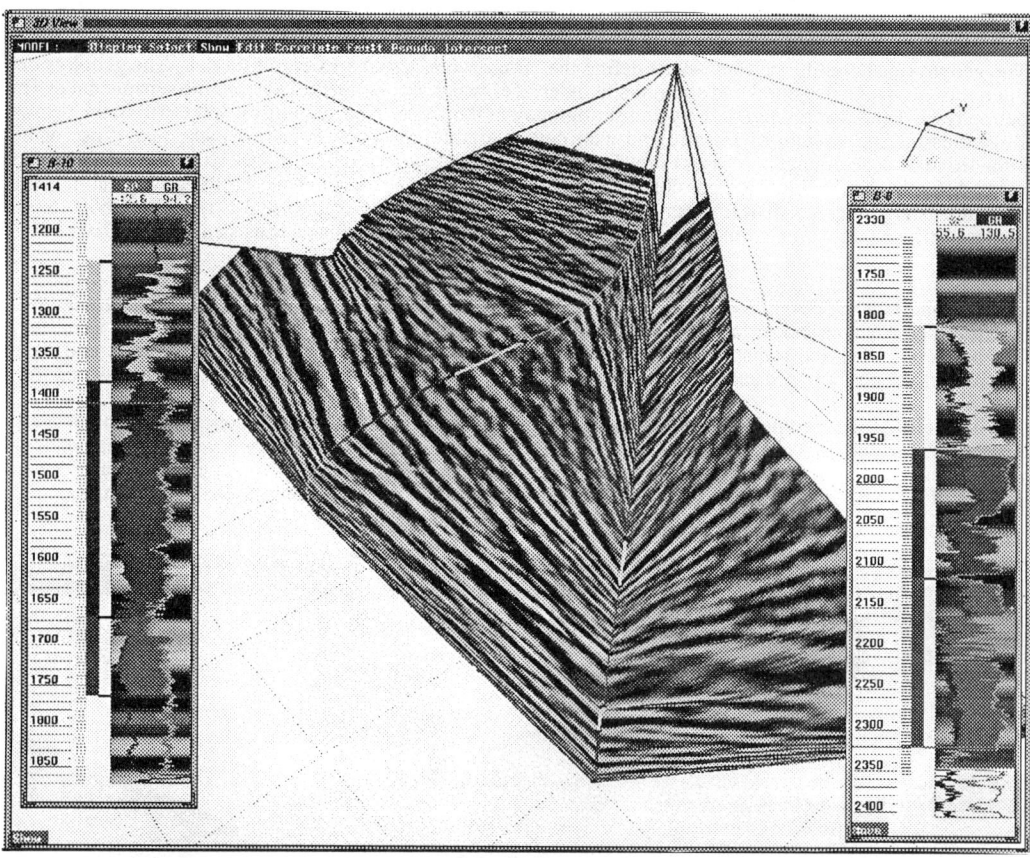

Fig. 1. 3-D seismic data display on ribbon sections between wells.

INTEGRATED PETROPHYSICAL INTERPRETATION

The prototype interpretation system allows the display of log and zone information in the well windows as was shown in Figure 1. Additional displays which would significantly improve integration and interpretation include lithological zones, core data, dipmeter data, production test data, biostratigraphic information, and general annotation. With these displays, the petrophysicist, petrologist, and geologist can develop consistent and detailed interpretation of rock structure, lithology, stratigraphy, and properties.

Computer-processed interpretation of petrophysical data is typically done on a well-by-well basis with the same procedure being applied to the entire well. From the processed logs a manual or automatic interpretation of lithology is typically performed. Stratigraphic zones are then identified and correlations made. At this point, the computer-processed petrophysical interpretation should be repeated, not on a well-by-well basis, but on a stratigraphic unit-by-unit basis. This approach should provide a more accurate "model-based" petrophysical interpretation. The improved lithological interpretation should in principle improve the geophysical and geological interpretation of the model.

INTEGRATED GEOLOGICAL INTERPRETATION

The geophysicist and the petrophysicist define the "framework" for the reservoir model. The geophysicist interprets the major structural and stratigraphic interfaces which can be identified from the seismic data. The petrophysicist can determine lithology and stratigraphy at the wells. The geologist is left to fill in the details. Regardless of the amount of seismic and well data available, the geologist must make heavy use of his knowledge of depositional processes to render any kind of detailed geological model. It is often the case that many different "realizations" are possible given the same data. As new data comes in, the geological model is often drastically affected. These observations emphasize the need for easy-to-use 3-D modeling tools for the geologist which allow him to quickly construct and edit a model which remains consistent with the data he has available.

Figure 2 shows a salt dome overlain by layers which have been faulted by the rising dome. This model uses 3-D parametric curves and surfaces which can be easily manipulated and modified. When a surface is edited, it can be recomputed and redisplayed in real time. This model was built with very little effort and again emphasizes the performance and interactivity which can be achieved with current 3-D graphics hardware.

Having defined the geometry of the model, the geologist must assist in the determination of property distributions within each unit which has been identified. Here again the data density is very low relative to the density desired. Furthermore, property measurements from core, logs, and seismic data are often suspect for a number of reasons. Permeability in particular is very difficult to derive from conventional data sources.

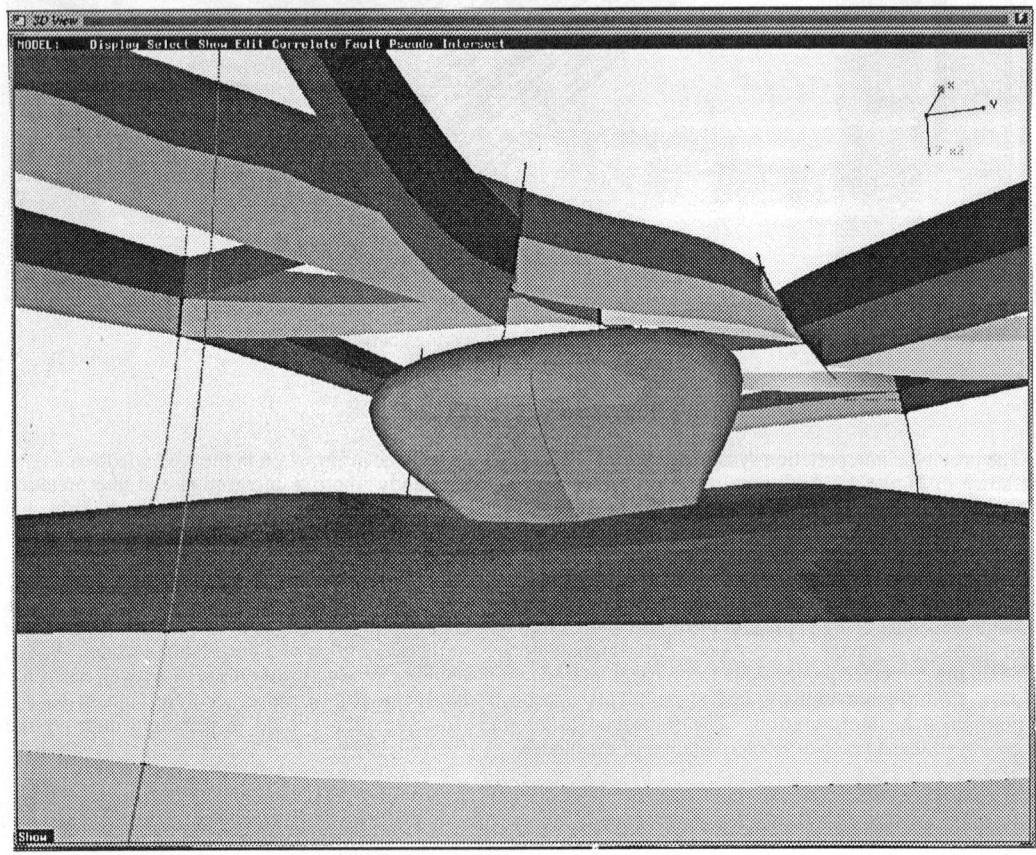

Fig. 2. Salt dome overlain by layers which have been faulted by the rising dome.

The property determination procedure begins with a display of the property interpolated over the units defined as is shown in Figure 3. The geologist often has a model for porosity and permeability distributions within each depositional type: fining upward with decreasing porosity and permeability for channels, for example. The available data and resulting interpolation should be evaluated in light of the appropriate model and interactively adjusted as necessary. Here again a number of "realizations" are possible and the tools should make the rendering of a number of realizations efficient.

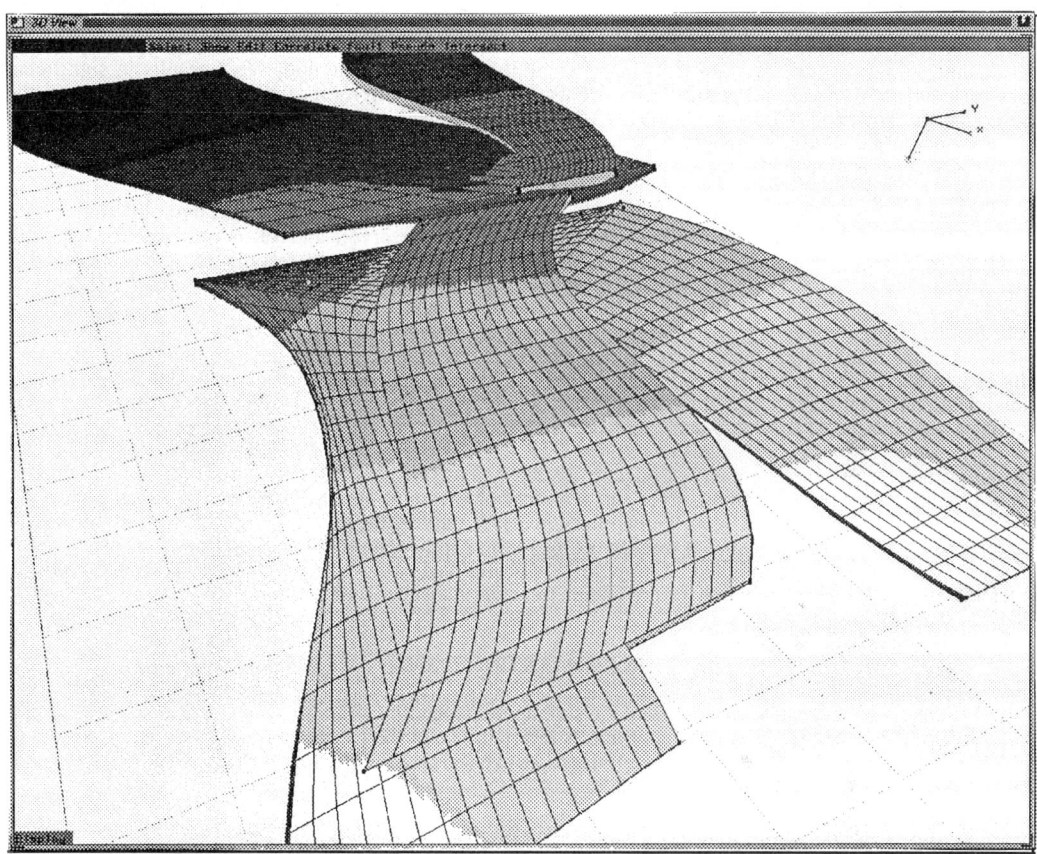

Fig. 3. Braided-channel sequence with interpolated permeability displayed.

INTEGRATED ENGINEERING INTERPRETATION

The model initially defined by the geophysicist and petrophysicist, and "filled-in" by the geologist arrives at the reservoir engineer. His responsibility is to identify the major "flow units" controlling the "hydraulics" of the reservoir and simplify the geometry of the model as much as possible while retaining its essential topology. Given this model his task is to determine the fluids-in-place, the reserves (how much can be recovered), and the producibility (how fast can it be recovered).

Conventionally, reservoir engineering models are almost always represented as layer-cakes. Producing horizons are typically modeled as single continuous layers when in fact they are a complex, partially disconnected network of flow units. This oversimplification of the geology often results in significant overestimates of reserves and producibility. It also is the major reason for lack of integration between the geologist and engineer. Once the engineer has converted the geological model to layers of grid blocks and subsequently modified grid-block properties in the reservoir history-matching process, the relationship between the engineering model and the geological model is often completely lost.

As was shown in Figure 3, a series of braided channels can be gridded for subsequent input into a reservoir simulation model. This model preserves the topology of the reservoir and yet is practical for reservoir simulation. As the geometry of the model is edited, the grid is automatically recomputed and redisplayed in real time.

The results of simulations run by the reservoir engineer can be easily shown as part of the unit display. The analysis of the results often leads to significant modifications of the geometry and/or property distributions. By preserving a single consistent model of the reservoir, the engineer can effectively communicate his model modification recommendations to the other interpretation team members and work in collaboration with them to define a new, consistent and presumably more accurate model.

An important task of the engineer is to determine the volume of hydrocarbons in place. For the modeling approach we have been describing, given the geometry of each unit, its associated porosity distribution, the geometry of the fluid contacts and the capillary pressure curves, the fluids-in-place for each unit can be computed. This is an improvement over the conventional approach in which fluids-in-place for the entire reservoir as a whole are computed. Computing on a unit basis allows the engineer to associate recovery factors with each unit depending on the production facilities associated with the unit, etc.

MODEL CONSTRUCTION AND EDITING TOOLS

The system we have been describing (IREX) is being developed by an industry consortium. It runs on the Silicon Graphics series of computers and makes efficient use of their high-performance three-dimensional graphics capabilities. Curve and surface definitions utilize parametric representation allowing for multivalued surfaces such as folds and reverse-thrust faults.

In the IREX prototype development, focus has been on evaluating interactive 3-D modeling functions. In addition, a high-performance flexible data management system has been developed which provides methods for allocating, storing, retrieving, and deleting data objects.

The graphics interface consists of five basic window types:

3-D. A 3-D view of the grid which shows an outline of a 3-D seismic data volume, wells (if highlighted), sections, maps, and parametric patches; the 3-D window is where all editing other than normal well definition and initial zoning takes place.

Map View. A map view of the model; well x-y values are defined in this view; the top window is also used for highlighting/unhighlighting wells and sections.

Well. A scalable and scrollable display of the well TVD, zoning, log curves, and amplitude extracted from the 3-D seismic.

Color Definition A window for selecting and editing color spectra associated with seismic, wells, zones, stratigraphy, and lithology.

Message. A window for data input and diagnostic message output.

The current system has the following basic model construction and editing tools:

Data loading. 3-D seismic data, well deviation surveys, well logs, and maps can be loaded from ASCII files.

Data management. The model and associated data can be saved at any time in a compact binary representation which can be subsequently reloaded.

Seismic display. Seismic amplitude data from a 3-D data set can be displayed on any arbitrary section as a series of Gouraud-shaded triangles offering smooth color interpolation without display artifacts.

Pseudo well definition. The 3-D path of a pseudo well is defined using the well window to pick time or depth and the map view window to pick x and y.

Zone definition. Individual well zones are interactively defined and edited in the well window. Once defined, zones can be edited in the 3-D window also.

Fault definition. Fault lines (both normal and reverse) can be defined. When added, up- and down-thrown block zones are automatically generated. The user can edit the displacement, slope, and curvature at the fault. Correlations can then be made along the fault surface on both the up- and down-thrown blocks. Faults crossing faults can be easily modeled with the system.

Correlation. In the 3-D window or in two well windows, the interpreter picks two zones in different wells belonging to the same stratigraphic or structural unit. Edges consisting of one or more parametric cubic curves are drawn between the respective tops and bottoms of the two zones.

Edge editing. Each parametric curve on an edge is defined by two end-points and two "handles" which control slope and curvature; the user can pick and move the points and/or handles.

Pseudo wells. Additional detail is added to a section between wells by specifying the path of a pseudo well on the section. Correlations crossed by the pseudo wells are automatically added to the pseudo well description as zones. Additional zoning can then be added to the pseudo wells.

Path editing. The paths of pseudo wells and fault lines defined on sections can be edited; a fault can be easily moved anywhere on a section for example.

Subunit definition. The geometry of uncorrelatable subunits can be specified by defining pseudo wells on existing sections, adding additional zones on the pseudo wells, and then correlating between these zones.

Fault surface definition. A series of fault lines can be connected by correlating zones on both sides of the fault surface. The cross-sections of units crossing the fault surface can be described on both sides of the fault. Faults crossing faults can be described by inserting fault lines on both sides of the fault surfaces with the appropriate vertical and lateral displacement.

Vertex definitions. A vertex can be inserted anywhere on an existing parametric curve, effectively breaking it into two curves each of which can be further edited.

Surface patch definitions. Three or four vertices can be selected on the same surface and a parametric patch or series of patches will be constructed. The patch is then computed along with normals so that a lighted display of the surface can be made.

Correlation display. Correlations on sections between wells can be shown as both wireframe or solid color fill; the latter uses a lighting model which gives the eye important visual cues to surface shape.

Surface display. The parametric patch surfaces can be shown as lighted and smooth-shaded surfaces.

As the model is defined and edited, an associated model topology is constructed and modified. Correlations and patches are formed into layers. Layers are collected into stratigraphic units as well as being organized into structural blocks. The interpreter can then display and edit the model by editing layer geometries and properties on a unit basis or on a block basis.

AN INTEGRATED INTERPRETATION CASE STUDY

Figure 4 shows an actual data set consisting of 3-D seismic data, three interpreted maps, and eleven wells with associated log curves. The maps are time maps interpreted on a seismic workstation. Velocity data was used to convert the well deviation surveys to time.

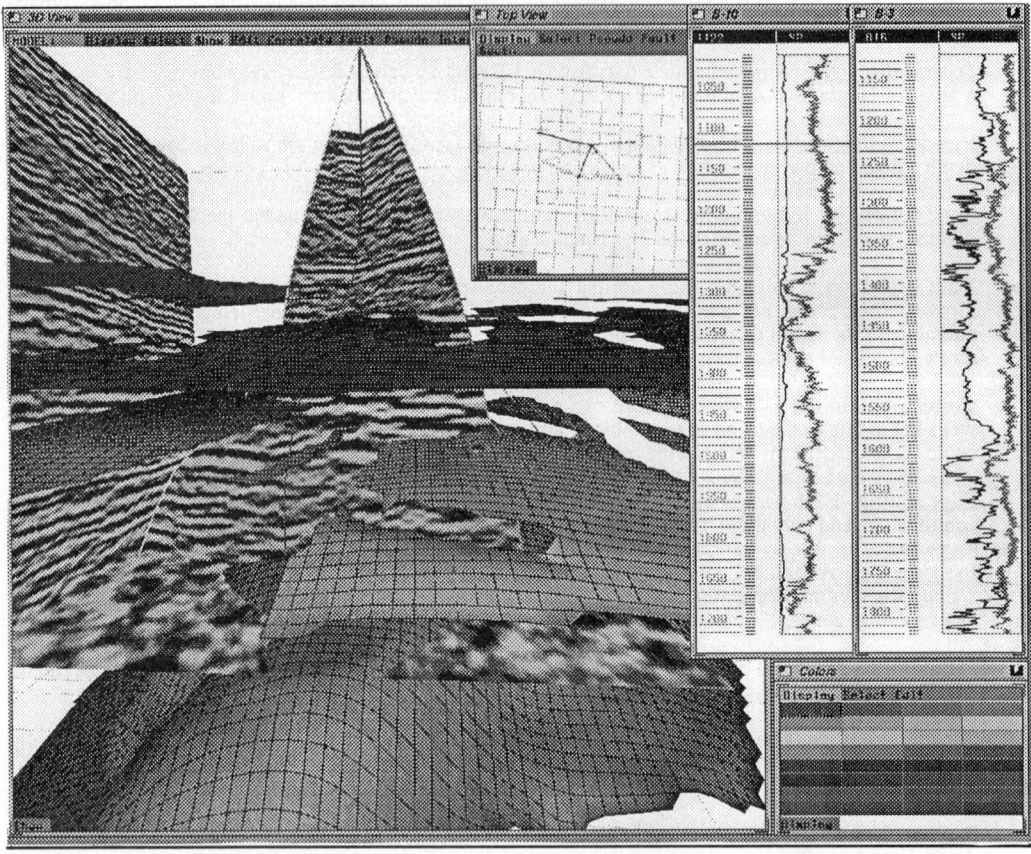

Fig. 4. Actual interpretation data set: 3-D seismic displayed as vertical, time-slice, and ribbon sections; well surveys and log data; interpolated maps.

Picking three or four wells, three- or four-sided patches are automatically cut from the intersected maps. Zones and correlations between mapped surfaces are automatically generated. Parametric patches approximating the map surface patches are also automatically computed. Ribbon sections are automatically built allowing the map patch edges to be displayed against the seismic data to validate the map interpretation. Where faults have been defined on the maps, the interpreter is required to build faults on the ribbon sections. The up- and down-thrown sides of the faults are then correlated thus defining individual fault blocks.

The zones that have been automatically built can be compared with the well log data again to validate the interpretation and can be edited as necessary. The log curves associated with each zone can be examined for subzones which can then be interactively defined and correlated. This would allow the geologist to construct a series of braided channel-sands within a bounding channel-sand unit as was shown in Figure 3.

In this case study, a log-derived permeability curve has been computed arbitrarily from the gamma-ray curve. In practice, the interpreter should be able to use a suite of property generation functions as well as generate customized functions. In Figure 5, the permeability curves for two of the wells are shown in the associated well windows along with the zone averages. The interpolated permeability map for the top layer is shown in the 3-D window. The user can interactively edit the permeability curves or the zone-averaged values. The interpolation is then automatically recomputed and redisplayed. Since we are currently using a simple interpolation scheme, the update and redisplay occurs virtually instantaneously.

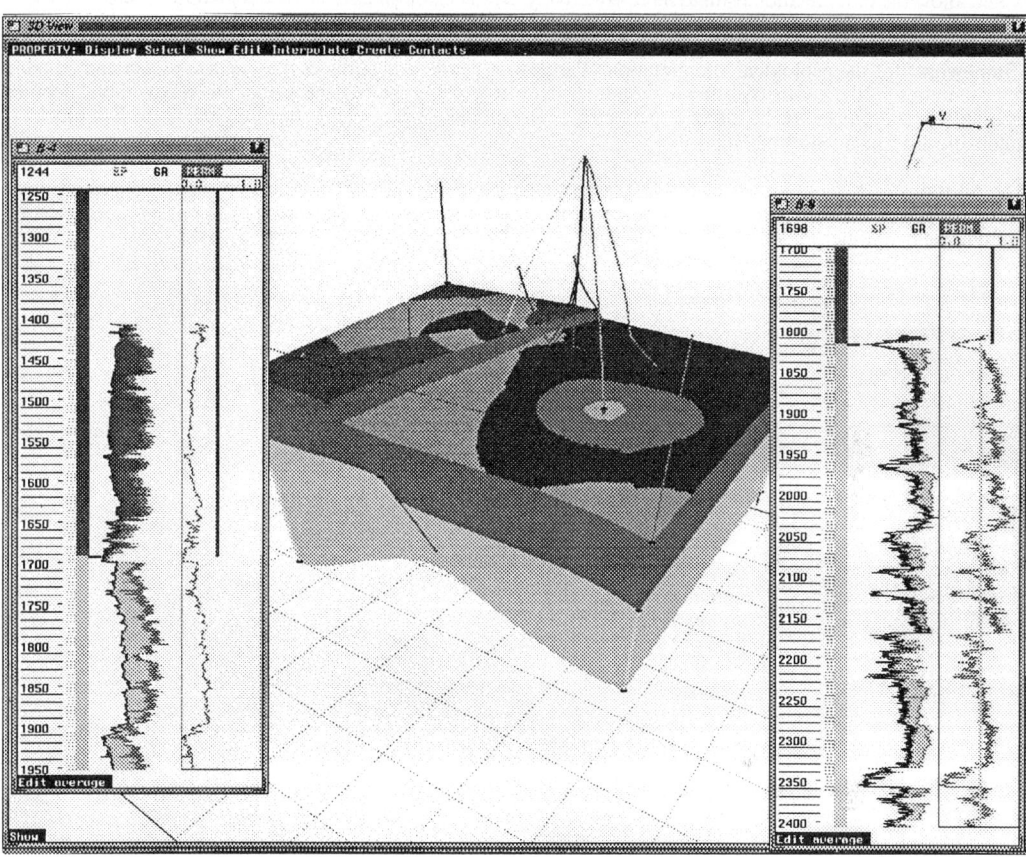

Fig. 5. Permeability display: curves and zone-averages in well window; interpolated distribution in 3-D window.

SUMMARY AND CONCLUSIONS

Integrated interpretation requires the representation of the reservoir as a single consistent model which can be displayed and interactively edited in three dimensions. To be practical, high-performance 3-D graphics systems and the requisite support software are essential. Integrated data display and editing tools allow interpreters from the various disciplines to view and edit the model from their unique perspectives. Facilitating the display of geophysical and petrophysical data in conjunction with reservoir simulation output as an example would allow the interpretation team to more effectively evaluate and ameliorate model inconsistencies.

Surfaces represented by detailed grids derived from maps and seismic interpretation are used to form the "framework" for the geological interpretation. Parametric curves and surfaces provide a compact and editable representation of the geometry allowing the interpreter to quickly modify the model when new data becomes available or alternative "realizations" are required. The model topology in terms of layers, units, and blocks is automatically generated ad maintained.

The oil industry is facing more difficult interpretation problems as reservoir development focuses on more complex geological environments. The development of interactive 3-D interpretation systems used by multidisciplinary interpretation teams should help offset the increased uncertainties and risks.

DIRECT GEOSTATISTICAL ESTIMATION OF IRREGULAR 3D VOLUMES

S.W. Houlding and M.A. Stoakes

LYNX Geosystems Inc, 1199 West Pender St.
Vancouver, BC   V6E 2R1   Canada

I. Clark

Geostokos Ltd.
36 Baker Street, London, W1M 1DG

ABSTRACT

Those who have had an occasion to apply geostastitics to the estimation of ore reserves are familiar with the conventional approaches of estimating average grade of quality values for arrays of regular grid squares or blocks. Further, the technology to perform geostatistical analysis and prediction for irregular two-dimensional areas has been available in a practical, readily useable form for some time. However, the generalization of this technology into three dimensions has been hampered by lack of a suitable geometrical characterization for realistic, three dimensional shapes and volumes.

The extension of these estimation techniques into three dimensions has now been made possible by a successful integration of the volume modelling capabilities of 3D Component Modelling with proven 3D Geostatistical Techniques. The result is a powerful new technique for mineral ore reserves estimation in a practical, useable form. The technique allows direct geostatistical estimation of grade and quality values for precisely defined, irregular, realistic, geological, and mining shapes or volumes.

This paper describes the two technologies, their integration and their application in practice to the estimation of mineral ore reserves. The paper closes with a discussion of the benefits provided by the new approach, in terms of accuracy, efficiency and practicality.

KEYWORDS

Geostatistics, 3D modelling, estimation, ore reserves, volumetrics.

INTRODUCTION

Mineral ore deposits exist in three-dimensional geological space in varying degrees of complexity. For detailed evaluation and mine-planning purposes we need to be able to represent deposits, and simulate their extraction, within a computer environment. For successful and efficient representation, the modelling process must be able to accommodate ALL aspects of three-dimensionality to a level of precision which is compatible with an acceptable degree of mining and financial risk. Traditional modelling methods are generally incapable of fulfilling this requirement. This paper describes a new modelling technology which satisfies the precision criteria in all aspects. The technology is based on the successful integration of new 3D

Geostatistical theory, which allows the direct estimation of irregular three-dimensional volumes, and 3D Component Modelling, which provides the necessary precision in geometrical representation of three-dimensional shapes and volumes.

BACKGROUND

A few deposits, like thin uniform tabular coal seams, can be approximated by gridded surface representation. A few others, like simple disseminated base metal deposits, can be approximated by traditional block modelling techniques. The majority of deposits exhibit a degree of complexity which renders their representation by traditional modelling methods unacceptable. The complexity may be due to structure (faulting, folding, intrusions, etc.), irregularity in shape and form, variation in grade or quality, or combinations of these and other three-dimensional effects. In all cases the complexity is increased by an order of magnitude when the need to represent three-dimensional excavation limits, and to simulate the extraction of the deposit, is superimposed on the modelling requirement.

The traditional computer modelling methods all involve approximations which introduce errors when applied to complex three-dimensional ore deposits. These errors affect the end results of any deposit evaluation or mine-planning exercise. Even less acceptable, most traditional methods are incapable of providing any measure of the magnitude of the estimation errors introduced. To elaborate, any geostatistical estimate of grade or quality is incomplete (and misleading) unless it provides not only an *average* value, but also a *range* and an associated *confidence level* (or probability) for the volume of concern.

Thus the reliability of an estimate produced by traditional modelling methods is unknown and cannot be related to an acceptable degree of risk. Considering the amount of capital required to implement a mining project, the high cost of operation, and the associated risks of today's volatile commodity markets, the elimination of these deficiencies by the new technology presented below is indeed timely.

Three essential features are required in order to successfully model complex three-dimensional deposits, and to simulate their extraction, within a computer.

An ability to define and represent irregular, realistic, geological and mining shapes with precision.

A precise method of determining the volumes of these irregular three-dimensional shapes, and their volumes of intersection.

A method for direct geostatistical estimation of grade or quality within these irregular three-dimensional volumes which accommodates the geometry of geological and mining shapes.

The first two requirements have been met in THEORY AND PRACTICE by the recently developed and proven capabilities of 3D COMPONENT MODELLING technology, as conceived by LYNX Geosystems Inc and implemented within the LYNX Mining System.

The third requirement has previously been met IN THEORY ONLY by 3D GEOSTATISTICS technology, as conceived by Geostokos Ltd.. The implementation of this theory in practice has been hampered by lack of a suitable geometric characterization for irregular three-dimensional volumes with which it can be integrated. This last requirement has now been provided in a compatible and complementary form by 3D Component Modelling.

The product of combining and integrating these technologies is a computer modelling methodology which provides a practical, logical and precise approach to estimation and evaluation of complex ore deposits. It also provides a stable platform on which to base mine-planning and production scheduling activities. The methodology eliminates the use of traditional modelling methods and their inherent approximations, constraints and errors, and considerably reduces the amount of computer storage and processing required.   Finally, it provides a measure of reliability (associated standard error) for any estimate produced.   The results of its application can therefore be related directly to an acceptable degree of mining and financial risk.

The two technologies, their successful integration, and their application in practice to the evaluation and mine-planning of mineral ore deposits are described below.

THEORETICAL BASIS

The theory, process and application of 3D Component Modelling technology have been described in detail elsewhere (1,2).   The key features which distinguish this technology from traditional modelling methods are the following.

  The ease and precision with which complex irregular shapes are defined
  and represented in the computerized geological environment. This is
  achieved by means of a geometric characterization of shapes termed 3D
  Solids Modelling; based on analytical geometry and solids of
  integration theory.

  The precision with which the enclosed volumes of complex irregular
  shapes, and the volumes of intersection of two or more shapes, are
  determined. This is achieved by means of an analytical procedure
  termed 3D Volumetrics; based on the geometry of intersection of any
  plane with an irregular shape, and volumetric integration.

  The precision with which spatially variable characteristics, such as
  grade and quality, associated with the volumes of irregular complex
  shapes are estimated and represented. This function was provided in
  the previous implementation of 3D Component Modelling by an
  integration of accepted geostatistical procedures with the features
  described above. This development in itself represented a significant
  advance in modelling technology.

In the present implementation the accepted procedures are replaced by the 3D Geostatistics technology described herein.   This technology is in all ways more compatible and complementary to the geometrical characterization of shapes used by 3D Solids Modelling, and to the analytical procedures of 3D Volumetrics.   The technology is an extension of existing geostatistical methods for the estimation of irregular two-dimensional polygonal areas into the third dimension.   Its implementation is based on the ability to readily determine the polygonal geometry of intersection of planes with an irregular shape.   The basic requirements are therefore similar to those of 3D Volumetrics.

Within 3D Component Modelling, the complex irregular shapes of realistic geological and mining units are modelled by sets of one or more 3D Solids components, as described below.   It is the unique way in which these components are defined and represented within the computer, and the ability to determine the geometry of intersection of a plane (or section) at any orientation, which make possible the integration of the 3D Geostatistics technology.   In order to present a meaningful description of the new technology and its integration, a brief preliminary description of these features is required.

1. Solids Modelling of Irregular 3D Shapes

A typical 3D Solids component, representing an irregular three-dimensional shape (or part of a shape), is illustrated by Fig. 1.

The orientation of the component is controlled by that of the viewplane from which it is defined.   This is specified in terms of N,E,Z,Azimuth and Inclination values in the Global Coordinate System.   The shape of the component is controlled by Mid-plane,

Fore-plane and Back-plane boundaries, defined interactively, and by the links between boundary points. The Fore-plane and Back-plane are assumed parallel to the Mid-plane and are therefore fully defined by a specified Fore-thickness and Back-thickness respectively.

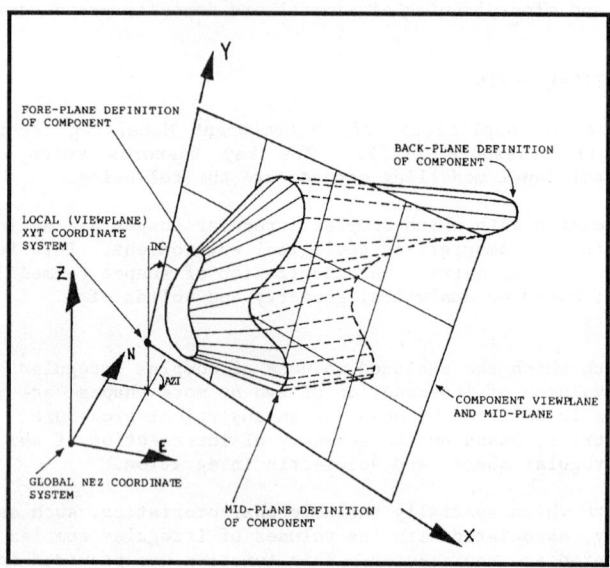

Fig. 1.   The definition and geometric characterization
          of a 3D Solids Modelling Component representing
          an irregular volume.

The points comprising the boundaries are defined in terms of Local (XYT) System coordinates. The volume of the component is that enclosed by the implied polygonal facets formed by the boundaries and links of the component. This completes the geometrical characterization of the component.

The component definition is completed by specification of several additional items, including unit and component identification strings. These are specified according to a selected convention and are referenced by the wild-card retrieval facilities of the modelling database. In this way *sets* of contiguous components can be used to represent more complex geological and mining units.

Despite the simplicity of the geometrical characterization of components, it has been successfully and efficiently applied to the representation of all forms of complex geological and mining shapes. Figure 8 at the end of the paper illustrates these capabilities.

2. Sections at Arbitrary Orientations

A feature of 3D Solids Modelling, based on three-dimensional analytical geometry and the geometric characterization of components, is the ability to readily determine the two-dimensional polygonal geometry of intersection of a plane with a component, irrespective of their relative orientation. This capability is illustrated by Fig. 2. It allows the display of sections at any orientation through irregular three-dimensional shapes represented by sets of one or more contiguous components. This capability also provides the basis for volumetric integration and 3D geostatistical estimation, as described below.

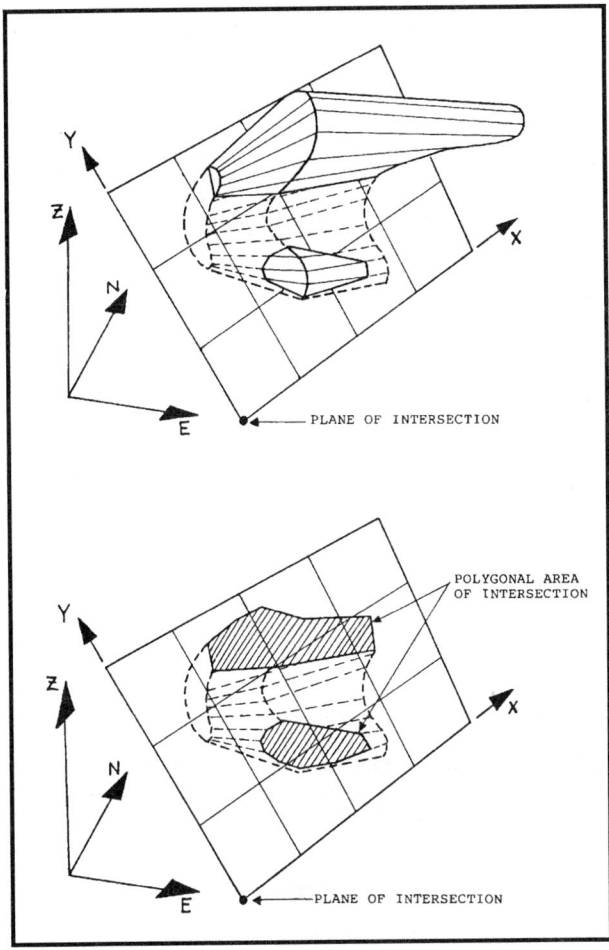

Fig. 2. Top: Component intersected by arbitrary
                 plane (refer Fig. 1.).
          Bottom: Polygonal geometry of intersection.

3. Volumetric Integration of Irregular Shapes

The process of determining the two-dimensional geometry of intersection of a plane
with a component can readily be extended to determination of the area of intersection
by means of simple analytical geometry.   If the concept of a thickness, measured
normal to the plane, is added then the result is an intersection *slice* with an
associated volume.   If the plane of intersection is replaced by a set of parallel
planes at a constant spacing, then a volume of intersection can be obtained for each
of the planes.   Integration (accumulation) of these volumes throughout a component
provides a precise measure of the volume of the component, providing that the
spacing, or integration increment, is sufficiently small.   The process is graphically
illustrated by Fig. 3.

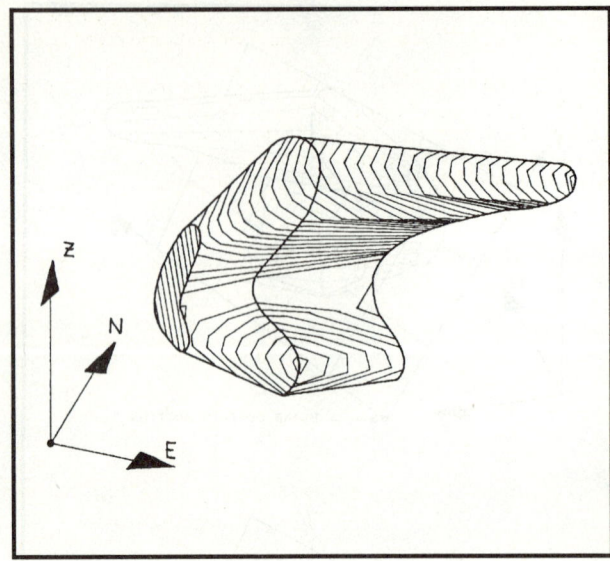

Fig. 3. Graphical representation of parallel intersection
through a component, used for volumetric integration
(refer Fig. 2.).

The volumetric integration process is extended from single components to the analysis
of complex irregular shapes by accumulation of results for the complete set of
components representing the shape. The process is independent of the orientation of
the slices relative to the components. Thus there is no requirement for all of the
components representing a geological or mining unit to be similarly oriented.

4. Geostatistical Estimation of Irregular Volumes

The primary objective of the 3D Geostatistics technology is to produce directly a
single estimate, and its associated standard error, for irregular realistic
geological and mining volumes. The requirement for this capability is based largely
on the following.

The estimation of the average value over a volume differs in concept, theory and
practice from the estimation at a single point. It is a fact that the average value
of all points within a mining stope, or a panel, has a different distribution of
values from that of the original sample data. In particular, average values tend to
vary less than sample values. Thus the variance of stopes or panels tends to be
lower than that of relatively small samples. In a similar way, the relationship
between a sample and a large volume is not the same as between the sample and (for
example) the centre of mass of that volume. Kriging is possibly the only estimating
technique which incorporates the geometry of the *volume to be estimated* into the
estimation process.

For this reason, the representation of geological and mining volumes by multitudes of
blocks and sub-blocks (as in traditional block modelling) not only introduces errors
of approximation at the volume boundaries but also introduces errors of estimation
throughout the volume. In the latter case the estimation error increases as the
blocks are made smaller. This situation is compounded by an inability to produce any
useful measure of the estimation error caused by sample distribution and volume
geometry. All of these approximations and deficiencies are eliminated if irregular
volumes are treated as *whole* units, as presented below.

Since the new technology is an extension of the existing approach for estimation of
two-dimensional irregular areas, it is instructive to summarize the latter first (3).

The major difference in technique between kriging the value at a point and obtaining
the average value for an area (or volume) is in the setting up of the system of
simultaneous equations which yield the kriging estimator and its associated standard
error. These equations are dependent on the size and shape of the area and its
relationship to the known samples. This factor becomes obvious on inspection of the
kriging estimation equations, and the variance equation, in their generic form (6),
as follows.

The kriging estimation equations,

$$\sum w_i \, \bar{\gamma}(s_i, s_j) + \lambda = \bar{\gamma}(s_i, v) \quad \text{[for each example]} \tag{1}$$

$$\sum w_i = 1 \tag{2}$$

and the kriging variance equation,

$$\sigma^2_\kappa = \sum w_i \bar{\gamma}(s_i, v) + \lambda - \bar{\gamma}(v, v) \tag{3}$$

where,

w_i is the sample weighting factor

s_i is the sample factor

$\gamma(\ , \ )$ is the semi-variogram function

v is the volume (area) of concern

λ is the associated standard error

In order to set up these equations the following semi-variograms must be calculated
and their contributions to the equation coefficients determined.

(i)   Between each pair of samples; represented by the term (s_i, s_j) in the
        equations.

(ii)   Between each sample and every point within the irregular area; represented by
        the term (s_i, v).

(iii)   Between every possible pair of points within the irregular area; represented
        by the term (v, v).

In (ii) above, instead of the semi-variogram between two points, ie. a sample and the
point being estimated, the requirement is to calculate the average semi-variogram
between the sample and EVERY point in the specified area.

The necessary contribution to the kriging equation corresponding to the sample can
then be made. This process is repeated for each available sample.

To perform this process in the case of an irregular area (or polygon) a numerical
approximation is used. This involves placing a rectangular grid over the polygon and
selecting every point which falls inside the polygon. These discretized points are
assumed to represent the whole area. Much investigation has been done on determining
the number of grid points required to obtain a suitable level of precision (4),(5).
It has been found that 100 (approximately) points falling inside a polygon provides
an acceptable result for most shapes; the grid density can be adjusted to suit the
size and shape (or degree of irregularity).   It should be emphasized that this
discretization is used only to represent the contribution of the area itself to the
kriging equations. Figure 4 illustrates the discretization concept.

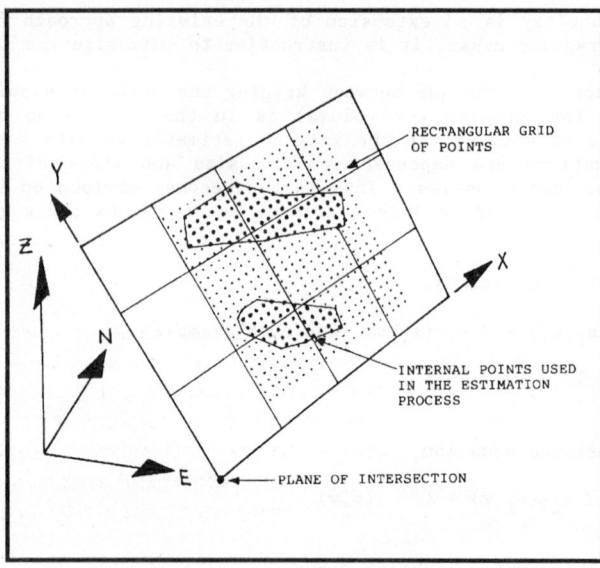

Fig. 4. The rectangular grid of discrete internal points
used in the estimation of a two-dimensional
polygonal area (refer Fig. 2.).

The final quantity of concern is the estimation (or kriging) variance. This is
partially dependent on the variability of values within the area being estimated, and
therefore is also sensitive to the size and shape of the area. The average
variability within the area is obtained by calculating the semi-variogram value
between every possible pair of points within the area and taking the average value
(refer (iii) above). The necessary contribution to the kriging variance equation can
then be made. The same numerical approximation grid as above is used in this case.

Solution of the set of kriging equations resulting from this process yields the
sample weighting factors, and hence a single estimated (kriged) value, and the
associated standard error for the irregular area.

The extension of the estimation process into three dimensions follows the same steps
as those outlined above for the two-dimensional process. In fact the only real
difference is in the mechanics of the discretization process itself.

Assume, as in the case of volumetric integration of irregular shapes (refer 3.
above), that the polygonal boundary of the area of intersection of a plane with a
component is representative of the geometry of a slice of uniform thickness. Assume
further that the discretization of the two-dimensional area by a uniform grid of
points is equally representative of the volume of the slice. If the plane of
intersection is replaced by a set of parallel planes at a constant spacing, equal to
the *slice* thickness, and the two-dimensional discretization process is repeated for
each plane of intersection, then the result is a three-dimensional discretization of
the volume of the component. The spacing of the grid points in the third dimension
is equal to the slice thickness and the set of contained points is representative of
the volume. Figure 5 illustrates the slice concept.

The slice thickness in this case is independent of the thickness used for volumetric
integration purposes and depends solely on obtaining a sufficient number of internal
points to provide adequate discretization of the volume.

In this implementation of the technology the grid density and the slice thickness are
variables which can be adjusted to suit this requirement.

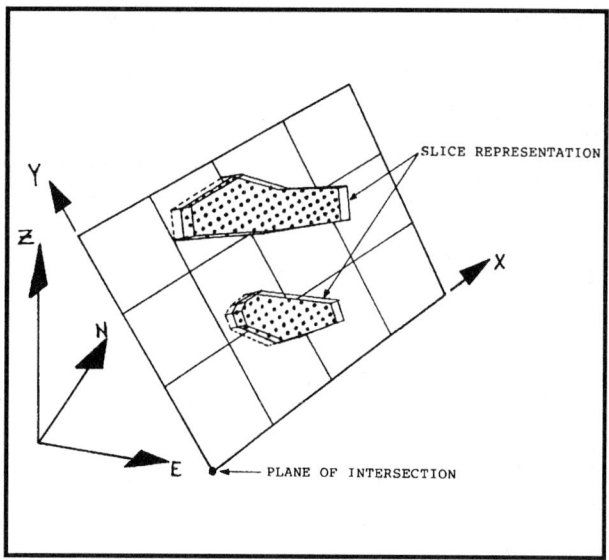

Fig. 5. The transition to three-dimensions - an
equivalent "slice" and its representative
discretization and geometry (refer Fig. 4.).

The semi-variogram values between samples and internal points are accumulated for the
entire volume during the slice discretization process, averaged at the end, and
contributed to the kriging estimation equations. The semi-variogram values between
all possible pairs of internal points are, of necessity, only determined and
accumulated once discretization of all slices is complete, ie. once the final form of
the representative three-dimensional set of internal grid points is known. The
average variability within the volume is then determined and contributed to the
kriging variance equation.

As before, solution of the equations yields a single estimated kriged value for the
irregular volume, and its associated standard error.

Orientation (of planes, slices and grid), grid spacings and slice thickness are all
variables which can be adjusted to suit the characteristics (geometry, anisotropy,
etc.) of the problem at hand. In the case of a more complex irregular shape, the
mechanics of the three-dimensional discretization process are readily extended to
include the concept of representation of the shape by a set of contiguous components.

5. Extention to Volumes of Intersection

The volumetrics and geostatistics procedures described above (refer 3. and 4.) are
readily extended to handle volumes of intersection between complex irregular shapes.
Consider volumetrics first and the case of two independent components which intersect
each other (refer Fig. 6.).

Select a plane which intersects both of the components, each of which produces a
polygonal area of intersection on the plane (refer Fig. 6.). The area of
intersection common to both components is readily determined by geometrical analysis
of the intersection of polygons. The result is a new set of *intersection* polygons.
As before, application of the *slice* concept, replacement of the intersection plane by

a set of parallel planes, and accumulation of the slice volumes provides a precise measure of the volume of intersection of the two components. A graphical presentation of this process is provided by Fig. 7.

Similarly, application of this intersection slice technique to the geostatistics procedures outlined above (refer 4.) leads to a direct kriged estimate, and its associated standard error, for the volume of intersection of any two irregular shapes.

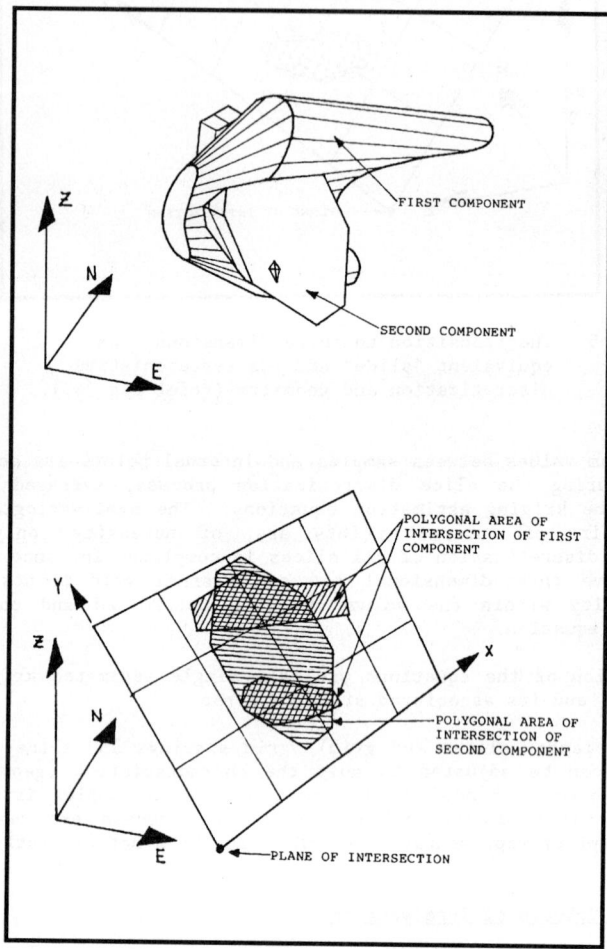

Fig. 6. Top: Two components which intersect each other
        Bottom: A common intersection plane and the
                overlayed polygonal areas of intersection

TECHNICAL ADVANTAGES

The obvious technical advantages of the 3D Geostatistics technology and its integration with 3D Component Modelling are many, these are listed below (there are no perceived disadvantages at this stage).

The boundary approximations inherent to most traditional modelling
methods are eliminated in the estimation process.

The size and shape of irregular volumes, and their relationship to
associated samples, are taken into account in the estimation process;
ie. the form of the resulting kriging equations is geostatistically
correct.

The methodology provides direct geostatistical estimation of
irregular, realistic volumes, and intersections of volumes; there are
no intermediate steps such as volume kriging of a multitude of
discrete rectangular blocks and subsequent intersection of these with
complex shapes and averaging of results.

The methodology provides a direct measure of the reliability of
estimation for any irregular volume, in terms of the associated
standard error.

The procedures are efficient in terms of computer processing and array
storage requirements, to the extent that realtime interactive
estimation of irregular volumes is now possible.

The methodology offers full flexibility in terms of orientation of the
estimation process to suit the geometry of the irregular volume and/or
the anisotropy of the grade or quality which is estimated.

The approach is flexible and provides a generic methodology for the
application of virtually any estimation technique; ie. it is not
limited to the use of *ordinary kriging*.

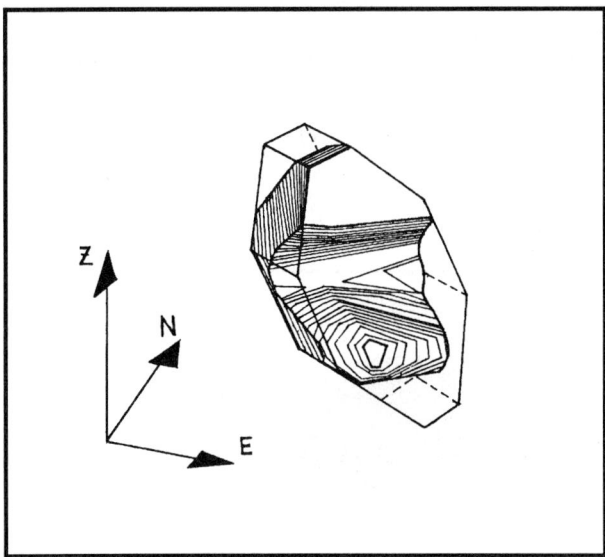

Fig. 7.   Graphical representation of parallel intersection
          planes through the intersection of two components,
          used for volumetric integration (refer Fig. 6.).

PRACTICAL APPLICATION

The technology described herein has been incorporated in the LYNX Mining System and
is available to the industry in a ready-to-use practical form for application to ore

deposit evaluation, extraction simulation, and the requirements of operating mines (refer Fig. 8).

The system, and the new technology, provide a logical, practical, precise approach to modelling. The conceptual philosophy throughout is one of *what you see is what you get*. In other words, the precision and detail provided by the user, and displayed graphically on the screen, is that used by the system for analysis and estimation. There are no hidden, internal approximations.

Figures 8a, 8b and 8c.   Practical Application of the New
Modelling Technology

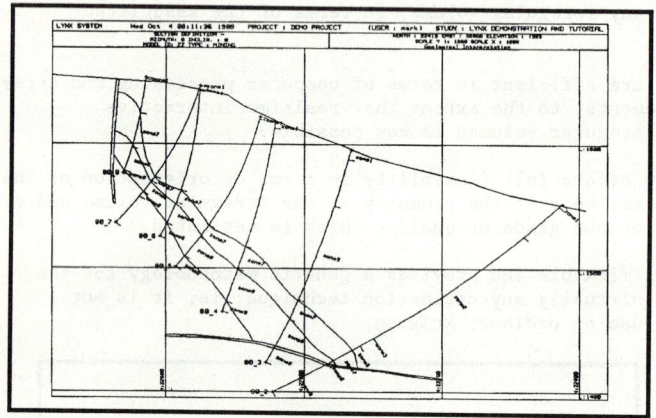

Fig. 8(a).   Geological Interpretation from Boreholes on
Section

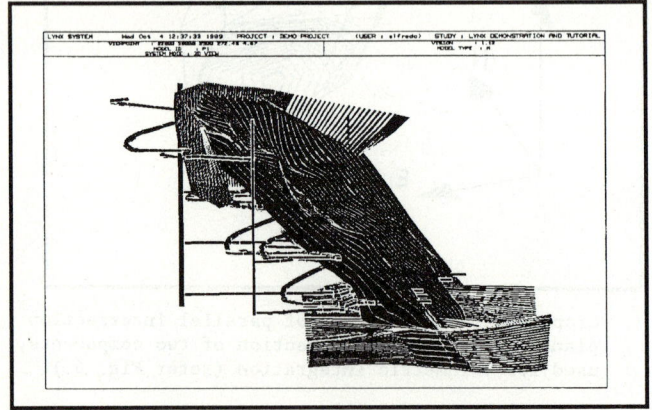

Fig. 8(b).   3D Perspective of Deposit Ore Zones and
Fault Structure.

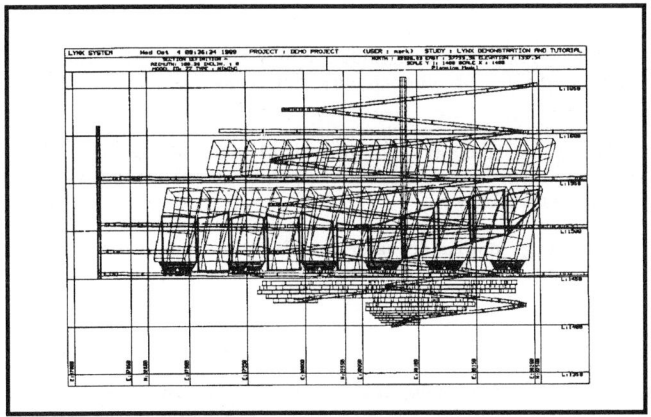

Fig 8(c). Design of Underground Mine Layout -
Stopes & Mine Development on Section

BENEFITS OF THE NEW TECHNOLOGY

The gross approximations and errors of most traditional modelling
methods, when applied to complex deposits, have been eliminated.

From the user's viewpoint, the whole approach to estimation is
simpler, more logical and intuitively more correct in its application
with the new technology. The interactive estimation capabilities
encourage an improved appreciation of the three-dimensional geological
and mining environment.

The direct estimation of irregular volumes means that the "true"
shapes of complex geological and mining units are accounted for in the
modelling process.

The ability to provide a more precise estimate, and a measure of the
reliability of the estimate, allows management to relate the results
to an acceptable degree of mining and financial risk.

The interactiveness of the estimation procedures encourages the
comparison of many design alternatives, and hence leads to
optimization of the mining extraction process.

Finally, the new technology provides a suitably precise platform for
detailed planning and scheduling of mining operations.

CREDITS

3D Component Modelling Technology is the intellectual property and copyright of LYNX
Geosystems Inc, Vancouver, Canada.

3D Geostatistics Technology is the intellectual property of Geostokos Ltd., London,
U.K.

AUTHORS

Dr. Isobel Clark, FIMM, C.Eng. is a founder and principal of Geostokos Ltd., London, UK. She is also responsible for conceptual development of the new 3D Geostatistics Technology.

Simon Houlding, P.Eng. is a founder and principal of LYNX Geosystems Inc, Vancouver, Canada. He is also responsible for the conceptual development and implementation of 3D Component Modelling Technology.

Mark Stoakes is a mining engineer and holds the position of Senior Technical Consultant with LYNX Geosystems Inc, Vancouver, Canada.

REFERENCES

Clark, I. (1976). Some Practical Computational Aspects on Mine Planning. *Advanced Geostatistics in the Mining Industry*, D.Reidel, Dordrecht, Holland, pp. 391-399.

Clark, I. (1977). Practical Kriging in Three Dimensions. *Computers & Geosciences*, Vol. 3, 173-180.

Clark, I. (1979). Practical Geostatistics. *Applied Science Publishers*, Vol. 3, United Kingdom, pp. 129.

Geostokos Ltd. (1988). *Geostokos PC Toolkit*. London, UK.

Houlding, S.W. (1987). 3D Computer Modelling of Geology and Mine Geometry. *Mining Magazine*, March ed., 226-231.

Houlding, S.W. and M.A. Stoakes (1989). Mine Activity and Resource Scheduling Using 3D Component Modelling. In: *Transactions of the Institute of Mining and Metallurgy Conference on Computer-Aided Mine Planning and Design*, UK, A53-A59.

MODELLING OF GEOLOGICAL DISCONTINUITIES FOR
RESERVE ESTIMATION PURPOSES AT NEVES-CORVO, PORTUGAL

E.J.Sides

Somincor, Apartado 12, 7780 Castro Verde, Portugal

ABSTRACT

The structure of a geological discontinuity modelling system, and its implementation as part of an integrated geological database and reserve estimation system, is described. The application of this system at the Neves-Corvo mine in southern Portugal is discussed.

KEYWORDS

Neves-Corvo, geological databases, 3-D geological modelling, reserve estimation, boundary representations, geological discontinuity model, volumetric modelling, block models.

INTRODUCTION

3-D geological modelling plays an important role in the generation of reserve models for use in planning the exploration, evaluation and exploitation of most mineral deposits. The objective of such modelling is to provide three dimensional representations of the variation of material types and their physical properties, such as density and metal content, within a deposit. By enabling the evaluation of tonnages and grades within defined mining volumes, such models permit the technical and financial evaluation of different mining scenarios. Results of such studies are used to determine the initial viability of mining projects, as well as for production control planning on existing mining operations.

The Neves-Corvo copper-tin mine in Portugal provides a good example of the importance of such geological modelling in the case of a folded and faulted stratiform orebody. Being one of the largest underground mining projects brought into production in recent years a strong emphasis has been put on the installation of computerised reserve estimation and mine planning systems. The development of a new approach to 3-D geological modelling, and its implementation as part of the reserve estimation systems at Neves-Corvo, is discussed below.

THE NEVES-CORVO DEPOSITS

The Neves-Corvo deposits consist of a series of massive sulphide lenses which are mined for copper and tin by underground methods.   They are located in southern Portugal, 220km south of Lisbon.   The deposits fall within the Iberian Pyrite Belt (Fig. 1), a metallogenic province within which several other massive sulphide deposits occur (Strauss and Madel, 1974, Barriga and Carvalho, 1983).

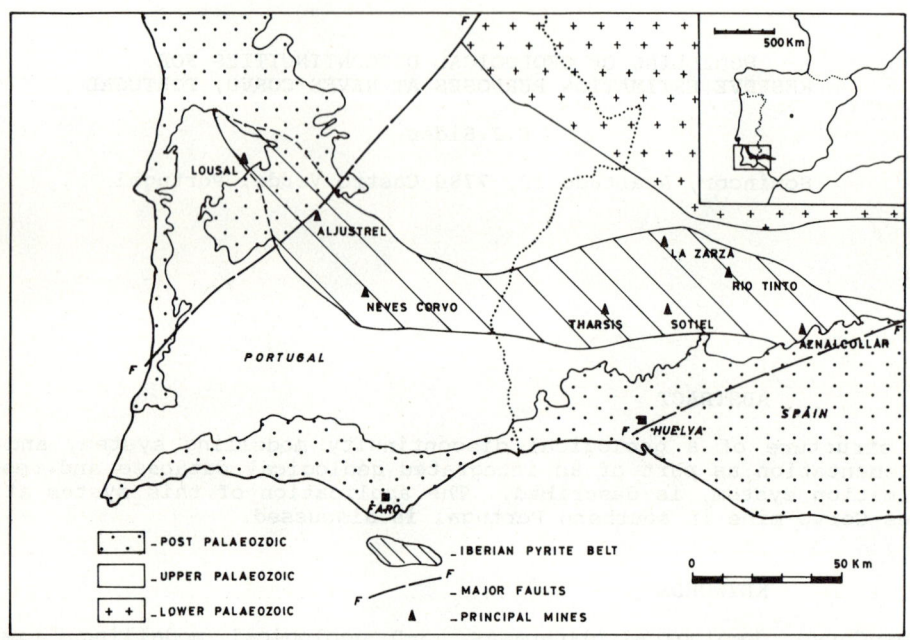

Fig. 1.   Simplified geology of the Iberian Pyrite Belt

The deposits were discovered in 1977 as a result of an exploration programme based on geological mapping and gravimetric surveying followed up by surface diamond drilling (Leca, 1985).   Mineralisation present includes both copper, copper-tin, and lead-zinc ores.   Subsequent to the discovery a major underground drilling and geological evaluation programme was completed to enable definition of reserves before the start of production at the end of 1988.

The mine currently produces and treats 1.3Mt of copper ore, and 0.3Mt of copper-tin and tin ore per annum.   Currently quoted geological reserves (Real and Murray, 1990) are as follows:

```
Copper ore    30.25 Mt averaging 7.94% Cu, 1.41% Zn
Tin ore        2.68 Mt averaging 2.42% Sn, 13.62% Cu, 1.27% Zn
Zinc ore      44.38 Mt averaging 5.48% Zn, 1.02% Pb, 0.50% Cu
```

The Neves-Corvo deposits show many characteristics in common with other deposits in the Iberian Pyrite Belt.   Nevertheless they show several unique features, particularly the very high grades of the copper and tin ores present.   Five separate massive sulphide lenses occur, two of which (Corvo and Graca) contain most of the copper and tin reserves on which the current operation is based.

Sulphide mineralisation is localised at the contact between an underlying tuff-shale sequence, and an overlying shale-greywacke-tuffite sequence, both of lower Carboniferous age (Fig. 2). Strong vertical and lateral metal zonation is present within the massive sulphides, such that most of the ore contacts are geologically very well defined. Subsequent to its formation the orebody was affected by Hercynian folding, and several steep faults cut and displace the ore horizon.

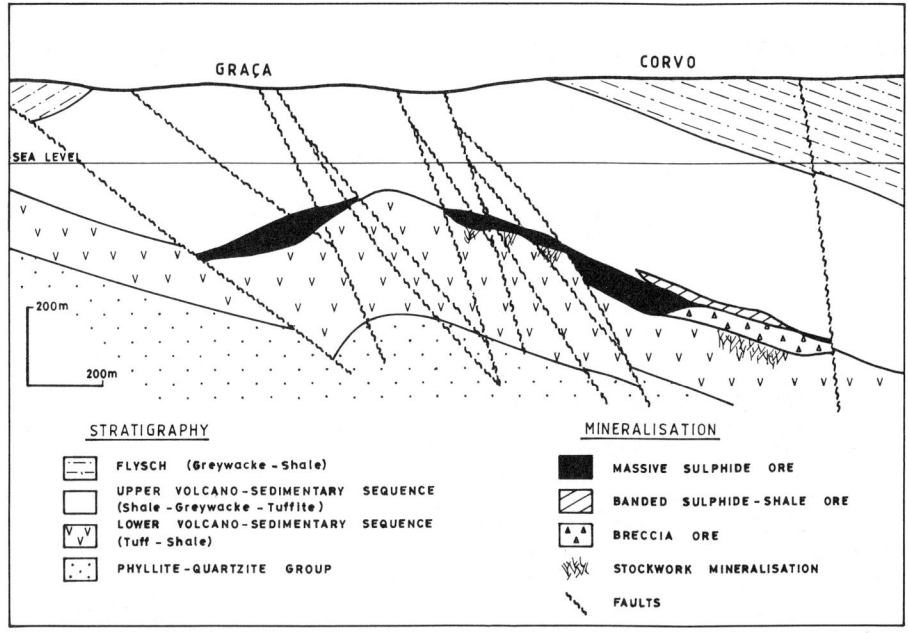

Fig. 2. Simplified geological cross section through the Corvo and Graca orebodies (modified from Fernandez-Rubio et al, 1988)

The geological features shown in Fig. 2 illustrate the importance of geological modelling to reserve estimation. The high grade nature of the mineralisation means that efficient mining must maximise recovery of the valuable ores, and minimise dilution from the barren country rocks. Accurate prediction of the position of the ore contacts, and important faults which may displace them, is therefore essential.

GEOLOGICAL MODELLING FOR RESERVE ESTIMATION PURPOSES

3-D geological modelling is of fundamental importance in the generation of computer models for mineral deposit evaluation and reserve estimation. The objective of such modelling is to provide accurate three dimensional representations of orebody characteristics. Such models are used in mine planning studies aimed at the selection of appropriate mining methods and extraction sequences.

Reserve Estimation Procedures

Over the past two decades several commercial software packages have been developed to for use in reserve estimation and mine planning (Gibbs, 1990). Several in-house packages have also been implemented, both within large mining companies, and at individual mines (Ashton and Harte, 1989). In addition to geological database handling, such packages provide the following functions to enable generation of geological, and mining, reserve estimates.

Geological Modelling:- the definition of boundaries between different lithological or mineralogical types. These define abrupt changes in one or more physical properties of the materials to be mined (eg. density, ore grades, hardness, etc.)

Grade Estimation:- the modelling of the distribution of ore grades within the different mineralised units present. Geological control is often imposed on the grade estimation procedures adopted, using the models developed in the preceeding step.

Mine Planning:- interrogation of the geological and grade models in order to estimate tonnages and grades within mining volumes corresponding to planning time periods. This enables planning of production to maximise the efficiency of an operation in both technical and financial terms.

Special Requirements

Geological modelling in the mining industry shows several differences in emphasis from similar modelling in other fields, such as petroleum geology, structural geology, etc.. A common requirement is the need to maintain models at different levels of precision for the same area. This reflects the progressive nature of information collection during evaluation and exploitation of a deposit, whereby a greater amount of more detailed information becomes available as a particular block of ground comes closer to being mined. There is also a need for more detailed information for short term planning than for long term planning. Such aspects are commonly reflected in the categorisation of reserve models into different confidence categories, either on a qualitative of quantitative basis.

Reserve Model Types

The first two steps involved in reserve estimation require two fundamentally different types of modelling procedure. Boundary modelling type approaches are generally used to define the geological discontinuities within a deposit. Subsequent volumetric modelling is used to represent continuously varying internal properties within broadly similar lithological, mineralogical, or other material types.

This distinction has been highlighted by other workers in the field of 3-D geological modelling (Raper, 1989), the former type often being regarded as analogous to vector-type computer graphics, and the latter to raster-type. Table 1 presents a summary of the main model types used for 3-D reserve modelling.

Many of the difficulties associated with geological modelling for reserve estimation purposes arise because most mine planning systems assume that both geological and volumetric information can be provided directly by the same model structure. In reality, however, geological features tend to be better represented by boundary representation type models, whereas volumetric type information is more readily handled in block model structures (Karonen, 1985, Bak and Mill, 1989). The work presented here will illustrate how this difficulty can be overcome by using different model structures for the two types of information, provided interfaces between the two are established.

Table 1. Modelling techniques used for reserve estimation

Method	Description
Gridded Seam	Multiple sub-parallel surfaces represented by grids of elevation points. Volumetric modelling is based on rectangular prisms between adjoining interfaces, precision being dependent on grid dimensions, and number of interfaces. Limited to use in tabular bodies (effectively 2.5-D).
Serial Slice	Geological interpretation of polygonal outlines on a series of regularly-spaced parallel sections or plans. Volumetric modelling is based on polygonal outlines assigned a prismatic volume of influence halfway to adjacent slices, precision being dependent on slice spacing and polygon size. Model precision can be improved by linking polygons on adjacent slices in 3-D. This is the most common method used for manually based geological modelling.
Block Modelling	Volume being modelled is divided into a series of regular rectangular blocks. Each block is normally assigned uniform geological and grade characteristics, precision being dependent on the minimum block size. Blocks are commonly equidimensional, but storage and handling efficiency can be improved through the use of blocks of different sizes (eg octree encoding). Regular block models are one of the most widespread structures used in computerised reserve estimation.
Solid Modelling	Based on modelling groups of geometric primitives defining volumes of common properties. Precision depends on model complexity, but can be very good for geological modelling. Such techniques are adapted from CAD/CAM applications, so that difficulties are often encountered in establishing links with pre-existing geological database and mine planning systems. This is a relatively new approach in reserve estimation whose full potential has still to be exploited.
Boundary Representations	Based on modelling discontinuity surfaces within a volume of ground. Precision depends on model complexity, but can be very good for geological modelling. Techniques are mostly adapted from applications in other fields, such as medical tomography. Commonly used for 3-D visualisation of geological and mining volumes, although handling multiple intersecting boundaries can be difficult. A lot of recent development work has concentrated on this type of model.

A GEOLOGICAL DISCONTINUITY MODELLING SYSTEM

In the light of the importance of accurate 3-D geological modelling in
the estimation of reserves at Neves-Corvo a review of available
techniques, both in existing mining software packages, and in 3-D
modelling applications in other fields, was carried out. Although
several packages of possible applicability were identified, all of these
showed limitations, particularly with regard to the feasibility of
establishing links with existing geological databases and volumetric
models.

Consequently the decision was taken to develop and implement a 3-D
geological modelling system based on using triangulated surfaces to
represent the different geological discontinuities (faults, contacts,
etc.) within the deposit. The system developed is termed a geological
discontinuity modelling system and is analogous to the use of boundary
representation models in other fields. An essential part of the system
is the ability to maintain links to other parts of the geological
database and reserve estimation systems.

The system described below has been implemented on a Sun 3/60 colour
graphics workstation, the programs used being written in Fortran 77.
Figure 3 illustrates some of the types of information, and geological
modelling requirements, which the system was designed to handle.

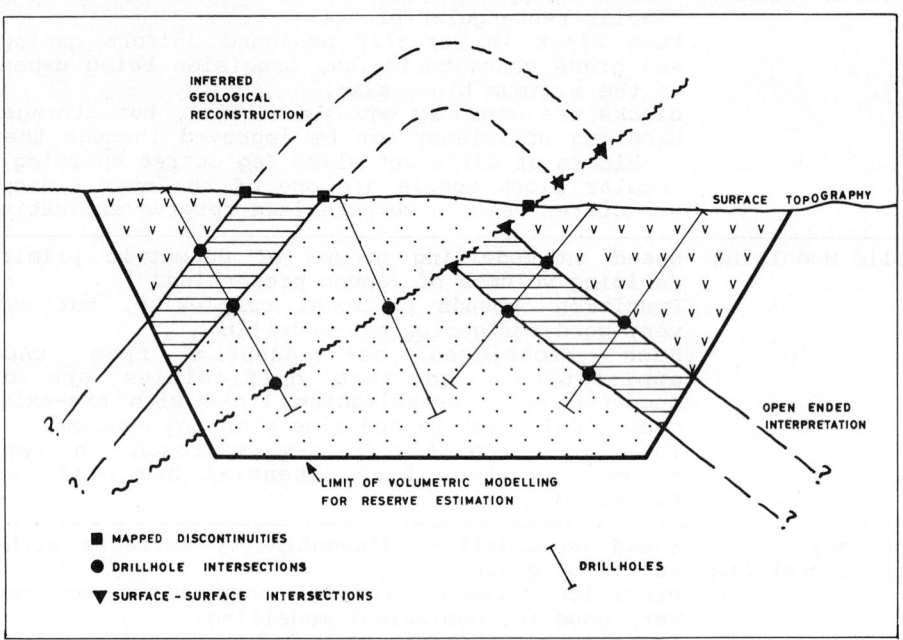

Fig. 3. Schematic cross-section illustrating some of the
data types, and interpretative requirements of a
geological modelling system

General Criteria

The model structure adopted is based on several general principles,
including the following:

- All model co-ordinate data is stored in mine co-ordinates to facilitate links with other databases.

- All data items included in the model are reduced to geometric primitives so as to facilitate handling and analysis.

- Links to geological databases are maintained by establishing groups of special line and point types.

- The basic system is designed around open ended surfaces, those bounding closed volumes being regarded as special cases.

- All surfaces are regarded as having unique positive and negative facing directions, with the possibility of assigning lithological codes to the material types on either side.

- Definition of volumes is done in a geological fashion by establishing a hierarchy of surfaces for intersection purposes.

Model Structure

Each individual discontinuity is modelled as a separate entity; several separate blocks of different information being stored for each, as summarised in Table 2.

Table 2. Geological Discontinuity Model File Structure.

Data type	Description
Header information	Name, associated codes, number of points, number of triangles, geometric limits, date generated, etc.
Point co-ordinates	X, Y, Z co-ordinates, plus up to 3 associated integer codes for all points in the model
Triangle pointers	Sets of three vertex pointers into the point array
Special points	Extra information about points derived from drillhole and mapping databases; such as drillhole name, depth, mapping location, orientation, etc.
Special edges	Triangle edges fixed by geological interpretation, to facilitate agreement with 2-D interpretations
Intersection lines	Intersection lines with other surfaces, used for special display features, and to control generation of volumetric models

Work on any particular project is controlled by a set of surface indices which allow grouping of discontinuities at two levels. The first level of association is normally defined on the basis of common geological type, such as surface topography, reserve category limits, late faults, early faults, stratigraphic, lithological and other contact types. Such groupings are arranged so as to correspond to the hierarchy of intersection priority used to control surface-surface intersections. A second (upper) level of association is also allowed in order to facilitate separation of models by deposit, user, or reserve category.

The geometric structure of an individual discontinuity model is shown in Fig. 4, which illustrates the main geometric primitives used to represent the model. Figure 4 also indicates the requirement to store the triangles in a common sense so as to enable the definition of positive and negative sides to a surface. This means that an edge shared by two triangles is incorporated as a vector in different directions in each.

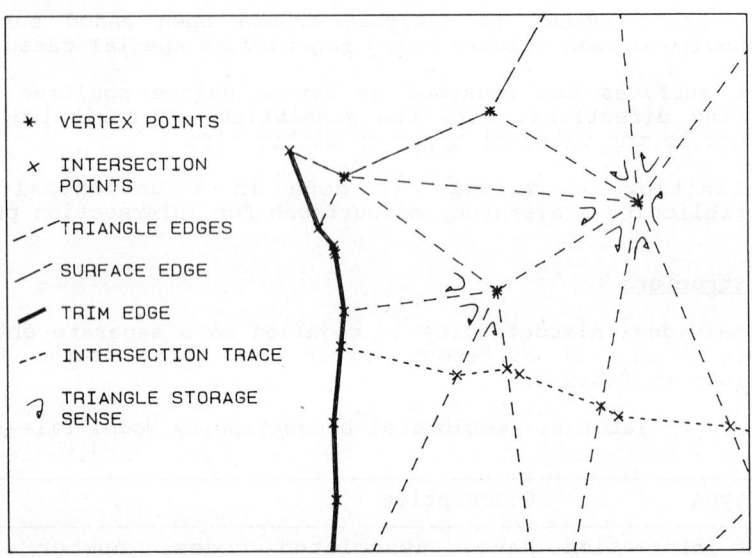

Fig. 4. Portion of a geological discontinuity model illustrating the model structure

Legend:
* VERTEX POINTS
x INTERSECTION POINTS
- TRIANGLE EDGES
- SURFACE EDGE
TRIM EDGE
-- INTERSECTION TRACE
TRIANGLE STORAGE SENSE

Model Creation

Before starting work on a new area an analysis is made of the discontinuities that are likely to be modelled and their geological relationships. Such information is used to establish the naming conventions, intersection hierarchies and file groupings, to be used in subsequent modelling work.

In the initial stages of a new project several options are available for generating new surfaces, including the following:

- initialise a blank file
- import ASCII file of point co-ordinates
- import ASCII file of triangulated points
- import digitised serial-slice data
- import digitised linked-slice data
- digitise isobath plans of a surface
- select contact positions from drillhole or mapping databases

Stored model parameters are initialised as zero before commencing an import, unless otherwise specified, and are then modified according to the input data. This allows a discontinuity model to be initialised using a list of co-ordinates; triangles being generated during subsequent editing.

Model Editing

Basic file handling operations for reading, writing and copying files are used for general file manipulation, allowing working copies to be stored in separate directories from the final models used for planning work. The other editing options implemented enable the user to generate and modify discontinuity models in an interactive graphics environment.

Global Editing allows changes to be made which affect the whole model simultaneously, including the following:

- Automatic re-triangulation (Delaunay)
- Check triangle storage sense integrity
- Remove all points of a selected type
- Remove all intersection lines
- Co-ordinate transformation
- Add another surface
- Extract part of a surface to form a new one
- Import additional points or triangles
- Surface-surface intersection options

Automatic re-triangulation, based on the algorithm of Watson (1982), is often the main editing technique used. However, at present its use is limited to simpler stratiform or stratabound deposits, since it cannot readily handle overfolds. The other global editing options facilitate general checking and verification of discontinuity models. Since surface-surface intersections are used to control several special display options, as well as in the generation of volumetric models, they are discussed separately below.

Interactive Editing allows modification of specific parts of the structure. This includes options to add, delete or modify the points, lines or triangles which make up the models. The use of interactive computer graphics enables the effect of changes to be seen directly on the screen. Links with other geological databases are used during the editing of special point and line types. In this manner editing of a drillhole point would require entry of a new downhole depth, thus forcing the point to be moved along the line of the drillhole trace.

Surface-Surface Intersections

The generation of coherent geological discontinuity models, that can be used to establish volumetric models, requires the modelling of cross-cutting surfaces in order to ensure proper closure of the volumes defined by intersecting discontinuities. Such aspects can be avoided whilst doing general geological modelling during exploration, or when modelling simple deposits where the main geological boundaries do not intersect (eg thick continuous coal seams). However, in order to set up a system of more general applicability procedures for generating and handling surface-surface intersections are essential.

The calculation of a surface-surface intersection line is fairly simple in geometric terms, being based on repeated triangle-triangle intersections. Nevertheless, a lot of programming effort was needed to set up appropriate software to handle this, mainly in order to deal with the special cases that arise, such as coincident vertices and edges.

Definition of volumes for display and evaluation purposes is done by defining a hierarchy of intersection priority. On the basis of the relative rankings of the pairs of intersecting surfaces, two categories of surface-surface intersection are recognised, namely;

<u>Trim lines</u>, where the intersected surface truncates the current surface. Such lines are used to define visible and invisible parts of a surface (Fig. 5). This procedure of using one surface to truncate another is termed <u>surface trimming</u>.

<u>Intersection traces</u>, where the current surface truncates the intersected one, and the intersection is used for display purposes only (Fig. 4). This is used, for instance, when a representation of surface geology is generated by intersecting all modelled discontinuities with a model representing the surface topography.

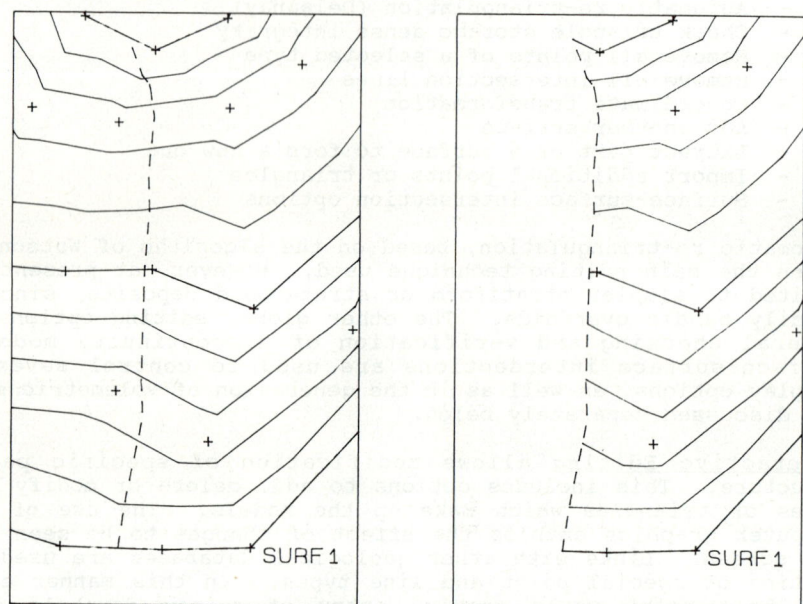

Fig. 5.   Untrimmed (left) and trimmed (right) displays of
          part of a contoured discontinuity model

Both intersection line types are generated and stored with the rest of the information in a discontinuity model file, rather than recalculating them each time a new display is to be generated. This is done mainly because user interaction is required to check and verify the intersection lines generated. This approach also reduces the amount of processing required for display generation.

Graphic Displays

The intersection of stored models with a plane in any orientation is fundamental to a complete 3-dimensional modelling system. For the discontinuity models used here, this is based on fairly simple geometric principles, namely the intersection of a triangle with a plane. Such aspects are discussed in many texts on computer graphics and geometric modelling (Angell and Griffith, 1987, Mortensen, 1985, Bowyer and Woodwark, 1983).

Intersection of a surface with a plane involves checking each triangle in turn, and generating a line point-pair whenever an intersection is encountered. The main difficulty is in establishing rules for dealing

with the special cases which occur when one or more of the triangle
vertices are co-planar with the plane of interest.

2-D graphic displays: Generation of such displays requires the
selection and/or generation of a set of points and lines to be
displayed, along with appropriate display symbols, colours and
labelling.   2-D displays are always generated with reference to a
particular viewplane, which can be zoomed inwards or outwards, and moved
forwards or backwards. The data displayed can also be selected within a
defined depth of field specified by minimum and maximum projection
distances onto the viewplane. Different combinations of display colour,
line types, and symbols can be used for different surface features,
allowing fairly complex displays to be built up.

3-D visualisations: These include 3-D graphics representations or
isometric views of the discontinuity models. Since many sophisticated
3-D display packages are available commercially, the approach adopted
has been to provide interfaces to existing display systems, rather than
to develop such options from scratch. This requires the modelling
system to output the stored surfaces in a form which can be used by
other systems, such as a set of polygons with associated colours. 3-D
displays are currently generated using the SUNCORE system provided on
older Sun graphics workstations.

Data Output

An important aspect of any modelling system is the ability to output the
model data for reporting or documentation purposes, use in other
applications, further analysis, etc.. Output options available include
the following;

- ASCII file dump of total file
- ASCII file of point co-ordinates +/- triangle vertices
- Printer listing of file contents
- 2-D graphic displays as plot files
- Interface files for 3-D visualisation software
- Gridded seam file output
- Serial slice type file output

APPLICATION OF MODELLING SYSTEM AT NEVES-CORVO

The discontinuity modelling system described above plays an important
part in the generation of reserve models for medium-long term planning
at Neves-Corvo. Some examples of the application of this technique are
given below.

Geological Database Links

The ability to include information about the origin of the point data
used in a discontinuity model greatly facilitates checking and
verification of the geological interpretations on which such models are
based. This is illustrated by the isobath plan of the massive sulphide
footwall presented in Fig. 6, where the drillhole intersections used in
the model are identified. Such displays highlight variations in data
density, as well as allowing verification of the data points used to
construct the model.

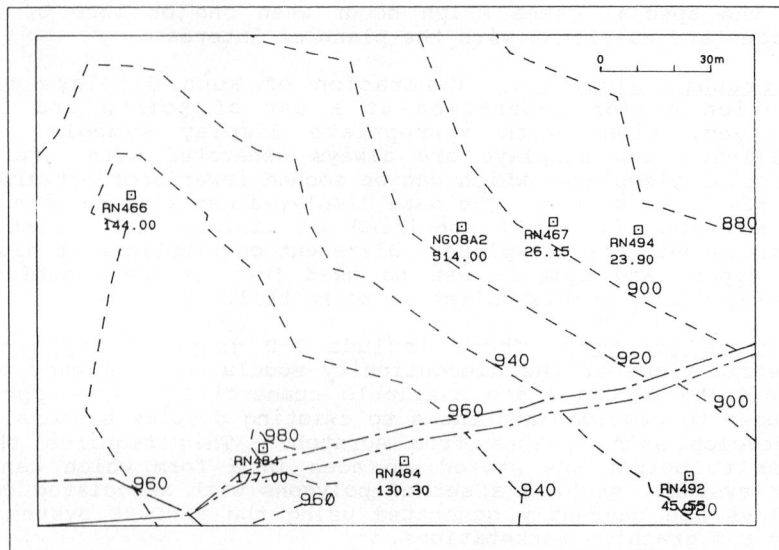

Fig. 6. Isobath plan of orebody footwall showing drillhole
intersections used and zone of fault loss.

Geological Modelling

Detailed geological evaluation of the Neves-Corvo orebodies prior to the
initiation of mining led to the definition of a series of major faults,
and oretype contacts within the mineralised units. These features have
important implications for reserve estimation and mine planning, and
were modelled as individual discontinuities as shown in Fig. 7. Four
main groups of discontinuities were defined, namely; reserve model
limits, major faults, main lithological contacts, and internal ore type
boundaries. These groupings correspond to different priority levels in
the intersection hierarchy used to control surface trimming.

Volumetric Model Generation

Using the modelled discontinuities (Fig. 7), oretype block models were
generated in order to control estimation of internal grade variations.
This was done by converting the stratigraphic contacts to gridded seam
type models, and then generating blocks at regular vertical intervals.
A comparison between the discontinuity models and a block model derived
from them is shown in Fig. 8. The loss of information about faulting
away from the mineralised horizon, as well the loss of internal detail
within the mineralised cupriferous sulphide (MC), should be noted.

Mine Planning Applications

Options to generate displays such as those shown in Figs. 6-8 are
available during mine planning using interactive graphics programs
(Howson, 1989, Teixera and Caupers, 1990). This allows planning to take
into account the fact that not all ore boundaries are the same. For
instance the approach adopted when mining up to a faulted contact will

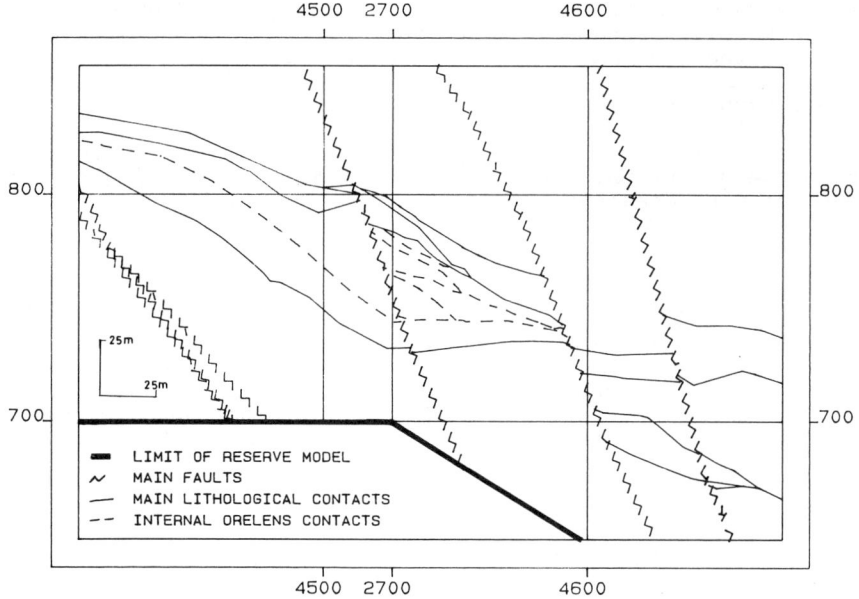

Fig. 7. Cross section through discontinuity models used for reserve estimation of the Corvo orebody.

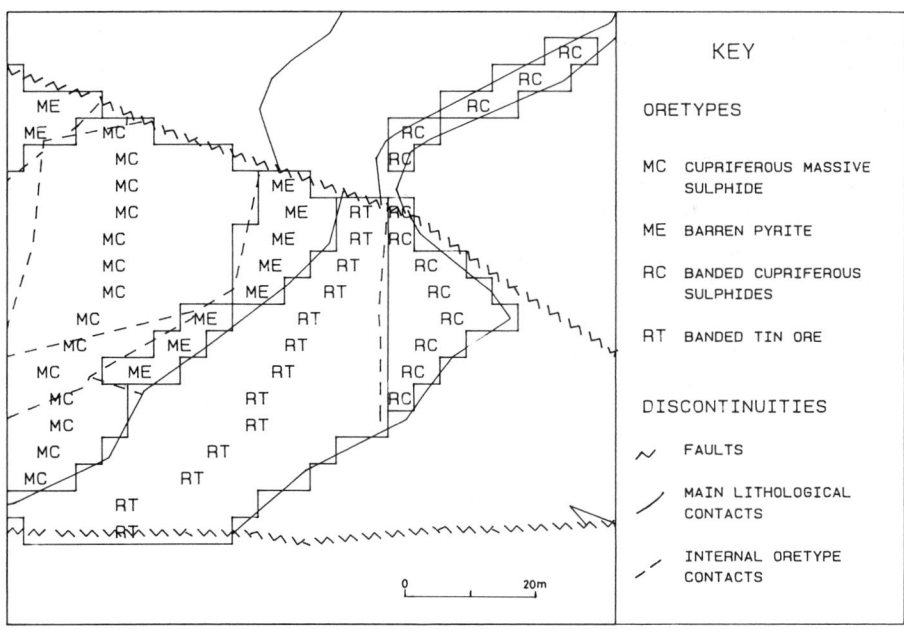

Fig. 8. Plan showing discontinuity models, and corresponding block model boundaries, for part of the Corvo orebody

differ from that involved when mining next to an internal ore contact within the massive sulphides. The availability of geological details of faulting in the footwall and hangingwall rocks (Fig. 7) is also of considerable importance when planning mine infrastructure development (eg access ramps and crosscuts).

A subsidiary benefit of the development of the discontinuity modelling system is the use of the same structure to model floor and roof surfaces of irregular mine openings, as described by Howson (1989).

DISCUSSION

The work described above has resulted in the development of a geological modelling system with several advantages over approaches used in other reserve estimation systems. Nevertheless, the present system suffers from several limitations and additional work is needed in order to improve the user interface, optimise file storage and access, and to fully implement its three dimensional modelling capabilities.

Advantages

The advantages of the model structure adopted, over other techniques of geological modelling, owe much to the fact that the structure has been set up specifically for 3-dimensional geological modelling. Most existing systems are based on approaches developed principally for use in grade estimation or mine planning, or borrowed from applications in other fields such as CAD/CAM (Karonen, 1985, Bak and Mill, 1989). Some of the main advantages of the system used include the following:

- links with drillhole and mapping databases facilitate initial interpretation and checking of models

- surfaces do not have to define closed volumes; those that do are treated as special cases

- storage of surface-surface intersections allows the generation of contact traces on faults, display of modelled geology on irregular surfaces, etc.

- the trimming mechanism applied for display and evaluation purposes facilitates proper closure of contacts for generating volumetric models and also allows extrapolated geological contacts to be modelled

- the structure used can handle storage and manipulation of complex overfolded surfaces

Limitations

One limitation of the system is that all surfaces are treated as sets of planar triangles, making some of the derived plots look artifically angular in character. This problem could be diminished by using curved patches, such as cubic splines or Coons patches, although this would make the handling of surface-surface intersections more difficult.

Although the structure developed can handle complex contorted surfaces with overhangs, and discontinuities made up of several segments, further work is needed to make visualisation and editing of such features practicable.

Options for the conversion of discontinuity models to different types of volumetric model are fairly limited at present. The procedure of establishing gridded seam type models as an intermediate step in the generation of regular block models, as used at Neves-Corvo, is limited to use in tabular or stratiform orebodies. It is envisaged that a more general approach based on initial generation of a regular series of sections or plans through the discontinuity models would allow creation of serial-slice type models by merging the individual discontinuities to form closed polygons. Such models could then be readily converted into block model structures.

Future Development

At present, conversion of geological discontinuity models to volumetric type models is seen as the best way of interfacing with existing reserve estimation systems. However, it is considered likely that in future planning programs may calculate volumes of different material types directly from the discontinuity models. Appropriate grade information could then be obtained from a series of separate block models for each material type. This would enable the common conflict between choice of a smaller block size to accurately model geological variations, versus the larger block sizes defined as being the optimum size from the point of view of geostatistical grade estimation, to be reconciled.

The approach adopted, in the work done to date, has been a deterministic one, in that the models generated have no way of handling fuzzy data, or incorporating measures of uncertainity. Work done by Mallett (1989) suggests that it is possible to use algorithms that can handle fuzzy data when modelling discontinuities, and this is obviously an aspect that should be investigated further.

Additional work is also needed on the problem of trying to associate locational uncertainities with the modelled discontinuities. Measures of surface irregularity, such as that suggested by Philip and Watson (1986), are one approach that warrants investigation. Studies of locational uncertainity would also need to take into account the fact that the uncertainities associated with many discontinuities in a deposit are interdependent. For instance a drillhole intersection may provide a very accurate estimate of the thickness of a particular seam, thus defining the separation between the two bounding discontinuities, however it may have quite a high locational uncertainity.

CONCLUSIONS

Although several software packages exist which a geologist can use for geological modelling in mineral deposits, their use often involves interpretative compromises and consequent loss of important geological information. It is considered that a clearly defined structure for modelling geological discontinuities, such as that established here, can eliminate many of the problems associated with existing systems. To facilitate widespread usage such a structure needs to allow for future modification and enhancement as geological ideas and modelling techniques evolve.

From the point of view of reserve estimation and mine planning such advantages are of little interest unless the models generated can be integrated with existing techniques for grade estimation and volumetric calculation. This has been successfully achieved in the case of Neves-Corvo, however further development work is necessary to allow the use of this approach in a wider range of deposit types.

ACKNOWLEDGEMENTS

The writer acknowledges the permission of Somincor and RTZ Technical Services Ltd. to publish this paper. The involvement of numerous colleagues in Somincor, RTZ Technical Services Ltd. and Riofinex North Ltd. in providing geological, programming, and other input to the work described is also acknowledged.

REFERENCES

Angell, A.O. and Griffith, G. (1987). High Resolution Computer Graphics using FORTRAN 77. MacMillan Education Ltd., England.

Ashton, J.H. and Harte, G. (1989). Technical computerization at Tara Mines, Ltd., Navan, Ireland. Trans. Inst. Min. Metall., 98, pA85-A97.

Bak, P.R.G. and Mill, A.J.B. (1989). Three dimensional representation in a Geoscientific Resource Management System for the minerals industry. In: Three dimensional applications in Geographical Information Systems (J. Raper, ed.), Chap.12, pp155-182. Taylor & Francis, London.

Barriga, F.J.A.S. and Carvalho, D. (1983). Carboniferous volcanogenic sulphide mineralizations in southern Portugal (Iberian Pyrite Belt). In: The Carboniferous of Portugal (M.J.Lemos and J.T.Olivera, eds.), Mem. Serv. Geol. Portugal, Lisbon, 29, pp99-113.

Bowyer, A. and Woodwark, J. (1983). A programmer's geometry. Butterworths, London.

Fernandez-Rubio, R., Carvalho, P. and Real, F. (1988). Mining-hydrological characteristics of the underground copper mine of Neves-Corvo, Portugal. Paper presented at the Third International Mine Water Congress, Melbourne, Australia, October 1988.

Gibbs, B. (1990). Mineral Industry Software. In: Mining Annual Review - 1990, pp219-235. Mining Journal Publications, London.

Howson, M.P. (1989). Mine design through the application of boundary modelling techniques. Mining Magazine, Sept. 1989, p198-204.

Karonen, O. (1985). Geometric Mine Modeling: Modeling of three-dimensional objects based on Incomplete Information. Acta Polytechnica Scandinavica, Mathematics and Computer Science Series, No.45, Helsinki.

Leca, X. (1985). The discovery of the sulphide deposits of Neves-Corvo (southern Portugal). (In French). Chron. Rech. Min., Orleans, 479, p51-62.

Mallett, J.L. (1989). gOcad: A computer Aided Design Program for geological applications. Paper presented at NATO Advanced Research Workshop on Three Dimensional Modelling with Geoscientific Information Systems, Santa Barbara, California, December 1990.

Mortensen, M. (1985). Geometric Modeling. John Wiley & Sons, New York.

Philip, G.M. and Watson, D.F. (1986). A method of assessing local variation among scattered measurements. Mathematical Geology, 18, p759-764.

Raper, J. (1989). The 3-dimensional geoscientific mapping modelling system: a conceptual design. In: Three dimensional applications in Geographical Information Systems (J. Raper, ed.), Chap.2, pp11-20. Taylor & Francis, London.

Real, F. and Murray, D.N. (1990). The Neves-Corvo Mine. Paper presented at Metal Bulletins 4th International Copper Conference Santiago, Chile, April 1990.

Strauss, G.K. and Madel, J. (1974). Geology of massive sulphide deposits in the Spanish-Portugeuse pyrite belt. Geol. Rundschau, 63, p191-211.

Teixera L.A. and Caupers D.J. (1990). Computerized planning at Neves-Corvo copper mine, Portugal. Trans. Inst. Min. Metall., 99, pA60-A64.

Watson, D.F. (1982). ACORD: Automatic Countouring of Raw Data. Computers and Geosciences, 8, 97-101.

THREE-DIMENSIONAL PREDICTIVE DEPOSIT MODELLING
BASED ON THE LINEAR OCTREE DATA STRUCTURE

RENÉ PRISSANG

Freie Universität Berlin, Institut für Geologie,
Geophysik und Geoinformatik, FR Geoinformatik,
Malteserstr. 74-100 (Haus D), D-1000 Berlin 46, FRG

ABSTRACT

The linear octree data structure is based on a spatial enumeration scheme to index
subcubes in a recursively decomposed universe. This structure allows to set up
efficient models of the geometry of a deposit and of the spatial distributions of
continuous properties, e.g. ore grades. Property modelling requires to discretize the
3D space. For this purpose a top-down decomposition is employed that takes into
account the number of samples per block. In addition to that, all blocks of a size too
large to represent fluctuations in the property under consideration will be subdivided
down to a user-defined level. The subsequent interpolation of properties can be
carried out by means of a number of different techniques. Due to the common data
structure it is possible to derive new geometrical objects from the information stored
in the property model. This way, regions in 3D space that may be assumed to be
homogeneous can be predicted. Connected component labelling will render number and
volumes of disjoint bodies. The technique is applied to evaluate the effect of a
cutoff on the extend and distribution of the mineable regions in an orebody.
Visualization of the results provides a vital aid for devising an appropriate stoping
strategy.

KEYWORDS

Octree, Deposit modelling, Property modelling, Mine planning, Visualization

INTRODUCTION

A vital objective of deposit modelling is the assessment of mineable reserves. This
comprises the calculation of the quantity of mineral as well as a prediction of the
geometry of economically viable regions in 3D space. Methods of Solid Modelling allow
to set up efficient computer representations of the deposit geometry and of the
spatial distribution of continuous properties (attributes), especially ore grades.
Based on this type of representation, tools can be provided enabling geologists and
mining engineers to evaluate the effect of a selected cutoff value on the extend of
the mineable regions in the orebody. Fast and convenient 3D visualizations of the
results, e.g. by means of the CA-DISSPLA-based rendering program by Tietze (1990) or
by means of the Linoct program (Atkinson et al. 1987), provide a vital aid for the
selection of an appropriate stoping strategy as well as for the initial design of the
pit layout.

The proposed way of interactive evaluation of deposits requires the use of a modelling
technique which allows to predict, analyse, and visualize regions in 3D space that can

be assumed to be homogeneous with regard to the distribution of continuous spatial attributes. These requirements are met by the octree based approach to deposit modelling described by Dunstan and Mill (1989) and Prissang and Skala (1990).

This approach is characterized by the following features :

- representation of deposit geometry, spatial distribution of continuous attributes, and the mine layout by separate models
- use of the linear octree data structure for all of the aforementioned models
- discretization of 3D space according to the density of information when modelling the distribution of continuous attributes
- availability of efficient algorithms for model generation, mass calculation, definition of homogeneous regions in 3D space as geometrical objects as well as for visualization of the resulting objects.

The programs are part of a prototype of an integrated CAD system for underground metalliferous mines, which has been developed in the framework of a European Community funded research project.

THE LINEAR OCTREE DATA STRUCTURE

Constructive Solid Geometry (CSG) is based on a hierarchical decomposition of complex objects into primitives. This decomposition can be carried out in a number of different ways. One of these ways is a recursive subdivision of a cuboid object space or universe into eight congruent disjoint subcubes ("octands") until it becomes possible to decide if a subcube is fully contained in the object under consideration or not. The process of decomposition can be represented by an octernary tree which is known as the octree (fig. 1). The subcubes that are actually constituting the decomposed universe are symbolized by the terminal nodes or leaves. They are called object elements or obels. In an octree the maximum depth is known to as the resolution. Nodes on this level represent the smallest object elements which have to be regarded as indivisible. Therefore, the term "volume element" or "voxel" has been coined to designate these subcubes.

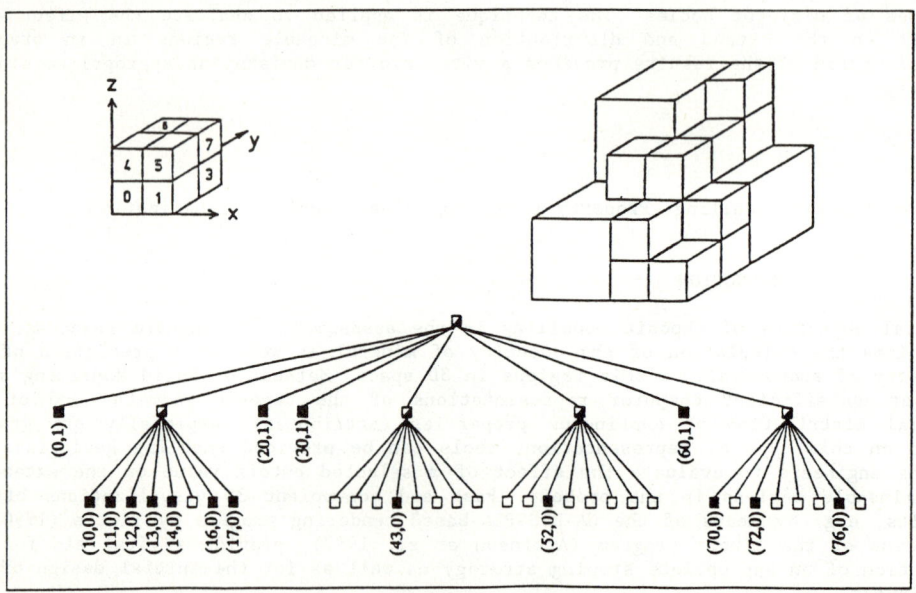

Fig. 1. Pointer-based and linear octree representation of a
three-dimensional object (Prissang and Skala 1990)

The pointer-based representation of an octree, the so-called regular octree, has the drawback that the root and all internal nodes have to be held in memory. Therefore, Gargantini (1982) developed a more efficient structure in which only those object elements are stored that are part of the object being modelled. This structure is based on a spatial indexing scheme. Octands are labelled by octal digits according to their position within their parent cubes. Hence, each object element can be encoded by a string of octal digits, representing the path from the root to the corresponding leaf. This string is called the key of the obel. The number of digits in the key is equal to the level of the node, i.e. to the number of subdivisions. Atkinson *et al.* (1984) further improved the efficiency of this structure by storing the entire key in a single integer variable. In order to use the key as a sorting criterion, it has to be expanded with trailing zeroes until its length is equal to the resolution of the tree. This made it necessary to introduce a so-called grouping factor which indicates the number of trailing zeroes and to set up an obel index consisting of the octal key and the grouping factor. The sorted list of obel indices associated with a number of attributes constitutes the linear octree. Figure 1 illustrates the relationship between both variants of octree representation.

STRUCTURE OF THE OCTREE-ENCODED DEPOSIT MODEL

When setting up a deposit model it is necessary to distinguish between of the geometry of an object and the spatial distribution of a continuous attribute. Hence, two types of octrees, geometry and attribute octrees, have been introduced. On the implementation level, this is reflected by the declaration of different data types. For the representation of the geometry, the obel index is associated with a fixed number of attributes, i.e. region (colour) and object identifications as well as two bit strings indicating the visibility of faces and the contact with objects belonging

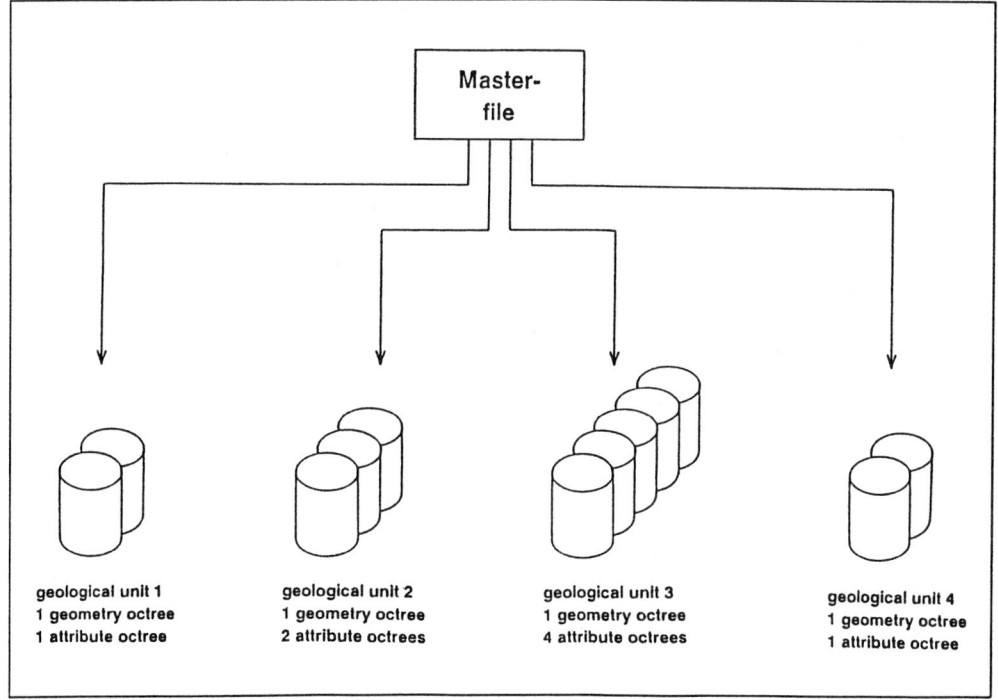

Master-file

geological unit 1
1 geometry octree
1 attribute octree

geological unit 2
1 geometry octree
2 attribute octrees

geological unit 3
1 geometry octree
4 attribute octrees

geological unit 4
1 geometry octree
1 attribute octree

Fig. 2. File structure of the deposit model

to other regions. In contrast to that, the main aim of setting up the attribute octree is to store the values of a number of user-defined properties. Therefore, a data type has been defined that associates the obel index with a pointer to a dynamic array of real values. Due to the different entities to be represented, the deposit model has been structured in a way that each geological unit is represented by a geometry octree in conjunction with a number of attribute octrees (fig. 2).

MODEL GENERATION AND PREDICTION OF ATTRIBUTE VALUES

Modelling starts with the delineation of the deposit geometry. For this purpose wireframe modellers acting as graphical user interfaces have proven to be most suitable. The resulting model can be converted to a linear octree encoded solid model using a suite of programs developed by Bak (1990). A survey of algorithms for the construction of octree-encoded geometry models is provided by Cheng and Huang (1988).

Setting up an attribute model requires to subdivide the 3D space into discrete blocks. Reasons of efficiency and accuracy lead to an approach which is based on a top-down decomposition of the 3D space guided by the sample density. This results in a model that allows to accomodate for detail. It furthermore prevents that information which has been aquired with high expenditures would be averaged away. A maximum number of samples per block as well as a user-defined minimum block size have been selected as stopping criteria for the decomposition process.

Blocks located fully outside the orebody will be removed afterwards from the decomposed universe. In addition to that, all blocks of a size to large to represent any grade fluctuations are subdivided down to a user-defined level. This level affects the accuracy of all geometrical objects which can be derived from the attribute model. Therefore, it has to be selected with care. The resulting octree can be regarded as the "empty model".

After the discretization of the 3D space the grade estimation will be carried out. This operation is based on the extention of sample data into the 3D space. For this purpose, different interpolation algorithms are available. In the current version of the estimation routine point and block kriging with corrections for geometrical anisotropy as well as inverse squared distance weighting have been implemented. The results are written on a grade octree, which acts as an inventory file for the given property.

The different steps are covered by respective programs (fig. 3) :

- deco (decomposition of the universe according to the sample density
- selecto (selection of obels fully or partly inside the orebody)
- vario (geostatistical structure analysis)
- esto (estimation of grades and other attributes)
- mass calculations in mining blocks.

VISUALIZATION AND ANALYSIS

The visualization of the spatial distribution of continuous attributes comprises several steps (fig. 4). At first disjunct classes of the attribute values as well as corresponding colour-scales have to be defined. By means of colour-coding (program coloro) new geometrical objects, that are characterized by a given range of attribute values, can be derived from the information stored in an attribute octree. The colour-coded octree is a geometry octree that can be displayed by a rendering program.

In case a high degree of geometrical accuracy is required, e.g. for the prediction of zones of similar ore grades, it is advisable to determine the intersection of the attribute octree with the geometry octree. The resulting "display octree" belongs to the class of attribute octrees. It allows to view the distribution of a spatial attribute confined in the boundaries of the orebody.

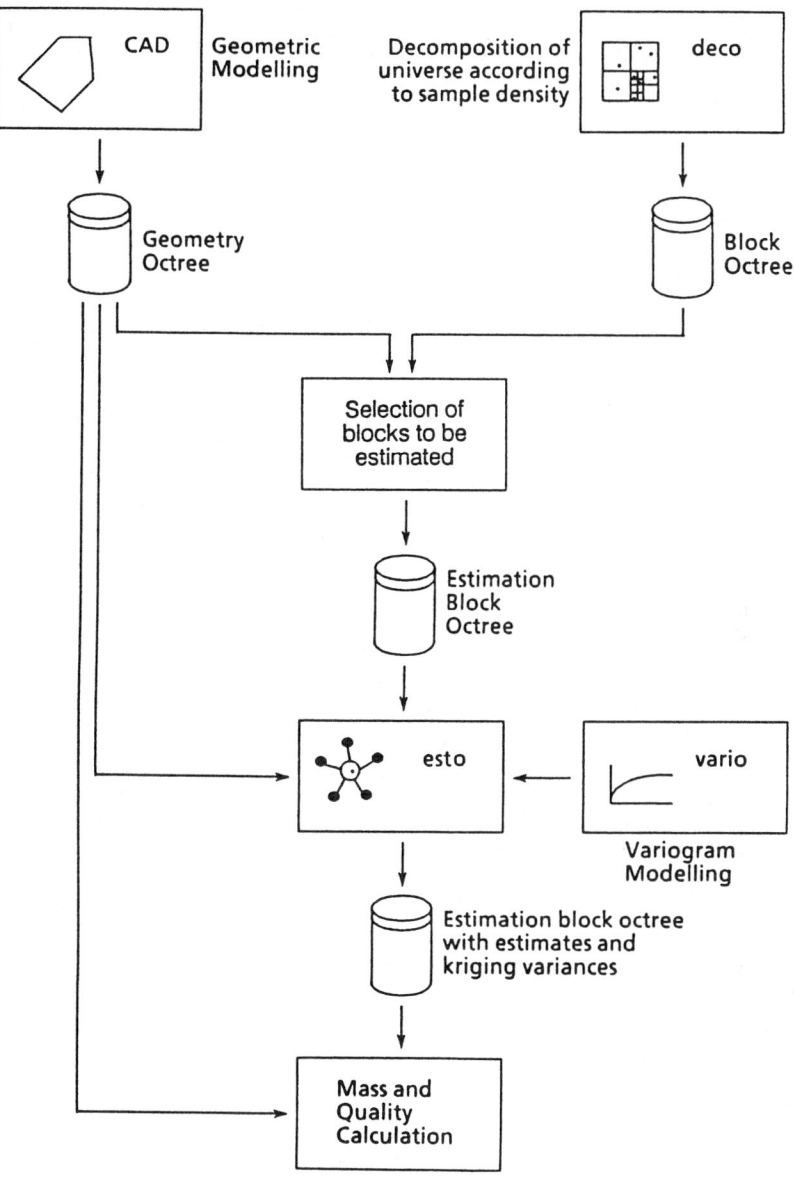

Fig. 3. Flow chart for modelling continuous spatial attributes (Prissang and Skala 1990)

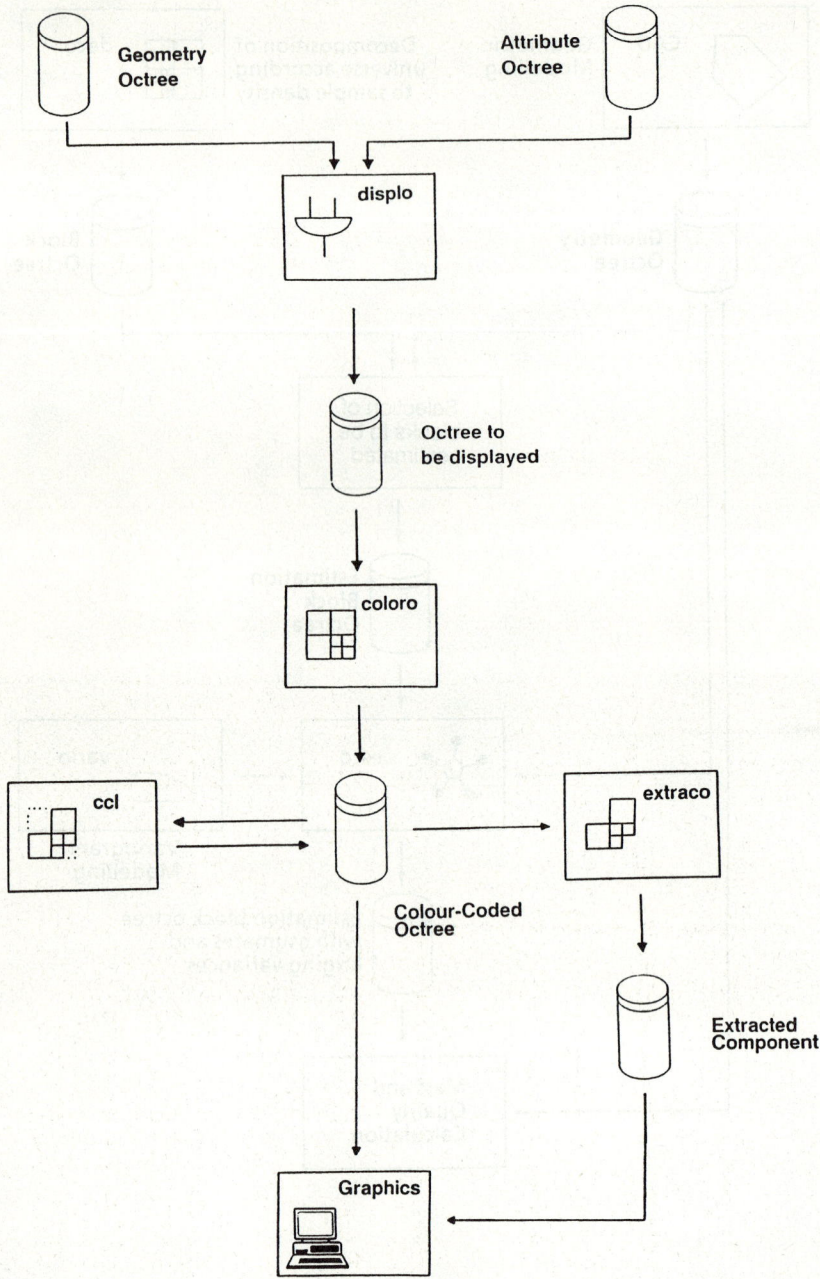

Fig. 4. Flow chart for the visualization of continuous spatial attributes and the analysis of the resulting regions in 3D space

Visualization is of paramount importance for the proposed approach to geological modelling, because the user is enabled to inspect all intermediate and final results. In order to arrive at high quality graphical presentations, we are currently using the CA-DISSPLA™ graphics library by Computer Associates International, Inc. (CA). The pictures shown in this paper have been generated with a program by Tietze (1990) which makes use of the object rendering facilities offered by this library.

All obels of the same colour form a so-called region in 3D space. Connected component labelling (program CCL) allows to analyse the regions in respect of the number of disjoint components or objects. The user can define, if the obels are checked for O(1)-, O(2)-, or O(3)-adjacency, i.e. face-to-face, edge-to-edge, or vertex-to-vertex connectivity. Components are ranked according to the enclosed volumes.

Unequivocal object identifications will be assigned to all obels that are part of a respective component. Like regions, single components can be extracted from the colour-coded octree (program "extraco") and visualized as individual geometrical objects.

PRACTICAL APPLICATIONS IN MINE PLANNING

A visualization provides the fastest way to convey information on spatial data. Therefore, rendering of a model showing the spatial distribution of zones of similar grades is more effective than reconstructing the grade distribution by going through a number of grid maps or sections.

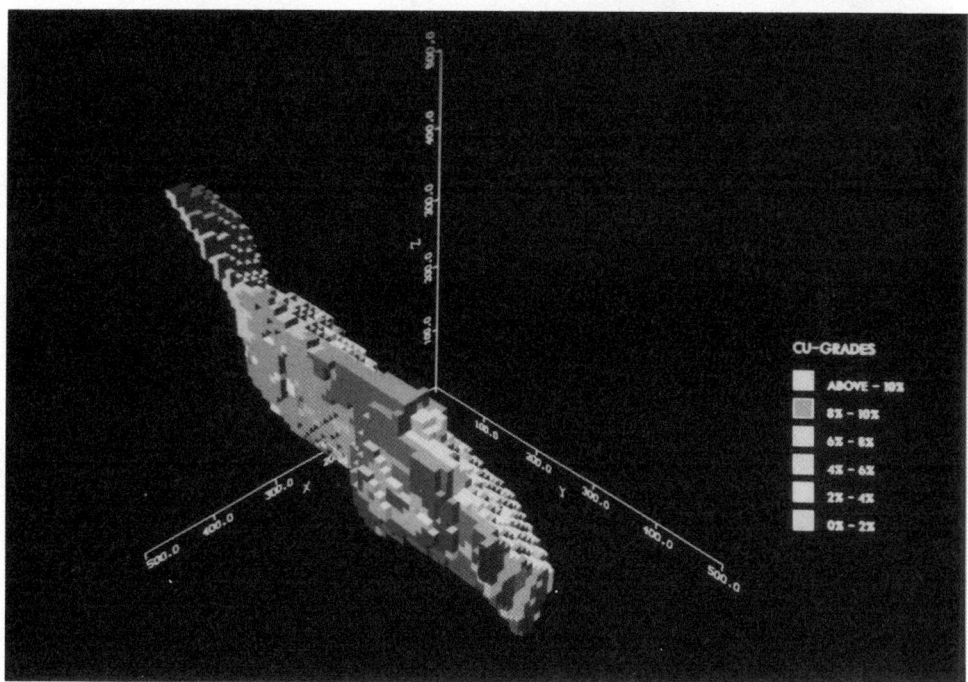

Fig. 5. Cayeli orebody colour-coded according to Cu grades

Furthermore, visualization can be used as a tool for geologists and mining engineers to inspect the effects of changes in class boundaries on the spatial distribution of zones of similar grades. Here, the possibility of evaluating the effect of a selected cutoff value on the extend of a deposit will be demonstrated. Data from the Cayeli Copper-Zink-Lead deposit in eastern Turkey is used to provide a testbed for the proposed approach.

The distribution of Cu-grades in the Cayeli deposit is illustrated by fig. 5. Classes of intervals of 2% Cu have been defined to set up the corresponding colour-coded octree.

Fig. 6 shows the Cayeli deposit from the same viewpoint. However, all obels with grades less than 2% are excluded from display. Here it becomes apparent that the SW part of the deposit does not contain any ore above a cutoff value of 2% Cu.

In fig. 7 only the richest parts of the deposit are shown. There are two clusters, separated by material of lower grades. This information provides a vital hint for the mining engineer when low and high grade ore have to mined out simultaneously to ensure a continuous flow of homogeneous material to the ore treatment plant.

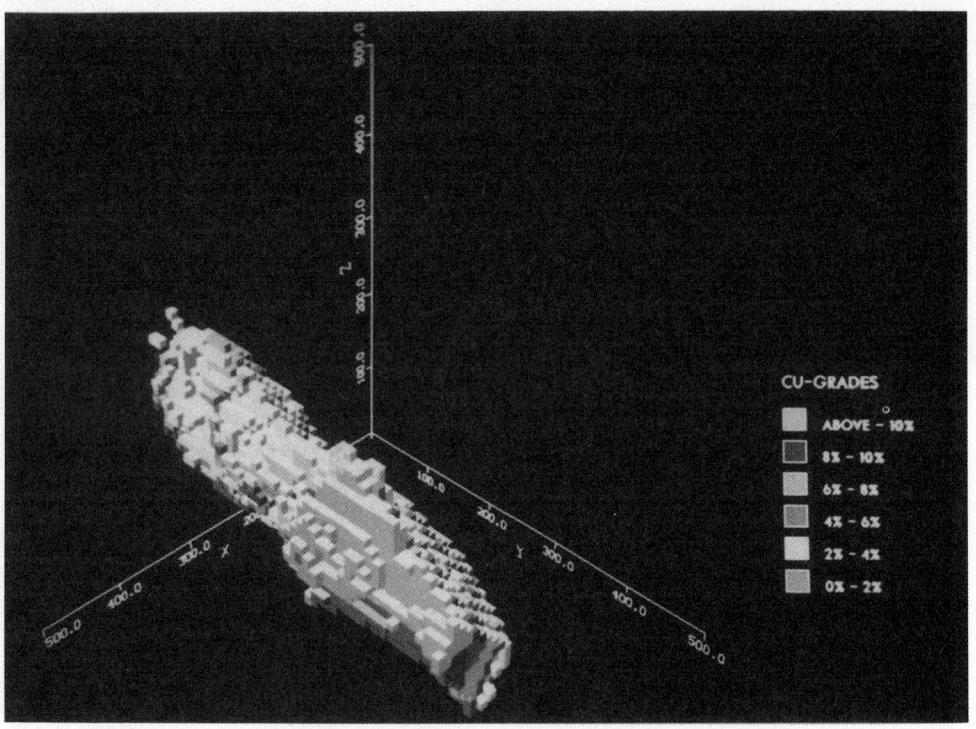

Fig. 6. Cayeli orebody, regions of Cu grades greater than 2%

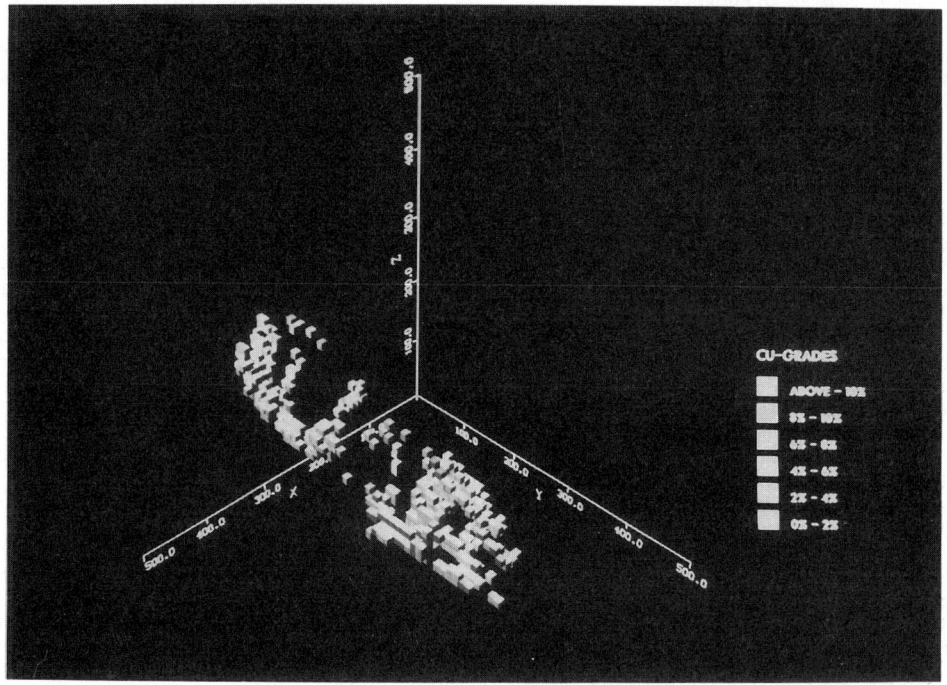

Fig. 7. Cayeli orebody, regions of high Cu grades (> 8%)

CONCLUSIONS

The linear octree encoded 3D variable block modelling technique provides a means to set up models that are more efficient than conventional 3D fixed block models in respect of the requirements for storage space and processing time. Generating the estimation block octree for Cayeli takes less than a minute. Due to the requested smallest obel size the presented attribute model contains 13380 object elements with edge lengths of 8, 4, and 2 metres. A 3D fixed block model of the same accuracy would comprise 606468 blocks. Based on 2136 samples an estimation of grades by means of point kriging took a HP9000/350 series workstation 33min user time. The estimation of the attribute values in the conventional type of model would take approximately 24h. The time requirements for colour coding and the auxiliary programs are negligible. Even the connected component labelling program, which is based on neighbour finding, does only require 14 seconds user time when applied to the full colour-coded grade octree of 13380 obels.

By the application of methods of 3D computer graphics, it is possible to gain new insights in internal structures of deposits. It therefore can be envisaged, that graphical evaluation of the information stored in property models will become a routine tool for geologists and mining engineers.

ACKNOWLEDGEMENTS

Research has been funded in part by the European Community Raw Materials Research Programme Grant No. MA1M-0026-D(B). The author likes to thank the Sachtleben Bergbau GmbH (Meggen, FRG) for their kind assistance.

REFERENCES

Atkinson, H.H., I. Gargantini and M.V.S. Ramanath (1984). Determination of the 3D Border by Repeated Elimination of Internal Surfaces. *Computing*, 32, 279-295

Atkinson, H.H., I. Gargantini and O.L. Wu (1987). Linoct (version 2.0) - viewing program. *University of Western Ontario, Dept. of Computer Science, Report No.* 176, 15 p.

Bak, P.R.G (1990). *Applied Solid Modelling in Mineral Resources Engineering*. Ph.D. Thesis, University of London, Imperial College, Dept. of Mineral Resource Eng., London (in preparation)

Chen, H.H. and T.S. Huang (1988). A Survey of Construction and Manipulation of Octrees. *Computer Vision, Graphics, and Image Processing*, 43, 409-431

Dunstan, S.P. and A.J.B. Mill (1989). Spatial Indexing of Geological Models using Linear Octrees. *Computers & Geosciences*, 15, 1291-1301

Gargantini, I. (1982). Linear Octrees for Fast Processing of Three-Dimensional Objects. *Computer Graphics and Image Processing*, 20, 365-374

Prissang, R. and W. Skala (1990). An Octree-Encoded 3D Variable Block Model for the Representation of Ore Grades in Underground Mining. *Proc. 22nd Int. APCOM Symp.*, Vol. 3, pp. 137-148

Tietze, J. (1990). Visualization of 3D predictive deposit models based on the linear octree data structure. *Freiburger geowiss. Beitr.*, 2, p. 103

3D - MODELING OF BIG STRATIFORM DEPOSITS

G.J. PESCHEL AND J.U. BERTHOLD

Department of Geological Sciences,
Ernst-Moritz-Arndt-University of Greifswald,
Friedrich-Ludwig-Jahn-Str. 17A,O-2200 Greifswald,Germany

ABSTRACT

A modeling strategy especially for big stratiform deposits was tested
which combines an only partly computed model data base with a set of
transformation procedures. It is based on a set of n+2 two-dimensional
models of every seam or layer, represented by digital maps or
matrices.Thereby is n the number of parameters, describing the
corresponding layer at a determined pixel, like for example the chemical
and physical properties. The additional two basis maps are the depth map
of the layer sole and the thickness map.The second part of modeling is
represented by a set of procedures, determining the true values of the
interesting parameters mentioned above for an arbitrary point or sector
of the tree-dimensional space. They execute the search of that layer
which is situated at the corresponding point and they support the
transfer of its properties from the two-dimensional maps to that
point.These procedures can be executed by means of the FORTRAN prototype
program REPRES.Some examples of 3D-modeling of a Miocene lignite seam in
Saxony have demonstrated the utility of the method.

KEYWORDS

Mathematical Geology, 3D-Modeling,Deposit-modeling, Stratiform deposits,
Coals, Lignites

TRADITIONAL MODELS

Models of geological bodies or deposits are stores of knowledge which we
can use to reproduce all the data and observations obtained by their
exploration. Geological models usually are made from paper. They consist
of a set of maps, sections, diagrams, tables etc. explained by
supplementary text. Such models are immediately understandable for men.
They can be used as basic documents for manifold construction and
calculation procedures in mining and other engineering application
fields. Therefore some CAD technologies are prefered for the modeling of
deposits, based on extended data bases and sophisticated methods of
geostatistical interpolation. They all intend to visualize immediately
the obtained primary data as vivid images of the geological bodies
(Hildebrand *et al.*,1984; Hinde *et al.*,1984; Peschel,G. *et al.*,1984;
Sukhendu *et al.*,1984; Cunliffe,J.J., 1989).This is the first way of
modeling (Fig. 1).

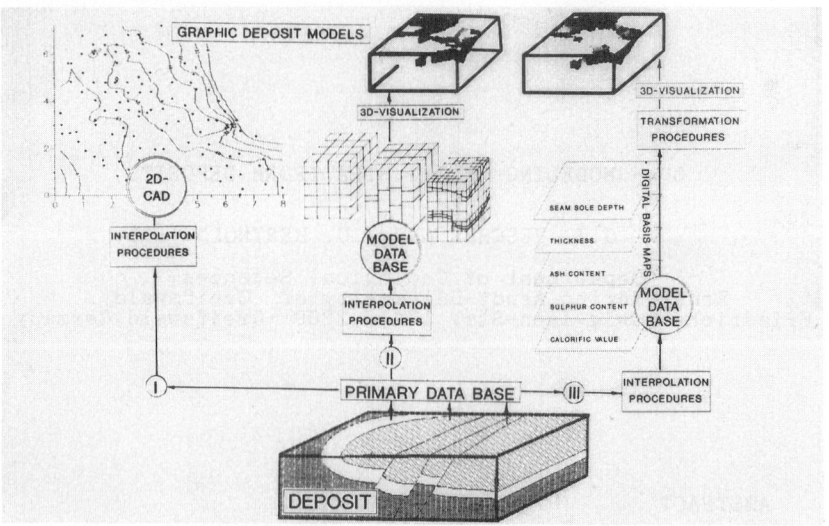

Fig.1. Three several conceptions for
deposit modeling

But we must point to some important disadvantages of this model type:

1. It reflects the three-dimensional object only by two-dimensional
sections.
2. These sections must not be consistent and it often appears for example
that maps and cross sections show a different seam thickness for the
same location.
3. The graphic deposit model reflects a temporary state of exploration in
a final manner. New results or changes of the object (e.g. changes by
mining) cannot be considered by an easy alteration of the model. The
model must be constructed anew.
4. The graphic deposit model is a deadlock for the automatization of the
geological exploration because the information, stored at the model
cannot be recycled into the computerized exploration process.
5. The graphic model contains only fragmentary the information which is
already obtained by the exploration of the object.So for instances the
construction of an isoline map demands the interpolation of an
extended surface covering data grid which disappears if the isoline
map is plotted out.

BLOCK MODELS

These are the reasons why was developed a second path of modeling during
the last decade (Fig. 1). Thereby the graphic models are supported by an
information model, represented by a model data base within the computer
memory and calculated out of the primary data base by means of
geostatistic and other interpolation or approximation methods. At this
strategy the visualized results can serve as an efficient interface
between the model data base and men, aiding the illustration and the
understanding of the intrinsic spatial nature of the deposits.
Traditional quality parameters of graphs as for example the scale and
exactness of maps lose their signification because all demanded
operations could be executed more precise and more efficient by
calculations with the data of the model data base.

Table 1. Estimation of the required memory
capacity for a 3D-voxel coal quality
model of a big lignite deposit.

```
┌─────────────────────────────────────────────────────────────────────┐
│                                                                       │
│  1. Deposit extension     : 10 000 x 10 000 x 100    m                │
│     Voxel   extension     :     50 x     50 x  0.25 m                 │
│     Total number of voxels :    200 x    200 x 400 = 16 000 000       │
│                                                                       │
│  2. Bytes per voxel       : Parameter              Bytes             │
│                                                                       │
│                             Northing                 8                │
│                             Easting                  8                │
│                             Depth                    7                │
│                             Stratigraphic code       5                │
│                             Rock (coal) type code    7                │
│                             Ash content              5                │
│                             Sulphur content          5                │
│                             Calorific value          6                │
│                                                                       │
│                             Total                   51                │
│                                                                       │
│  3. Required memory capacity              816 000 000                │
│                                                                       │
└─────────────────────────────────────────────────────────────────────┘
```

Three-dimensional model data bases of deposits or other geological
bodies consist usually of a set of regular blocks or cells with
interpolated or estimated data for each one (Abbachi,A., 1989; Benest,J.
and Winter,P.E. ,1984; Dowd,P.A., 1984,1988,1989; Folarin,A.B. and
Mill,A.J.B. ,1989); Galloway, D.W., 1986; Henley,S. and Stokes,W.P.C.
,1984; Nemec, V.,1971; Pflug,R.,1988).

Modeling which uses such a model type is characterized by the following
sequent mode. At first we must define the cell structure. Then we have
to compute the corresponding values for each cell and put them into the
memory. Thereby it is often necessary to change or to refine the cell
structure, depending from the real structure of the corresponding
deposit and its data. At last we can visualize the model with arbitrary
sections, maps and pseudo-tree-dimensional graphs or even stereo images.

This strategy is very favourite because of the short response time of
the interaction during vizualization. But it is not favourable for big
objects.

At first we need a very extended computer memory. So for example the 3D-
voxel model of a complicated endogenous disturbed subrosive type lignite
deposit with an area of more then 100 square kilometers and a small
block volume corresponding to the required exactness of the planning
documents for mining demands a memory region of approximately 800 MB
only for the geometrical coal quality model (Table 1.).

Furthermore the interpolation or estimation of values for all blocks
will take a long computing time. All this expenditure will be not
effectively because the visualization of the model allways is limited to
a few selected sections, maps and pseudo-spatial graphs which use mostly
less then 1 % of the computed information. Moreover the consideration of
additional drillings requires anew the computation of a huge information
garbage.

THE MAP-BASED 3D-MODEL

General features

For this reason a third way of modeling especially for big stratiform deposits was tested (Peschel, G.J., 1988) which combines an only partly calculated model data base with a set of transformation procedures (Fig. 1).

This strategy is based on a set of n+2 two-dimensional models of every seam or layer of the given deposit, represented by digital maps or matrices. Thereby is n the number of parameters, describing the corresponding layer at a determined pixel, like for example the chemical and physical properties, the rock type, the quality sort of the raw material etc.. The additional two basis maps are the depth map of the layer sole and the thickness map or alternatively the map of the upper seam border depth.

The second part of modeling is represented by a set of procedures, determining the true values of the interesting parameters mentioned above for an arbitrary point or sector of the tree-dimensional space. They execute the search of that layer which is situated at the corresponding point and they support the transfer of its properties from the two-dimensional maps to that point. This transfer is going on without any interpolation because the layers are considered to be vertically homogenous. For this reason the procedures work fastly.

Map model types

Parameters maps of stratiform deposits can be constructed for such layers or layer slices which can be regarded as vertically homogenous within given square elements. Supplementary to this assumption modeling allways is based on one of the following three conceptions.

1. The continual field conception. Such models can be constructed only for the spatial variation of a single quantitative i.e. measured parameter e.g. ash content, thickness, depth etc. Their typical visualizations are isoline maps.
2. The discreet field conception. Discontinuous models can be constructed for all parameter typs like qualitative observations (e.g. colour of rocks), estimated data (e.g. clay content of soils), classes of measured data and also for complex data sets representing rock types, facies - classes or sorts of raw material. Their typical vizualization is the geologic map with hard bordered areas of different colours and patterns.
3. The conception of the partial continuous field. It combines the first two model types and consists of a lot of sharply demarcated areas. The regarded single deposite parameter can be modeled as a continual field within each area but its spatial variation is disreet at the area borders. This model type must be applied if the deposite is disturbed by faults.

Algorithms for the construction of basis maps

The construction of basis maps using the continual field conception can be executed by means of geostatistical or similiar methods like invers distance weighted interpolation. The quality of the resulting two-dimensional model depends not only from the primary data but also from some decisions and estimations of controlling parameters of the used algorithms like for example the variogram model and its parameters, the definition of the interpolation neighbours, the weighting exponent etc.

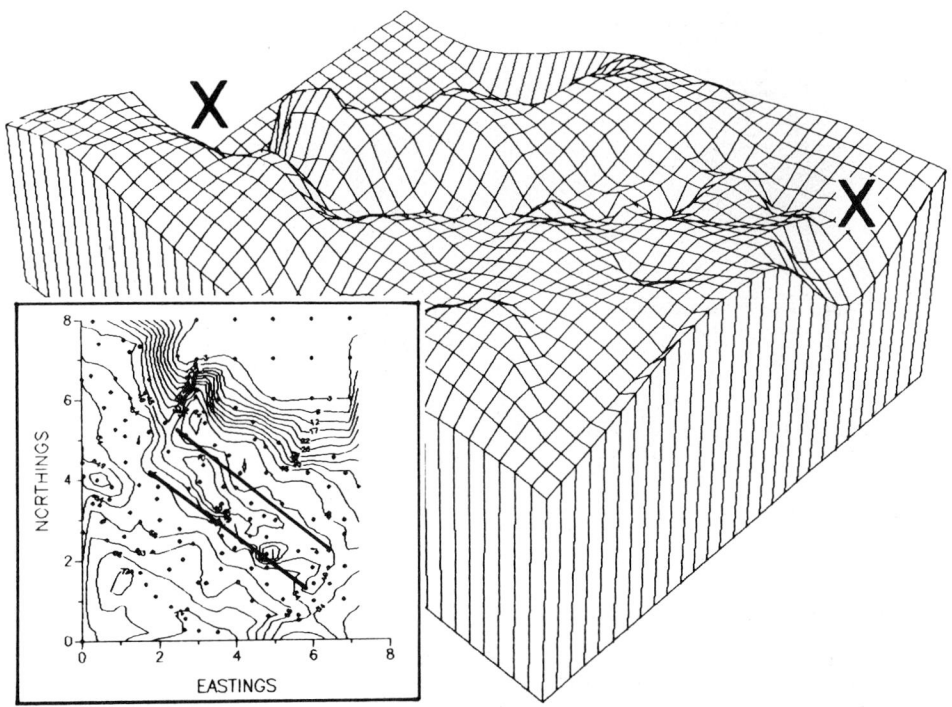

Fig. 2. 3D-seam surface model of a complicated
lignite deposit. View from south-west.
(X - X: Graben zone,bordered by step faults)

An extensive study using a comprehensive data base of a big lignite
deposit shows that the smoothing properties, the estimation error and
the computation time of the interpolation algorithms can be optimized,
whereby invers distance weighted interpolation using approximately 10
neighbours and a weighting exponent between 0.5 and 2.0 gives results,
very similiar to that of Kriging but with a computation time which is
only one half from that of Kriging.

For discreet field map constructions the VORONOI graph successfull was
used and also a new pixel by pixel classification algorithm.

All tools for the basis map construction have been united by a FORTRAN
program package which is named REPRES. It contains also moduls for
triangulation of primary data points and gives finally the complete
basis map model of n+2 two-dimensional data matrices of the regarded
deposit (Bittner,U. 1990).

The following figures show some examples of the basis map system of a
brown coal deposit in Saxony. This deposit has a very complicated
geometrical structure. The surface of the main Miocene lignite seam
sinks approximately 70 m over a horicontal distance of 5 km.But this
deeping is not uniformously.It is overlaid by some special structures
like for example a deep graben zone,bordered by step faults with a
mutual distance of about 1 km (Fig. 2).

Fig. 3. Digital depth map of the seam sole

Fig. 4. Thickness map of lignite seam 4380

The map of the seam sole depth (Fig. 3) and that of the seam thickness (Fig. 4) define the geometrical properties of the considered layer. The maps of the two-dimensional variations of the ash content (Fig. 5.),of the sulphur content (Fig.6.)and of the calorific value (Fig.7) determine together the coal quality of the regarded seam. These maps are interpolated from the data of 134 bore holes at an area of 67.5 square kilometers. The pixel spacing is approximately 60 m x 60 m.

Fig. 5. Digital map of the ash content

Fig. 6. Digital map of the sulphur content

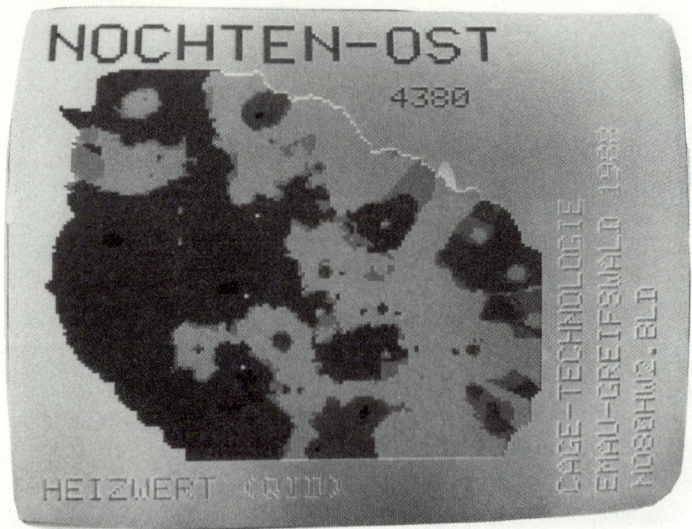

Fig. 7. Digital map of the calorific value

Fig. 8. Digital Voronoi map of some
coal quality sorts

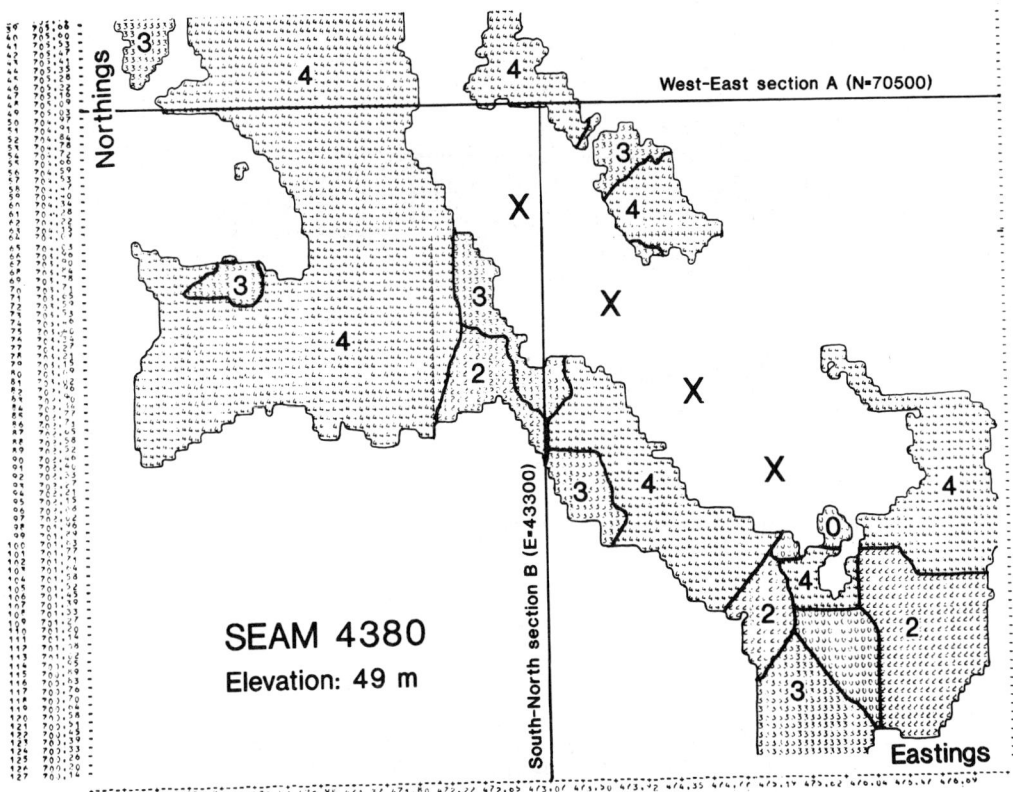

Fig. 9. Outcrop map of the discriminated coal
quality sorts on elevation of 49 m
above mean sea level.
(White:waste,0-2:Boiler coal,3:Gas coal
4:Coking coal, XXX: Graben zone).

Transformation procedures

Because of the mixed character of the basis map system there can be
executed different types of transformation procedures. At first we can
transform or combine such maps which represent the seam properties. So
for instance the common classification of the ash content, the sulphur
content and the calorific value determined by some quality criterions of
the refining industry gives a coal quality map, which is shown in Fig. 8
as an example of the VORONOI mapping algorithm. A further transformation
for example by computing the sum for all pixels of the thickness map
under the condition that the colour code of the same pixel of the coal
quality map has a determined value, gives the resource appraisal of a
given quality class if this sum will be multiplied by the square unit of
the pixels. This example already shows that the basis map system can be
used for calculations within the three-dimensional deposit model. This
will be yet more clearly if we consider such transformations which are
connected with the geometrical properties of the seam.

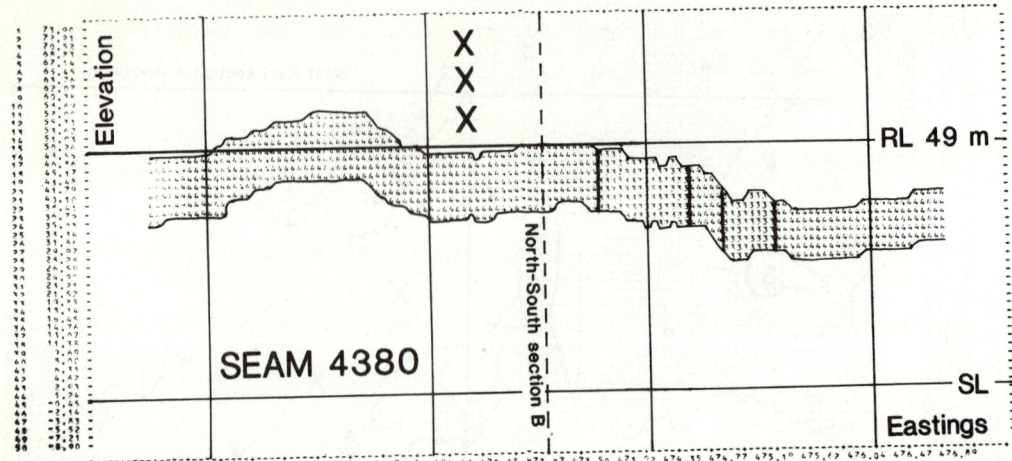

Fig. 10. West-East section along N=70500
(Legend see Fig. 9).

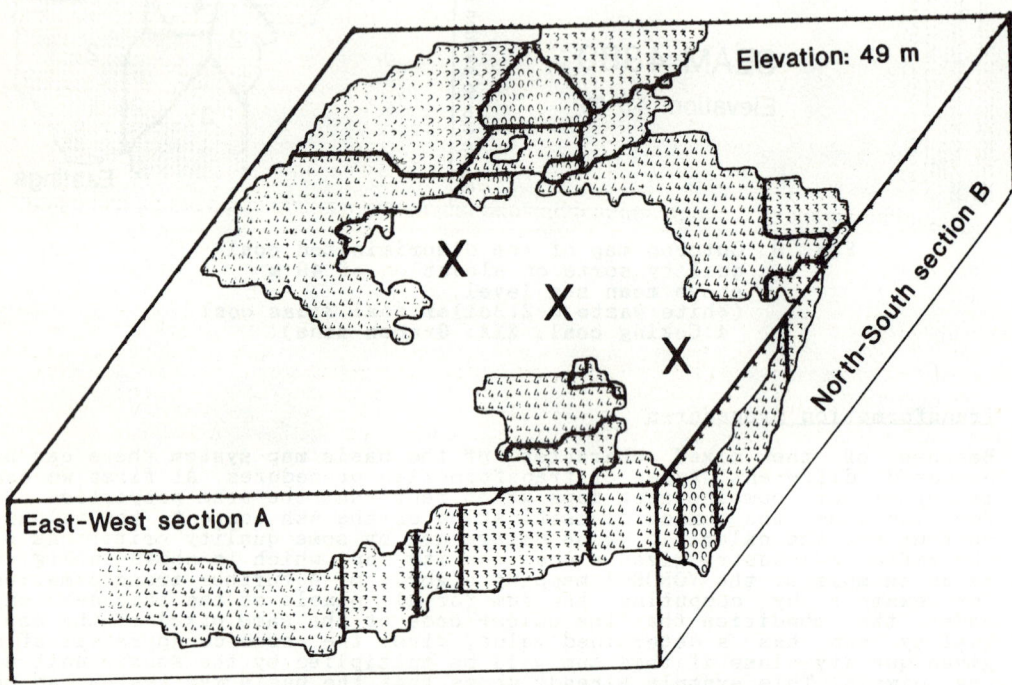

Fig. 11. Block diagram showing the spatial
arrangement of the coal quality sorts
of the seam 4380 in the eastern part
of the deposit Nochten (view from north).

For mine planning and engineering work it is very important to arrange the regarded geological object into some geometrical relations. So for instances the outcrop lines of a given layer at a given plain or pair of parallel plains must be constructed. This can be realized by a selection procedure which determines for each pixel of the basis map system if the seam sole depth is deeper and simultaneously if the depth of the upper seam border is less then the given level. Recording the results of this test by 0 or 1 we get an additional digital map. Multiplying this map pixel by pixel by the grey tone ore colour key numbers of the quality map (Fig.8) a detailed quality map on a given mining plain can be constructed . An example is shown in Fig. 9.

The completation of such maps by some cross sections can support the understanding of the spatial distribution of the raw material and its quality. So for instance the West-east section A,shown in Fig. 10 can help to clear if the seam within the white holes on the plain map is situated above ore below the considered plain.We can see that the seam can be mined on the considered plain by means of a combination of high cut and deep cut on a stope lenght of 2 km.Easterly from the beginning of the graben zone it can be won from the same plain by deep cuts using two benches also for a distance of two kilometers.Then, at the last part the mining plain must sink to an elevation of approximately 40 m above mean sea level.

The best visualization of the geometrical relation can give the combination of a level plain map with two cross sections to a block diagram (Fig. 11).But it is clearly visible that further decisions and constructions as mentioned above cannot miss the undistorted two-dimensional graphs which must be necessary consistent if they are derived from a 3D model data base.

The program package REPRES can dismember the regarded deposit into four blocks bordered by plains parallel to the co-ordinate axes of the basis maps whereby the disjunction can take place at an arbitrary point. Each block can be inspected from 8 directions.Moreover there can be constructed sections in arbitrary directions whereby the section trace may be a polygone, connecting some given points like drill holes etc (Berthold,J.U., 1990).

Visualization of the results

All results of the computations also the pseudo three-dimensional block diagrams are two-dimensional maps. They can be stored and in this way the informational part of the model, that is to say the model data base is growing up. The digital maps will be stored as matrices using regular lattices. After their transformation from floating point numbers to Byte normalized cardinals they can be immediately visualized by gray tone or coloured maps.For this an image processing unit with a 768 x 512 pixel array and an 8 bit information for each pixel has been used.

REFERENCES

Abbachi,A. (1989).Use of CAD Techniques to Control Geostatistical
    Estimates in Stratigraphic Orebodies: *LUMA,* Leeds, pp. 129-135.
Benest,J. and Winter,P.E. (1984).Ore-reserve estimation by use of
    geologically controlled geostatistics: *18-th APCOM,* London,
    pp.367-378.
Berthold,J.U. (1990). Ein Strukturmodell zur dreidimensionalen Modellie-
    rung stratiformer Lagerstätten: *Dissertation,* Ernst- Moritz- Arndt-
    Universität.

Bittner,U. (1990).Mathematische und methodische Beiträge zur rechner-
    gestützten Lagerstättenerkundung: Gitterbildung und Kartenherstellung
    aus unregelmäßig verteilten Punkten mit Merkmalswerten: *Disserta-
    tion*,Ernst- Moritz- Arndt- Universität Greifswald.
Cunliffe,J.J. (1989).The Application of Geostatistics in the Robertson
    Group: *LUMA*, Leeds, pp.143-157.
Dowd,P.A. (1984).Two Algorithms for Geometrical Control in Kriging and
    Orebody Modelling: *Science de la terre,Informatique Geologique,*
    21,pp. 189-208.
Dowd,P.A.(1988).Practical Aspects of Applying Geostatistics in
    Stratigraphic Orebodies: *Science de la Terre, Informatique Geologique,*
    27,2, pp.291-315.
Dowd,P.A.(1988).Practical Aspects of Applying Geostatistics in
    Stratigraphic Orebodies: *Science de la Terre, Informatique Geologique,*
    27,2, pp.291-315.
Dowd,P.A. (1989).Microcomputer Modelling of Orebodies for Interactive
    Planning: *Mining Pribram*,2, pp.383-396.
Folarin,A.B. and Mill,A.J.B. (1989).Application of Computer Graphics in
    Mineral Reserve Assessment: *Mining Pribram*, 2, pp.261-271.
Galloway,D.W., (1986).A Classification of types of rectangular block
    models: *LUMA*, Leeds,pp.116-128.
Henley,S. and Stokes,W.P.C. (1984).Improved geological modelling
    techniques for mine planning: *18-th APCOM*, London, pp.265-268.
Hildebrand, R.T., Carey, M.A.,Smith, R.J., Blake, D. and Culbertson,
    W.C. (1984).Graphic-enhanced coal-resource evaluation by use of the
    National Coal Resources Data System of the U.S. Geological Survey:
    *18-th APCOM,*London, pp. 89-95.
Hinde, C.G., Newton,M.J. and Rickus,J.E. (1984).Application of an
    interactive graphic mine planning system at Manlanjkhand open-pit
    copper mine, India: *18-th APCOM*, London, pp. 175-189.
Nemec,V. (1971).Space models of inclined limestone deposits: *Can. Inst.
    Min. Met.*, 12,pp.209-217.
Peschel,G.J.,Kolyschkow,P., Poppitz,H.H. and Bittner,U. (1987).
    Lagerstättenmodellierung am Dialogarbeitsplatz,in Peschel,
    G.(Ed.).Mathematische Lagerstättenmodellierung: *Wiss. -techn.-
    Informationsdienst d. Zentr. Geol.Inst.Berlin,* 28.A, 4, pp.55-63.
Peschel,G.J. (1988).The exploration and evaluation of low-rank- coals by
    means of an advanced CAGE-technology. *Science de la Terre,
    Informatique Geologique,* 27,pp.145-155.
Pflug,R. (1988).Solid modeling of geological objects with 3D rasters.
    *Geol.Jb.,*A 104, pp. 213-219.
Sukhendu L.B. and Young,C.K. (1984).Coal mine planning with interactive
    graphics: *18-th APCOM*, London, pp.191-200.

GEOSTATISTICAL MODELLING OF GEOLOGICAL LAYERS AND OPTIMIZATION OF SURVEY DESIGN FOR THE CHANNEL TUNNEL

Raymonde Blanchin and Jean-Paul Chilès

BRGM, BP 6009, 45060 Orléans Cédex 2, France

ABSTRACT

The geometrical modelling of the geological layers in the Channel tunnel area used geostatistical techniques for computing predictions and confidence intervals. The results were used to design the alignment of the tunnel. After two years of tunelling, the first 10 km of the service gallery driven off the French coast enables the reliability of the predictions to be assessed. The comparison between the predicted and actual levels of the strata, based on 48 probes, validates the geostatistical approach.

KEYWORDS

Geostatistics ; kriging ; survey design ; risk analysis ; validation ; Channel tunnel.

PROBLEM AND METHODOLOGY

The presence of a favourable layer for boring a tunnel beneath the Channel between France and Great Britain has been well known since the 1876-1882 surveys for such a project (see Fig. 1). This layer, the Cenomanian chalk marl, is composed of relatively soft and virtually impermeable chalk and clay, thus an ideal medium for tunnelling ; it is overlain by a layer of grey chalk, more fractured and permeable than the chalk marl, and is underlain by the Gault clay that cannot be penetrated without serious civil engineering problems.

The Channel tunnel consists of three galleries: a north gallery and a south gallery for transport, and a smaller central gallery for servicing. Although the optimization of the tunnel alignment had obviously to take into consideration the operational constraints imposed by a high-speed railway (slope, curvature), the primary constraints were geological and geotechnical : maintaining the tunnels sufficiently deep beneath the sea bed, and, above all, within the chalk marl formation. As the chalk marl layer is inclined and only 30 metres thick, a precise delimitation of this layer was essential. The various parameters (top, thickness, slope, permeability) were determined by means of a geostatistical analysis of bathymetric and seismic-reflection surveys (1,500 km of seismic profiles and 19 boreholes during the 1986 and 1988 surveys, plus 90 old boreholes since 1958). The main geological parameter used in this project was the top of the Gault clay.

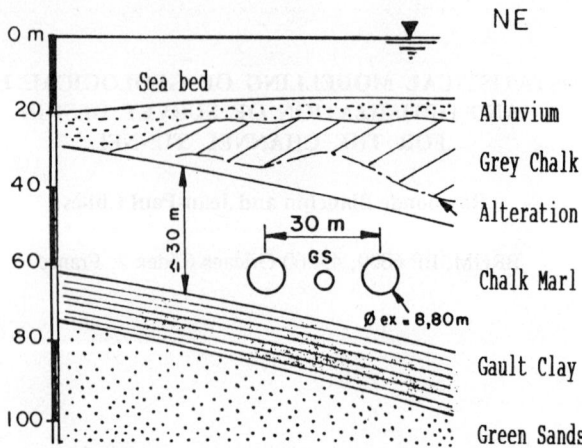

Fig. 1. Typical geological cross-section with the three galleries.

STUDY OF THE TUNNEL AS A WHOLE

Geostatistics (Matheron, 1971) was used to calculate a reliable digital model and produce representative maps and sections. The interpolation method, kriging, is based on a preliminary variogram analysis, which describes the variability (degree of continuity, trend, anisotropy) of the field under study. The originality of kriging is to provide at each interpolated point both the estimate and its confidence level, characterized by a standard deviation. This enables to quantify the risk of a discrepancy between prediction and reality.

Blanchin *et al.* (1989) details the geostatistical study of the overall survey. The interpretation was done in four phases:

1. estimation of the depth of the sea bed, $S(x)$, from the bathymetric profile data;
2. estimation of the two-way time from the sea-floor to the top of the Gault clay, $T(x)$, from the seismic profile data;
3. estimation of the average velocity, $V(x)$, from various geophysical data (retro-analysis at wells, well logs, fine-grain velocity analysis along profiles);
4. combination of the above results to provide the estimation of the depth to the top of the Gault clay from sea level, $G(x)$, by:

$$G(x) = S(x) + V(x) T(x) / 2$$

The estimates take into account:

- the local variability of each parameter in 2-D space (local variography);
- the various sources of measurement error (location uncertainty, tide correction, migration of seismic reflectors, velocity calculations), their order of magnitude, and their degree of spatial correlation (for profile data).

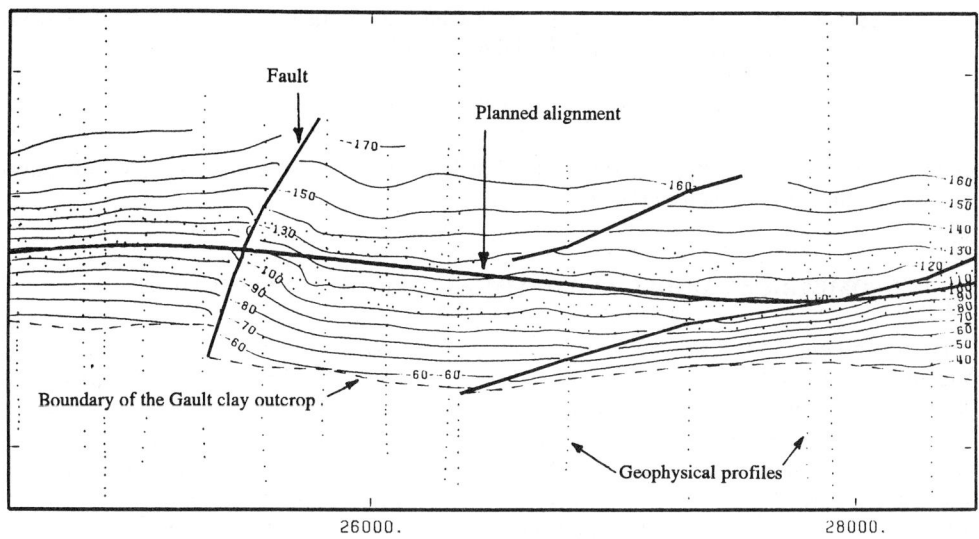

Fig. 2. Contour map of the elevation of the top of the Gault clay (metres).

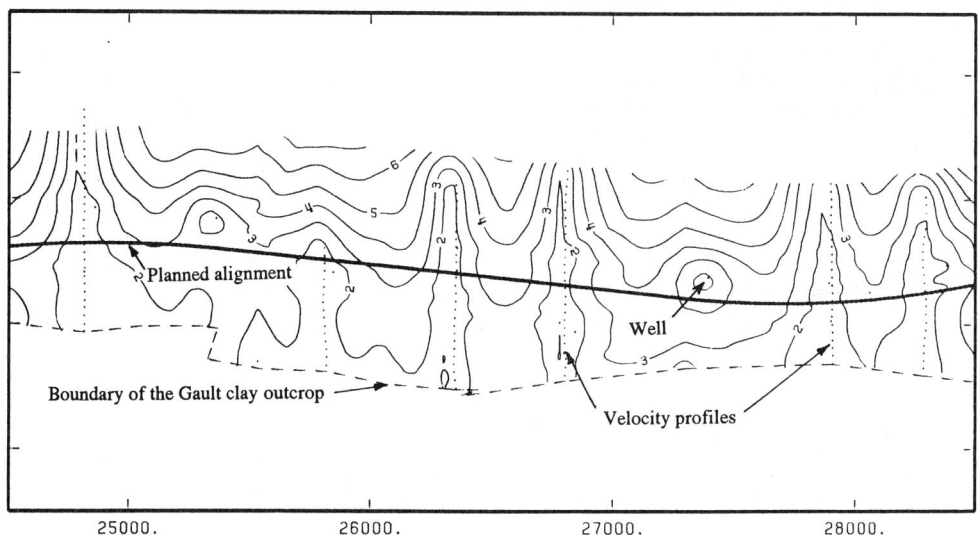

Fig. 3. Contour map of the kriging standard deviation of the Gault clay (metres).

The kriging standard deviation was computed for each interpolated field. As the various estimates are independent, the precision σ_G concerning the top of the Gault is given by:

$$\sigma_G^2 = \sigma_S^2 + (\sigma_V^2 . T^2 + V^2 . \sigma_T^2 + \sigma_V^2 . \sigma_T^2) / 4$$

In this expression, σ_S, σ_T, σ_V are the standard deviations attached to fields S, T, V, respectively, and all the values depend on the location x.

GEOMETRICAL MODEL AND OPTIMIZATION OF THE ALIGNMENT

The investigation area is a strip 500 m wide centred on the main axis of the first planned alignment. Figures 2 and 3 illustrate the results on a part of the French side of the map. The general pattern of the top of the Gault clay (Fig. 2) is a reflection of the regional geological trend - an anticline whose axis is parallel to the alignment. The boundary of the Gault clay outcrop is shown on the map as a dotted line. The standard deviation (Fig. 3) displays well-marked minimum values near the profiles (good knowledge of the seismic time) and boreholes (consistent knowledge of seismic time and average velocity) ; it increases from south to north, because the thickness from sea floor to the top of the Gault increases, so that the uncertainties concerning the velocity have a greater impact in the north. Throughout the whole studied zone, the standard deviation lies between 2 and 6 metres ; it does not exceed 4 metres along the marine part of the planned alignments, and usually lies between 2 and 3 metres.

Figure 4 shows the kriged results on a vertical section along part of the south gallery. The distorsion of scale that affects the axes (the horizontal scale is 20 times smaller than the vertical scale) amplifies the oscillations. The sea floor and the top of the Gault clay are represented with their 68% confidence interval ($\pm$ 1 standard deviation). When considering the estimated top minus one standard deviation, it can be seen that the first alignment could intersect the Gault clay at several places. Therefore a new alignment was designed from these results, as shown on the cross-section.

The optimization of the final alignment was made in such a way as to nearly always maintain the tunnels at least one standard-deviation above the estimated value of the top of the Gault. Thus the risk of having to penetrate the Gault clay is minimized, but of course not totally excluded.

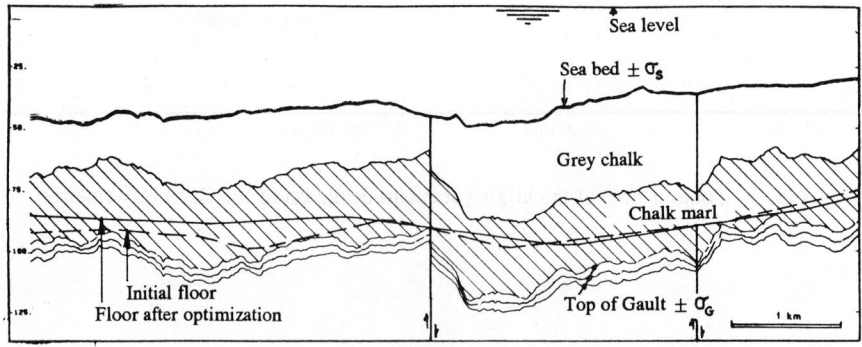

Fig. 4. Cross-section of the kriged results along the profile of the south gallery.

DESIGN OF COMPLEMENTARY SURVEYS

Cross-overs are planned at two locations to enable the trains to pass from one tunnel to the other if necessary. As these cross-overs are 21 m in diameter and 200 m in length, their precise location and design needs be based on more accurate geological predictions than the for primary galleries. For this reason complementary geophysical surveys were carried out, which were designed from a geostatistical study that determined the data density and accuracy necessary to obtain the required precision (standard deviation = 1 m). It was found that the transverse seismic profiles had to be 25 m apart on the French cross-over and 100 m apart on the British cross-over, and that the seismic velocity had to be defined by at least four boreholes located on either side of the tunnel.

COMPARISON OF THE ORIGINAL INTERPRETATION WITH THE NEW ONE

The complementary surveys enabled a new variography to be performed and the maps and cross-sections to be redrawn with a much greater precision (lower standard deviation) than for the original model. In comparison with the first one, the new estimate can be considered as the reality. Figure 5 shows a comparison between the first estimate with its 68% confidence interval (± 1 standard deviation) and the second estimate for the French cross-over. It can be seen that the new estimate usually lies within the 68% confidence interval of the first estimate, and only exceptionally departs the 95% confidence interval (± 2 standard deviations), thus proving the efficiency of geostatistics.

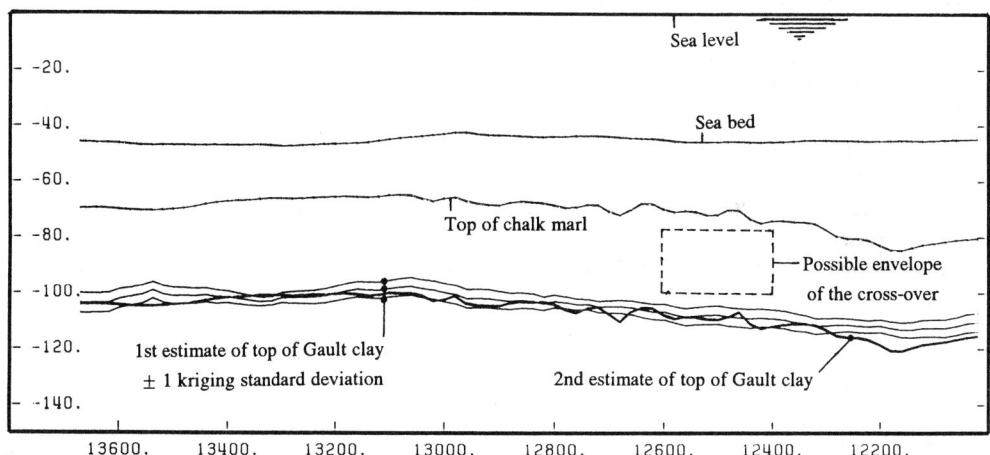

Fig. 5. Cross-section comparing the original interpretation with the new one
along the French cross-over.

COMPARISON OF THE GEOSTATISTICAL PREDICTION WITH REALITY

Subvertical boreholes are being drilled every 200 m during the boring of the service tunnel so as to check the level and the dip of the Gault clay. A comparison between the borehole data and the predicted values will provide a deeper analysis of the quality of the kriged estimates, and this study will be done after the link up of the English and French galleries. However, a preliminary analysis over the first 10 km of the French marine gallery has already been carried out, and the results concerning the difference reality-prediction are as follows :

- number of boreholes for checking :   48.
- minimum value of the difference :   -4.90 m below prediction.
- maximum value of the difference :   +5.10 m above prediction.
- average algebraic difference   :   +0.77 m.
- standard deviation from zero   :   1.98 m.

The average value of the kriging standard deviation in this zone is 2.93 m. The observed differences are usually lower than this value and nowhere exceed twice the average kriging standard deviation. The discrepencies are even slightly less than what could be expected, since the observed standard deviation (1.98 m) is smaller than the kriged one (2.93 m). This is probably due to the fact that the interpolation errors are correlated, due to correlations on the velocity data, and so 48 probes are not enough to derive a stable value of the standard deviation. This is confirmed by the probes available in the English gallery : they show a good agreement with the predicted levels, except three of them where the difference between prediction and reality exceeds twice the kriging standard deviation due to a lack of accuracy on the velocity. Such a result is not contradictory to the prediction : statistically speaking, it is normal that the discrepency exceeds twice the standard deviation in 5% cases.

CONCLUSION

The use of geostatistical techniques enabled an accurate geometrical modelling of the geological layers involved in the Channel tunnel project, and so optimized the design of the tunnel alignment. The observations in the service gallery are in a good agreement with the model and its predicted accuracy. The experience of the first 10 kilometres bored under the sea on the French side validates the efficiency of the geostatistical modelling, since the service gallery did not penetrate the Gault clay in this section.

REFERENCES

Blanchin, R., J.P. Chilès and F. Deverly (1989). Some applications of geostatistics to civil engineering. In: *Geostatistics* (M. Armstrong, ed.), Vol. 2, pp. 785-795. Kluwer Academic Publishers, Dordrecht, Netherlands.

Matheron, G. (1971). *The theory of regionalized variables and its applications*. Les cahiers du Centre de Morphologie Mathématique de Fontainebleau, Fasc. 5, 211 p., ENS des Mines de Paris.

POTENTIAL APPLICATIONS OF THREE-DIMENSIONAL GEOSCIENTIFIC MAPPING AND MODELING SYSTEMS TO REGIONAL HYDROGEOLOGICAL ASSESSMENTS AT YUCCA MOUNTAIN, NEVADA

A. Keith Turner
Department of Geology and Geological Engineering
Colorado School of Mines
Golden, Colorado 80401, USA

Kenneth E. Kolm
Department of Geology and Geological Engineering
Colorado School of Mines
Golden, Colorado 80401, USA

ABSTRACT

Three-dimensional Geoscientific Information Systems (GSIS) are being evaluated for supporting 3-D ground-water modeling activities required to evaluate the paleo-, present, and future hydrology at Yucca Mountain, Nevada, the proposed site of the nation's repository for high-level nuclear waste. The complexity of the regional ground-water system requires the use of a 3-dimensional ground-water modeling approach integrated with a true 3-dimensional geologic model. Integrated GSIS offers important capabilities for: 1) necessary data management and data audit trails; 2) the integration of diverse data sources; 3) rapid development, visualization, and testing of alternative model conceptualizations; and 4) integration with the numerical modeling steps.

The process of hydrogeological analysis may be considered in terms of four fundamental modules: 1) subsurface characterization; 2) three-dimensional GSIS; 3) statistical evaluation and sensitivity analysis; and 4) ground-water flow and transport modeling. The subsurface characterization module includes a variety of analytical techniques, including geological process simulation models that may combine both stochastic and deterministic elements, and Inverse Plume analysis. The statistical evaluation and sensitivity analysis module includes standard statistical screening methods, as well as more sophisticated techniques such as Kriging.

The ground-water flow and transport modeling module includes a variety of 3-D finite-element and finite-difference models. The 3-D GSIS module combines the use of the other three modules for data management and 3-D spatial display and visualization. A combination of these four modules will be used to develop regional ground-water flow models in southern Nevada and Death Valley, California.

KEYWORDS

Geoscientific Information Systems; GIS; Geohydrology; Modeling; Ground Water; Three-Dimensional; 3-D.

ACKNOWLEDGMENTS

The authors wish to thank their colleagues at the United States Geological Survey, especially Mr. Joe S. Downey, for their willingness to debate the concepts contained within this paper. However, the ideas and conclusions expressed in this paper are solely those of the authors, and do not necessarily represent the official positions of the U.S. Geological Survey.

INTRODUCTION

The correlation and synthesis of a variety of three-dimensional geoscientific data is a fundamental requirement in achieving the site characterization goal of the Yucca Mountain Project. This work must be accomplished within quality assurance and quality control guidelines that specify stringent documentation of all procedures and analysis steps.

Definition of Geoscientific Information Systems

Geographic Information Systems (GIS) are often used to map essentially two-dimensional land surface phenomena such as land-use, forestry, or soils. Most commercially available GIS products cannot accept true three-dimensional data, but can process topographic data, usually as a digital elevation model (DEM), and display isometric views or contour maps (Figure 1-a). Most DEM's use either gridded elevation matrices, or triangular meshes to represent the terrain. In these cases, the elevation, or z-coordinate, is a dependent variable.

An extension of the traditional two-dimensional GIS methods is required for geological applications. Geological problems require representation of the subsurface depth dimension in addition to the areal extent of geologic features, and linkages to various geological data manipulation procedures. As a result, the term "Geoscientific Information System", or GSIS, is used to differentiate these geologically oriented systems from the more common, two-dimensional, GIS products.

Some geological applications can be accomplished by reducing the three-dimensional representation to a quasi two-dimensional one through the use of surfaces (Figure 1-b). These surfaces, which can represent bedding planes, for example, can then be contoured or displayed as isometric views. However, in these cases, the elevation of the surface is not an independent variable, and so these systems are best defined as quasi- three-dimensional, or 2.5-dimensional systems. These 2.5-dimensional systems can only accept a single elevation (z) value for any surface at any given location. Accordingly, several important geologic structures, such as folded or faulted conditions, which cause repetition of a single horizon at a given location, cannot be represented by these systems. Nevertheless, many regional geological studies can be accomplished quite well in a 2.5-dimensional mode.

In contrast, true three-dimensional systems, containing three independent coordinate axes, can accept repeated occurrences of the same surface at any given location (Figure 1-c). The demands for detailed three-dimensional subsurface data, represented by a true three-dimensional system, are critical whenever the depth dimension is in the same general range as the surface dimensions, the true spatial relationships are important to the problem analysis, or increasingly quantitative and accurate rock-property characterizations are required within the three-dimensional subsurface environment. Ground-water flow and contamination modeling, geotechnical site characterization for increasingly complex construction projects, and petroleum reservoir characterization are examples of geological applications which typically can benefit from three-dimensional GSIS.

Until recently, affordable, fully-functional, three-dimensional GSIS products were not commonly available due to the lack of marketplace demand and costs. However, due largely to the rising interest in three-dimensional graphics systems for many uses, and the development of affordable new high-speed computer workstations that can support the rapid generation of three-dimensional graphical displays, new three-dimensional GSIS are being developed. The objective of this paper is to show conceptually how three-dimensional GSIS capabilities can be used, in combination with other modeling and analysis tools, to materially assist hydrogeological studies.

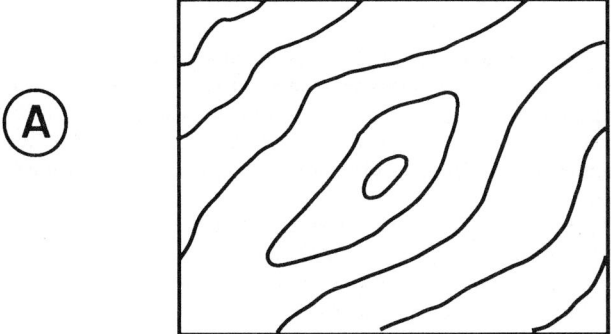

2-Dimensional Traditional GIS: a Contour Map

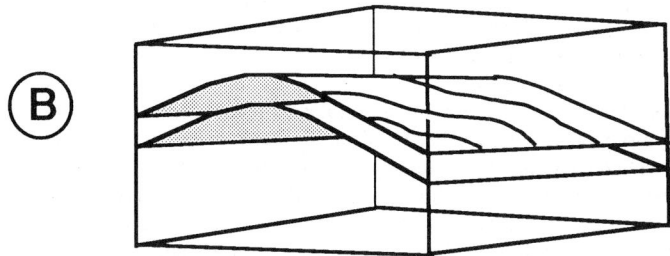

2.5-Dimensional GIS: Stacked Surfaces

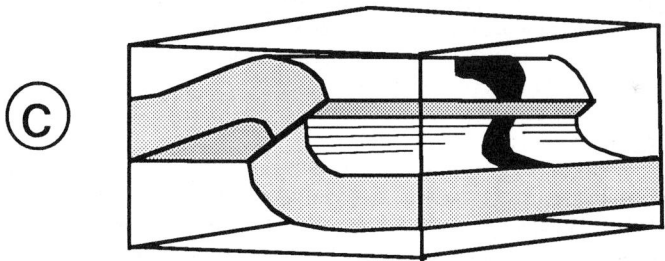

True 3-Dimensional GSIS Representation

Figure 1.   Comparison between 2-dimensional, 2.5-dimensional, and
3-dimensional spatial representations.

Importance of Spatial Visualization

Current hydrogeological applications require increasingly quantitative and accurate rock-property characterizations within the three-dimensional subsurface environment. These applications must address four major difficulties, which distinguish them from applications in most other fields (Turner, 1989):

1) normally incomplete and, sometimes, conflicting information is available concerning the dimensions, geometries, and variabilities of the rock units, at all scales of interest, from the microscopic to the megascopic;
2) the natural subsurface environment is characterized by complex spatial relationships;
3) economics prevent the sufficiently dense sampling required to resolve many uncertainities; and
4) the relations between the rock-property values and the volume of representative rock being averaged (the scale effect) are usually unknown.

The ability to rapidly create and manipulate three-dimensional images can materially assist the geoscientist's understanding of the subsurface environment. For example, typical calculations of three-dimensional ground-water flow using accepted, publically available models can be completed in only a few hours on a powerful "386-class" personal computer. However the interpretation and visualization of the results from each model analysis, achieved by contouring a series of two-dimensional surfaces and slices using available computer programs, may require more than a week to complete.

The latest generation of three-dimensional ground-water simulation models are capable of efficiently and accurately calculating the hydrodynamic flow characteristics of the fluids being evaluated, provided suitably accurate three-dimensional characteristics of the geological materials can be supplied. The geological characterization of the modelled volume in three-dimensions is often difficult to visualize and check. The model results are sensitive to the selection of input parameters, and traditional model calibration methods may fail to identify problems. In fact these models have outstripped our ability to supply the necessary data using traditional methods. A "parameter crisis" faces those who wish to use such models (Turner, 1989).

The use of true three-dimensional GSIS products may help solve both the spatial visualization and data management problems facing the users of these sophisticated ground-water models. This will require data and communications linkages between the three-dimensional GSIS programs and the ground-water models. However, three-dimensional GSIS techniques must also interact with a variety of other analytical procedures in order to solve the subsurface characterization problem, and hence offer a solution to the "parameter crisis".

Associated Analytical Procedures

The heterogeneities of an aquifer must be known in order to simulate or predict the transport of contaminants in a ground-water system, or the depletion of hydrological resources within an aquifer. Researchers have focussed on the inherent uncertainities associated with definitions of the subsurface, including the measurable properties and features at all scales of interest. In order to exactly determine these properties, every part of the region of interest would have to be tested. Stochastic modeling approaches have been used to solve the problem of subsurface uncertainties. A stochastic phenomena or process is one that when observed under specified conditions does not always produce the same result. Therefore, there is no deterministic regularity, but different outcomes may occur with statistical regularity. Stochastic methods have been used extensively for reservoir characterization methods in the petroleum industry (Augedal, and others, 1986; Haldorsen, and others, 1987).

Many deterministic and/or stochastic geologic-process-simulation computer models have been developed for a number of geological environments. These models combine deterministic components, often using empirical formulae, with stochastic components in order to introduce a suitable level of complexity, or uncertainity, into the results. Measures of statistical or geometrical properties have demonstrated that these models replicate actual systems. Use of these models can be considered a type of "Expert System" because the expertise of many geologists is incorporated in the model formulation, and the thought processes of experienced geologists are emulated to develop a conceptualization of subsurface conditions from limited data.

Domenico and Robbins (1985) and Domenico (1987) described methods, termed inverse plume analysis, which determined aquifer and contaminant source chartaceristics from the spatial distribution of contaminant concentration within a contaminant plume. Inverse plume analysis techniques determine the three orthogonal dispersivities, the center of mass of the the contaminant plume, and the contaminant source strength and dimensions from contaminant concentration data and assumed aquifer characteristics. The original technique was restricted to isotropic and homogeneous aquifers, but Belcher (1988) has extended the method to heterogeneous aquifers.

BUILDING THE GEOLOGICAL CONCEPTUAL MODEL

Development of geological conceptual models is the major process in subsurface characterization. The procedure of using multiple working hypotheses is a fundamental tenet of geology. Given the sparseness of geological data, geoscientists must develop one or more "most probable" or "equally likely" scenarios, or conceptual models, in order to expand the infrequent known observations to create a model of the entire subsurface.

Figure 2 illustrates the basic data flow when the conceptual model building process is supported by computerized methods. A variety of data types must be combined or synthesized. This requires a centralized data base capable of handling a great variety of data. The geoscientist using such a system desires to interact with the data base in ways that retain the spatial relationships, in order to visualize the subsurface in three-dimensions. The logical way to achieve this is to use three-dimensional GSIS as the interface.

Data Management Considerations

The combined use of several sophisticated numerical procedures to support the evolution of a geological conceptual model demands careful data management and quality assurance/quality controls. The data management procedures can be conceived of as having four stages, shown in Figure 3. The first stage, data capture, consists of both data gathering, the actual collection of new raw information, and data extraction, the selection of appropriate data items from existing data collections. The study purposes, goals, and objectives represent policy requirements and define the scope and type of data capture activities. Technological considerations may also constrain these activities.

Data-capture activities produce raw, original, data. Quality assurance and control procedures commonly require that: (1) such data be safeguarded, secure from any form of data modification, and (2) data must always retain "pedigree" information documenting their origins.

The second stage in data management involves data-edit preprocessing. The raw, original data usually cannot be used directly in building a coherent data base. Data collected at different times, by different people, using different methods, will not be consistent. Historical expert "knowledge bases" and data-base stucture design (the "data model") control this stage.

The data must first be validated. Validation includes both the identification of errors, defined as incorrect values due to instrument or equipment failures (such as "dropped bits" during electronic data transmission), and blunders, defined as incorrect values due to human mistakes, such as mis-labelling, mis-location, or mis-identification of samples.

Furthermore, descriptive data may not be consistent. Definitions of rock units may change over time, or different geoscientists may use different terms to describe the same thing. Data "parsing" involves the review and conversion of descriptive data to consistent, standard terminology and formats.

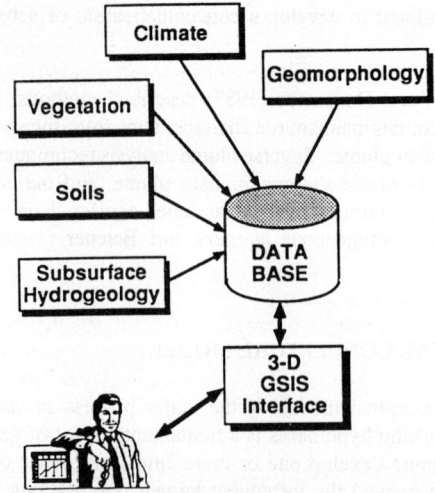

Figure 2. Data flow during conceptual model development.

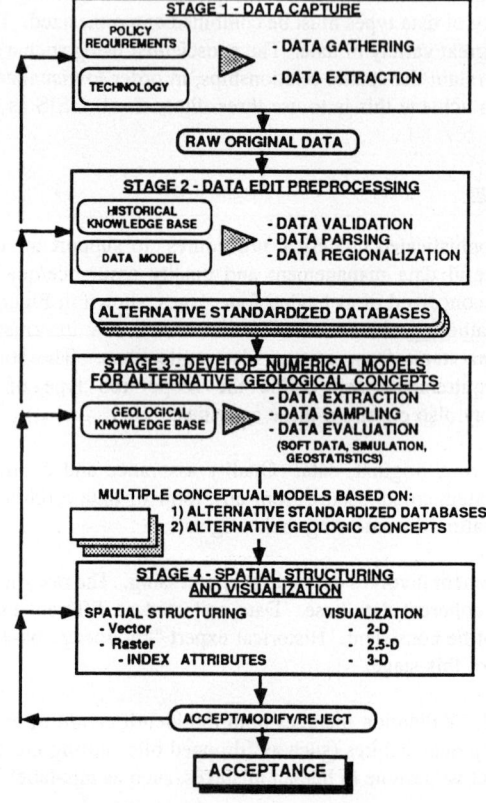

Figure 3. Data management considerations.

Finally data must be "regionalized". It must be adjusted to represent appropriate levels of detail in order to accomplish the purposes of the study. Data should be neither too detailed, nor too generalized. Excessively detailed data can be generalized by appropriate sampling, averaging, or other statistical methods. Data that are too generalized cannot be made more specific without new information. Such data should be identified, discarded and replaced by better data, if possible, or at least used with caution.

These data-edit preprocessing procedures may produce multiple, alternative, standardized data bases from a single set of raw, original data. Each data base is distinguished by the methods and assumptions used in its formulation.

The third stage involves the development of numerical models according to alternative geological interpretations. Different geoscientists may suggest different interpretations of subsurface conditions from the same standardized data base. By applying accepted geologic concepts and knowledge, and by using data extraction, sampling and evaluation methods, multiple conceptual geological models may be developed from each standardized data base.

The fourth and final stage involves the testing of these conceptual geological models. The major contribution of the GSIS technology occurs at this point, as the spatial structuring and display of the data allow the geoscientists using the system to evaluate, and then to accept, modify, or reject their conceptual models.

The Role for Three-Dimensional GSIS

Three-dimensional GSIS alone cannot solve the hydrogeological analysis problems. The process of hydrogeological analysis can be considered in terms of four fundamental modules:
1)   subsurface characterization;
2)   three-dimensional GSIS;
3)   statistical evaluation and sensitivity analysis; and
4)   ground-water flow- and contaminant-transport modeling.
The three-dimensional GSIS must, therefore, be used in combination with the remaining modules that use many developed analytical tools. Figure 4 summarizes these concepts by showing the dominant information flows and cycles among these modules.

The process starts when the geologist investigator combines geological experience with limited field data to begin the subsurface characterization process (Figure 4). The subsurface characterization module contains a variety of analytical techniques, including geological process simulation models that may combine both stochastic and deterministic elements.

The information generated by the subsurface characterization module is linked directly to the three-dimensional GSIS module. This feed-back loop is an important consideration in defining appropriate interfaces between the GSIS and the analytical tools within the subsurface characterization module. The interfaces must be designed to provide both data management and spatial visualization support. A number of iterations are expected before the most probable subsurface conditions are defined. In some cases, a unique solution may not be achievable, and two, or more, alternative characterizations may be used.

Once a suitable subsurface characterization has been defined, the analysis continues with a second cycle, where the three-dimensional GSIS is used to support appropriate ground-water or contaminant-transport models (Figure 4). This involves the creation of finite-difference or finite-element meshes by sampling from the data base. The definition of an optimal mesh has recently been studied by one of the authors and some colleagues (Stam, and others, 1989).

Figure 4 shows a strong linkage between the three-dimensional GSIS module and a module labelled "Statistical Evaluation and Sensitivity Analysis". This module contains methods for assessing the "usefulness" and "reasonableness" of the subsurface characterizations. The investigator can continue the analysis of the hydrogeological conditions only when the subsurface conditions are clearly defined and shown to be statistically acceptable.

The spatial visualization capabilities of the three-dimensional GSIS are one way of making such a "reasonableness" assessment. In addition, other more numerical approaches, including standard statistical screening methods and "geostatistical" techniques, are usually required. The term "geostatistics" has been used to describe several spatial interpolation methods using spatial autocovariance functions, the so-called "regionalized variables" (Olea, 1975; Clark, 1979; Lam, 1983). These techniques, often referred to as "Kriging", have been widely applied within the geosciences. They assume the data are time-invariant. Another method, Kalman filtering, allows both time and spatial variation in the data (van Geer, 1987). These methods have been used for optimizing sampling networks, but appear to have special utility in analyzing seasonally varying contaminant data. The use of such techniques, in conjunction with the analytical methods contained within the other modules, allow for a second level of information cycling and feed-back (Figure 4). An important question that is often posed in ground-water contamination modeling studies concerns the sensitivity of the answers to variations or uncertainties in the input parameters. This "sensitivity analysis" requires the combined use of all the modules shown in Figure 4.

REGIONAL GROUND-WATER FLOW MODELING AT YUCCA MOUNTAIN

Over the past decade the United States Geological Survey (USGS) has been studying the regional geohydrologic system around Yucca Mountain in support of the Department of Energy (DOE) efforts to design a high-level nuclear waste repository. As part of this effort, the authors are evaluating the effects of climate and tectonic changes on the regional hydrological system.

Descriptions of the current and paleo-geohydrologic framework and hydraulics, and anticipated future changes, will provide input to calibrate and validate three-dimensional numerical models of the paleo-, modern, and future ground-water flow systems. Given the complex nature of the regional ground-water system, including the aquifer geometry, hydrologic inputs and outputs, and sparseness of data needed for parameter estimation, the Geoscientific Information System (GSIS) presented previously will be used to facilitate the three-dimensional modelling efforts using data management and spatial visualization techniques.

Conceptual Model of the Ground-Water Flow System

The regional ground-water flow system underlying Yucca Mountain may be conceptualized as two types of connected aquifers:

1)   an intermontane, shallow ground-water flow system, and
2)   a regional, deep ground-water flow system.

The intermontane, shallow ground-water basin, composed of unconsolidated to poorly consolidated sediments of Tertiary to Holocene age and layered Tertiary volcanic rocks, is unconfined to partially confined. The dominant flow direction in this aquifer is from north, northeast, and east to the south (fig. 5). Ground-water flow in this system may be either matrix or fracture controlled (see potentiometric surface anomalies along 40-Mile Wash and the lower Amargosa River on Figure 5).

The regional, deep ground-water flow system, composed of faulted and fractured carbonate rocks of Paleozoic age and volcanic rocks of Tertiary age, is a multiple-layered group of confined aquifers. The dominant flow direction in these aquifers, as indicated by the arrows in Figure 6, is from the north, northeast, and east to the south and southwest. Ground-water flow in this system probably is fault and fracture controlled.

In general, the deeper, regional aquifer may have greater heads than the shallow, intermontane aquifer in the area south and southwest of the Yucca Mountain region (Sinton, 1987). As a result, ground-water leaks vertically from the deeper systems to the shallower aquifers, and ground-water mixing is observed frequently at regional discharge areas (Figure 7). In addition, flow within the shallow aquifer may be restricted to the

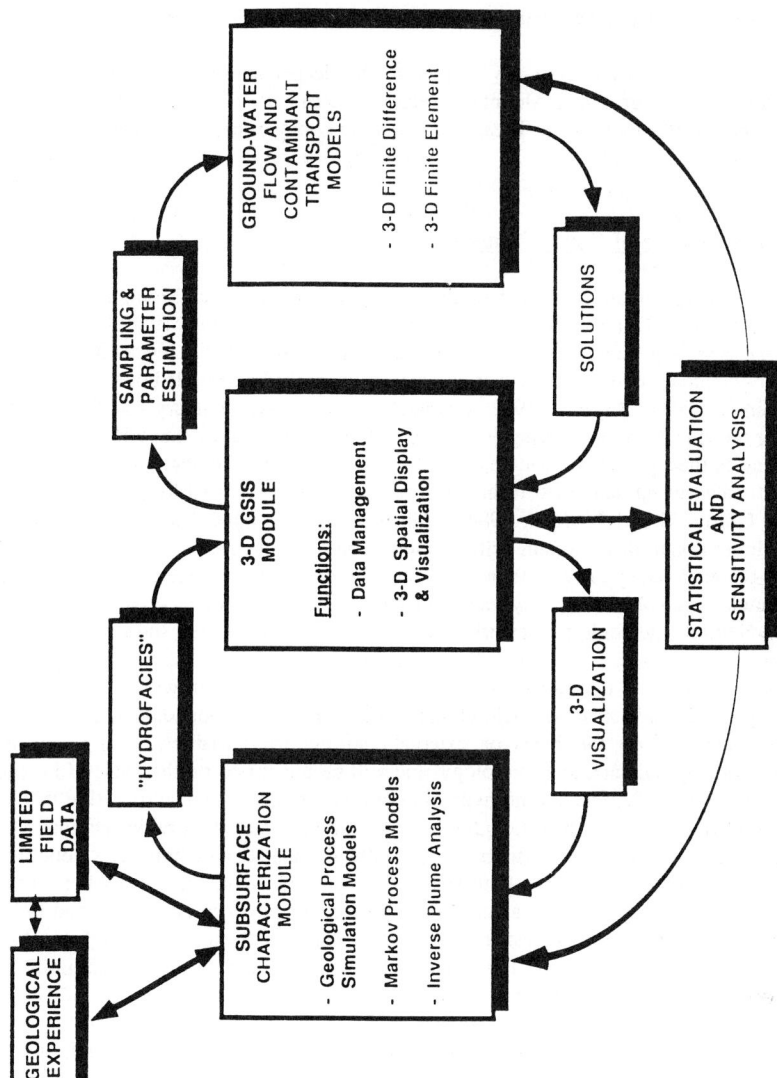

Figure 4.   Information flow and the role of three-dimensional GSIS for hydrogeology.

intermontane basin, whereas the deeper aquifer may show transbasinal flow (Figure 7).

By comparison, downward leakage from the shallow to the deeper aquifers has been observed in areas northeast, east, and southeast of Yucca Mountain, such as Yucca Flat, Frenchman Flat, and the Pahrump Valley (Winograd and Thordarson, 1975). These areas, therefore, may be conceptualized as regional recharge areas.

Three-Dimensional Simulation of Ground-Water Flow

Data defining the hydrologic boundaries of the regional aquifers in the Yucca Mountain region are lacking (Winograd and Thordarson,1975; Harrill and others, 1988; Plume and Carlton, 1988). As a result, most of the existing numerical models simulating the ground-water flow system in the Yucca Mountain region use estimated values for boundary conditions to approximate the unknown hydrologic boundaries (Czarnecki and Waddell, 1984; Sinton, 1987; Waddell, 1982; Czarnecki, 1987). All boundary conditions used in prior models, except for the Ash Meadows spring line and playas of Death Valley, have been selected based on general geology and limited potentiometric-head data. Many of the hydrological parameters used in model simulations are estimated using numerical techniques, such as inverse analysis for estimating transmissivity (Czarnecki, 1987; Czarnecki and Waddell, 1984). All of the models have estimated boundary conditions based on a two-dimensional flow system with homogeneous and isotropic, matrix ground-water flow. However, this two-dimensional approach does not accurately represent the present-day regional system, which is important to simulate paleohydrologic systems and for studies of the future ground-water flow system, where boundary conditions, geologic framework, and climate may change with time.

A three-dimensional approach to boundary selection and parameter estimation will be developed for use in this region of complex geologic structure, arid climate, and sparse data. Boundary conditions and aquifer framework will be selected on the basis of external and internal variables, such as geology, soils, geomorphology, geobotany, climate, and hydrology, and will be quantified on the basis of known relations to ground-water flow (Figure 8). The aquifer hydraulics will be simulated by incorporating this information with other internal variables, such as heads, hydraulic conductivities, and storativities (Figure 8). By using this approach, the ground-water fluxes can be numerically calculated and boundary conditions can be estimated using three-dimensional models, both steady-state and transient, for areas where information is unavailable or difficult to collect. These calculations can be compared with laboratory and field data, and the three-dimensional ground-water flow system can be visualized and adjusted accordingly until steady-state calibration is achieved (Figure 8). Each additional data set collected after steady-state calibration is achieved may be used for model validation (Figure 8).

This approach is most applicable to simulating the paleo and present-day ground-water flow systems. In these situations, evidence of past and present-day geological, pedological, geomorphological, botanical, and hydrological conditions can be obtained, and thus numerically modeled. The models may then be applied in studies of future ground-water flow systems that may result from changes in climate, geologic framework, or water use.

CONCLUSIONS

Three-dimensional GSIS technologies are important tools for future quantitative assessments required for hydrogeological applications. Three-dimensional GSIS cannot be used solve these problems without closely interfacing with a variety of existing analytical techniques for subsurface characterization, ground-water modeling, and statistical assessment.

Future three-dimensional GSIS products equally support both spatial visualization and data management. Until now, the visualization of three-dimensional subsurface geologic features has been a major constraint. The ability to rapidly create and manipulate three-dimensional images can materially accelerate the

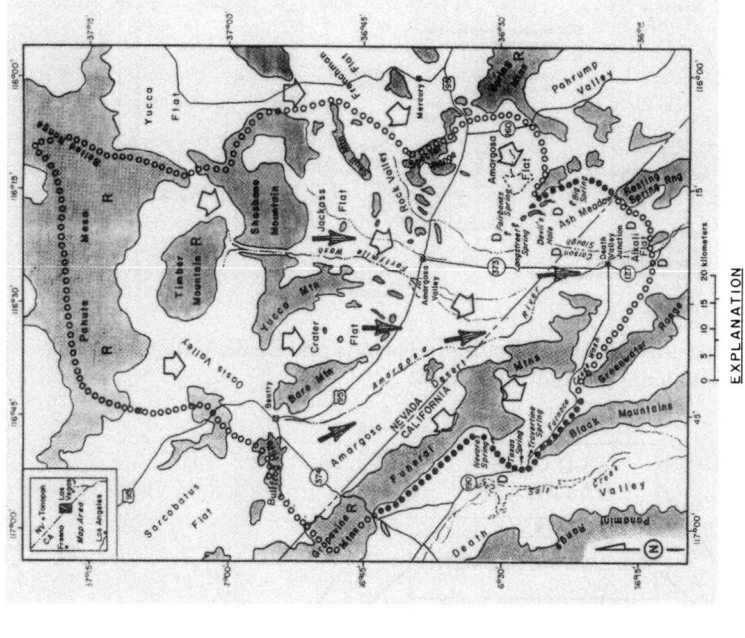

EXPLANATION

⇨ Generalized deep, regional ground-water flow system.

⇧ Generalized shallow, intermontane ground-water flow system.

D Discharge zone (estimated flux).

R Recharge zone (estimated flux).

BOUNDARY CONDITIONS

○ Constant head or unknown flux.

● Constant head or specified flux.

Figure 6. Generalized hydrogeology and flow paths of the deep, regional ground-water flow system.

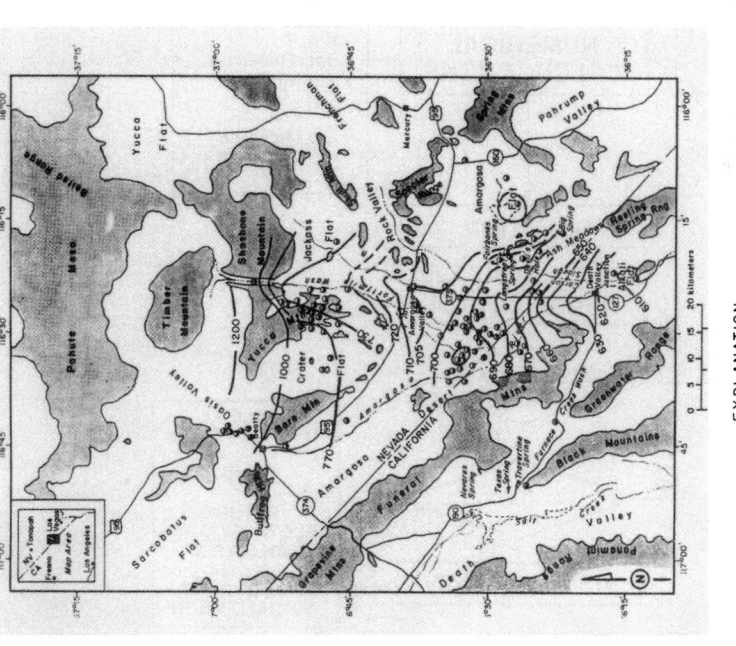

EXPLANATION

—700— Manually contoured potentiometric head, shallow intermontane aquifer. Contour interval, in meters, is variable. Datum is sea level.

Figure 5. Potentiometric surface of the intermontane, shallow ground-water flow system.

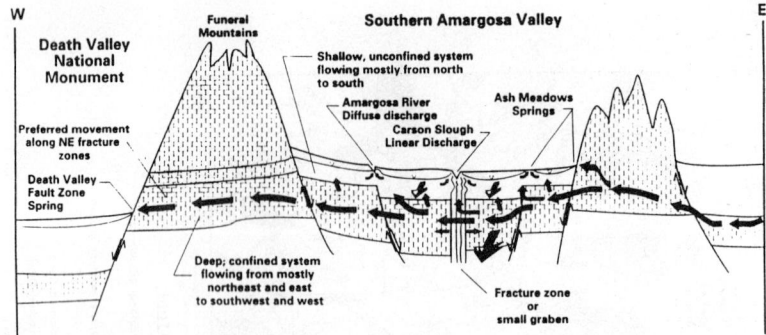

Figure 7.   Generalized cross-section of Southern Nevada and Death Valley, showing three-dimensional ground-water flow.

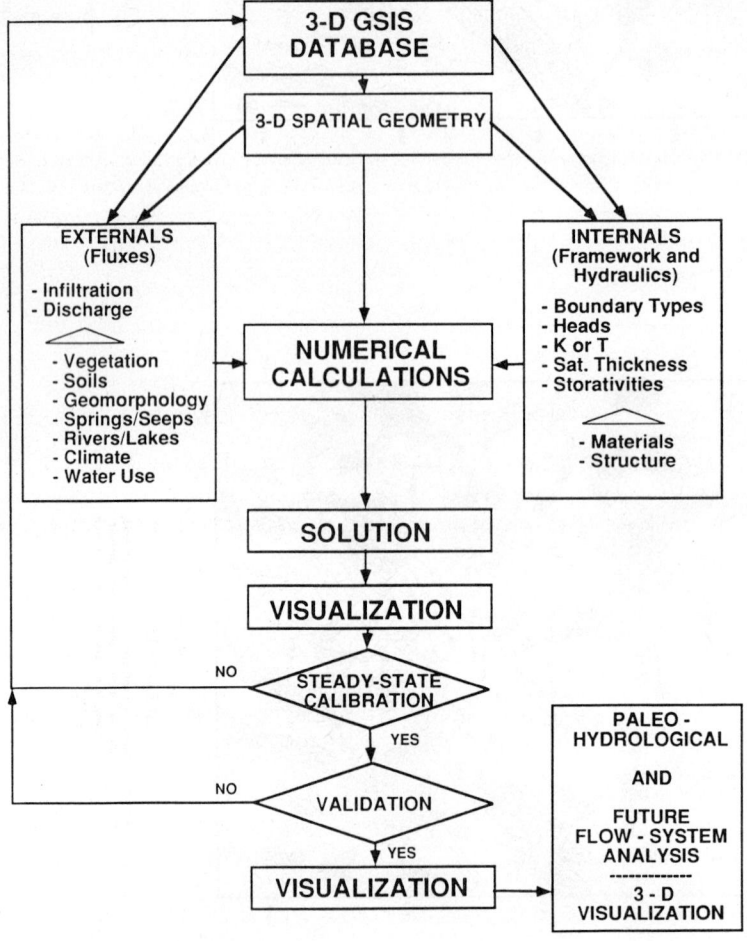

Figure 8.   Data flow during ground-water numerical model calibration.

geoscientist's understanding of the subsurface environment. Many types of geoscientific modeling require the extraction of information from large multi-parameter data sets, and the representation and modification of complex and uncertain geological objects of interest.

GSIS techniques offer important capabilities to the hydrologic modeling activities required by the Yucca Mountain Project:

1) data management and data audit trails suitable for quality assurance and contol documentation requirements;
2) the integration of diverse data sources;
3) rapid development, visualization, and testing of alternative model conceptualizations; and
4) integration with subsequent numerical modeling steps.

A three-dimensional spatial geometry will be conceptualized and tested based on external and internal variable input and resulting aquifer framework and hydraulics. Numerical calculations will be performed and the resultant solutions visualized using GSIS until a suitable steady-state calibration is achieved.

The conceptual, three-dimensional ground-water model of the Yucca Mountain area includes two aquifer systems:

1) a shallow, intermontane, mostly unconfined aquifer composed of unconsolidated or poorly consolidated sediments of Tertiary to Holocene age and layered, Tertiary volcanic rocks, and
2) a deep, regional, multiple-layered, confined aquifer system composed of faulted and fractured carbonate rocks of Paleozoic age and volcanic rocks of Tertiary age.

The hydraulic heads in the aquifer systems indicate that ground-water generally leaks vertically from the deeper to the shallower geologic units in the Amargosa Desert, and from the shallower to the deeper geologic units in Frenchman Flat and Yucca Flat. Water in the shallower aquifer of the Amargosa Desert may not flow beyond the intermontane subbasin, whereas water in the deeper aquifers may indicate transbasinal flow to the playas of Death Valley. Most of the hydrologic boundaries of the regional aquifer systems in the Yucca Mountain area are geologically complex and probably cannot be determined accurately with currently available data.

Most of the existing numerical models simulating the ground-water flow system in the Yucca Mountain region are based on limited potentiometric-head data, elevation and precipitation estimates, and simplified geology. These models are two-dimensional, and are not adequate to represent past and future changes in the ground-water flow system. A three-dimensional approach to simulating the regional ground-water flow system and estimating unknown boundary conditions for the Yucca Mountain area should provide a reasonable simulation for the paleo and present-day regional hydrologic system, and will be useful for studies of future ground-water flow systems that may result from changes in climate, geologic framework, or water use.

REFERENCES

Augedal, H.O., Omre, Hans, and Stanley, K.O., 1986, SISABOSA- A Program for Stochastic Modeling and Evaluation of Reservoir Geology, in Proceedings, Reservoir Description and Simulation. Institute for Energy Technology(IFE) Norway, Oslo, Norway, September 1986.

Belcher, W.R., 1988, Assessment of Aquifer Heterogeneities at the Hanford Nuclear Reservation, Washington, Using Inverse Contaminant Plume Analysis. Colorado School of Mines Master of Engineering Report ER-3594.

Clark, Isobel, 1979, Practical Geostatistics. (London: Applied Science Publishers), 129p.

Czarnecki, J.B., 1987, Characterization of the Subregional Ground-Water Flow System of a Potential Site for a High-Level Nuclear Waste Repository: PhD Dissertation, Univ of Minnesota, Graduate School, 317 p.

Czarnecki, J.B., and Waddell, R.K., 1984, Finite-element simulation of ground-water flow in the vicinity of Yucca Mountain, Nevada-California: U.S. Geological Survey Water-Resources Investigations Report 84-4349, 38 p.

Domenico, P.R., 1987, An Analytical Model for Multidimensional Transport of a Decaying Contaminant Species. Journal of Hydrology, Vol. 91, pp49-58.

Domenico, P.R., and Robbins, G.A., 1985, A New Method of Contaminant Plume Analysis. Ground Water, Vol.23, pp.476-485.

Haldorsen, H.H., Brand, P.J., and MacDonald, C.J., 1987, Review of the Stochastic Nature of Reservoirs, presented at Seminar on the Mathematics of Oil Production, Robinson College, Cambridge University, July 1987.

Harrill, J.R., Gates, J.S., and Thomas, J.M., 1988, Major ground-water flow systems in the Great Basin region of Nevada, Utah, and adjacent States: U.S. Geological Survey Hydrological Investigations Atlas HA-694-C, 2 sheets.

Lam, N.S., 1983, Spatial Interpolation Methods: A Review. American Cartographer, 10, pp.129-149.

Olea, R.A., 1975, Optimum Mapping Techniques Using Regionalized Variable Theory. Kansas Geological Survey Series on Spatial Analysis no. 2, 137 p.

Plume, R.W., and Carlton, S.M., 1988, Hydrogeology of the Great Basin Region of Nevada, Utah, and adjacent States: U.S. Geological Survey Hydrologic Investigations Atlas 694-A, 1 sheet.

Raper J.F., 1989, The 3-dimensional Geoscientific Mapping and Modelling System: a Conceptual Design, in Three Dimensional Applications in Geographic Information Systems (J.F. Raper, editor), Taylor and Francis, London, pp.11-19.

Sinton, P.O., 1987, Three-dimensional, steady-state, finite-difference model of the ground-water flow system in the Death Valley ground-water basin, Nevada-California: Unpublished Master of Engineering Thesis, Department of Geology and Geological Engineering, Colorado School of Mines, Golden, Co., 145 p.

Stam, J.M.T., Zijl, Waulter, and Turner, A.K., 1989, Determination of Hydraulic Parameters from the Reconstruction of Alluvial Stratigraphy: in Computers and Experiments in Fluid Flow, Carlomagno, G.M., and Brebbia, C.A., editors, Proceedings of the 4th International Conference on Computational Methods and Experimental measurements, Capri, Italy, Springer-Verlag, New York, pp. 383-392.

Turner, A.K., 1989, The Role of Three-dimensional Geographic Information Systems in Subsurface Characterization for Hydrogeological Applications, in Three Dimensional Applications in Geographic Information Systems (J.F. Raper, editor), Taylor and Francis, London, pp.115-127.

Waddell, R.K., 1982, Two-dimensional, steady-state model of ground-water flow, Nevada Test Site and vicinity, Nevada-California: U.S. Geological Survey Water-Resources Investigations 82-4085, 72 p.

Winograd, I.J., and Thordarson, William., 1975, Hydrologic and hydrochemical framework, south-central Great Basin, Nevada-California, with special reference to the Nevada Test Site: U.S. Geological Survey Professional Paper 712-C, p. C1-C126.

van Geer, F.C., 1987, Applications of Kalman Filtering in the Analysis and Design of Groundwater Monitoring Networks. TNO Institute of Applied Geoscience, Delft, The Netherlands, 130p.

THREE-DIMENSIONAL MODELLING AND GEOTHERMAL PROCESS SIMULATION

Kerry L. Burns

Division of Earth & Environmental Sciences,
Los Alamos National Laboratory,
Los Alamos N.M. 87545 U.S.A.

ABSTRACT

The subsurface geological model or 3-D GIS is constructed from three kinds of objects, which are a lithotope (in boundary representation), a number of fault systems, and volumetric textures (vector fields). The chief task of the model is to yield an estimate of the conductance tensors (fluid permeability and thermal conductivity) throughout an array of voxels. This is input as material properties to a FEHM numerical physical process model.

The main task of the FEHM process model is to distinguish regions of convective from regions of conductive heat flow, and to estimate the fluid phase, pressure and flow paths. The temperature, geochemical, and seismic data provide the physical constraints on the process.

The conductance tensors in the Franciscan Complex are to be derived by the addition of two components. The isotropic component is a stochastic spatial variable due to disruption of lithologies in melange. The deviatoric component is deterministic, due to smoothness and continuity in the textural vector fields. This decomposition probably also applies to the engineering hydrogeological properties of shallow terrestrial fluvial systems. However there are differences in quantity. The isotropic component is much more variable in the Franciscan, to the point where volumetric averages are misleading, and it may be necessary to select that component from several, discrete possible states. The deviatoric component is interpolated using a textural vector field. The Franciscan field is much more complicated, and contains internal singularities.

KEYWORDS

3D GIS, geological information system, subsurface geology, geophysics, FEHM, Finite Element Heat and Mass Flow, lithotope, textural vector fields, geothermal regimes, conductance tensors.

EXPLORATION FOR GEOTHERMAL RESOURCES

Northern California Geothermal Province

Exploration for geothermal resources is costly, due to the lack of surface expression of deep-seated physical processes, and the high costs of deep exploratory drilling. One method of optimizing exploration costs is to obtain maximum information from exploration data, and one promising method is to construct a numerical model of the geothermal regimes.

A geothermal regime is the physical process of transport of heat and fluids through the Earth's crust in a define area. Construction of a numerical process model requires formulating a description of the material properties, physical boundary conditions, and phase of migrating fluids, followed by trial and error adjustment to fit those conditions.

A large geothermal anomaly occurs in northern California, marked by a cluster of hot springs (Fig. 1). At the center of the cluster is the Geysers steamfield, the world's largest geothermal power producer, with a nameplate capacity of about 1800 MW(e).

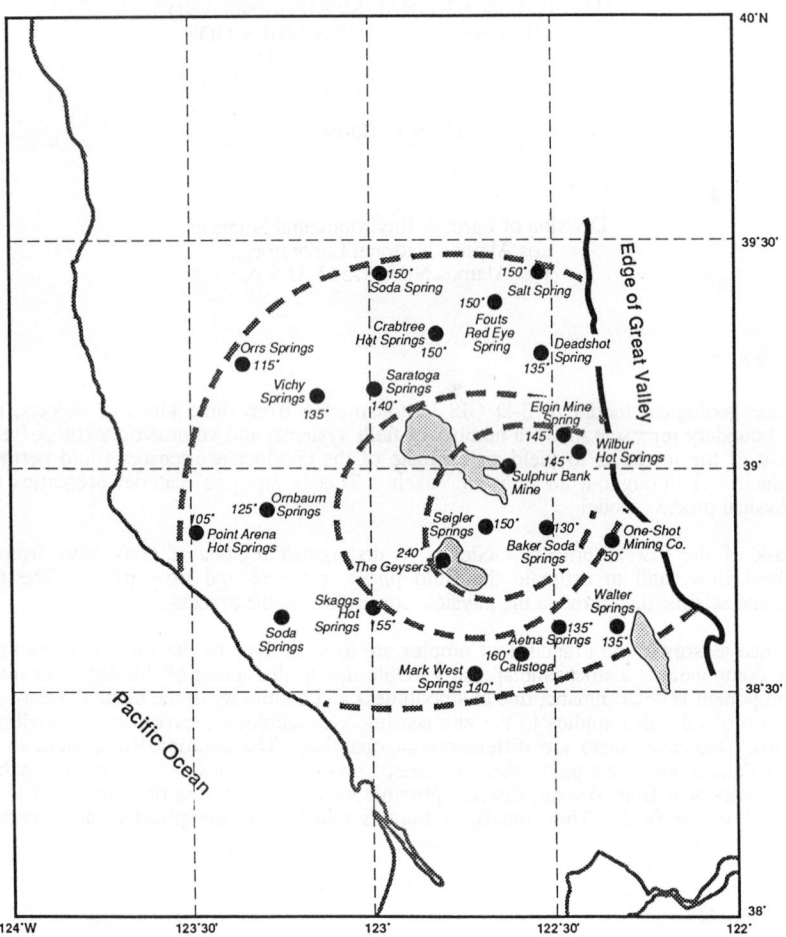

Fig. 1: The geothermal anomaly in northern California, showing the cluster of hot springs (filled dots), steamfield at the Geysers (shaded), and approximate temperature contours at a depth of 1.5km.

Inhomogeneous Geothermal Regimes

Geothermal resources are heterogeneous. Hydrothermal regimes include steamfields, hot springs and hot water aquifers, in which heat is carried mainly by convection in fluids. Petrothermal regimes, in which heat is carried mainly by conduction in rock, are the potential Hot Dry Rock resources. The framework stresses and pore pressure differ within the four regimes, so different production methods are required in each. The most important result of the numerical modelling for resource assessment is the determination, from place to place, of first, the sources of heat and fluids; second, the method of heat transport (convection or conduction); and third, the relative stresses and pressures. The FEHM (Finite Element Heat and Mass flow) model, developed by George Zyvoloski on the supercomputers at Los Alamos National Laboratory, will be used for the physical model.

Concept of an L^2T^1 Transect

A linear belt of high heat flow extends from the Geysers steamfield (bottom left of Fig. 2) to Clearlake (top right), as shown by the location of deep exploratory wells. The subsurface geological, geophysical and geochemical information is assembled, by interpolation and extrapolation, onto a cross-section along the line BB. This has two dimensions in space (L), none in time (T), so the dimensionality is L^2T^0. The FEHM model is constructed as a slice through the crust, along the section line BB, with a thickness of 1km (front slice of Fig. 3). This has dimensionality L^3T^0. Current conditions along this transect can then be modelled.

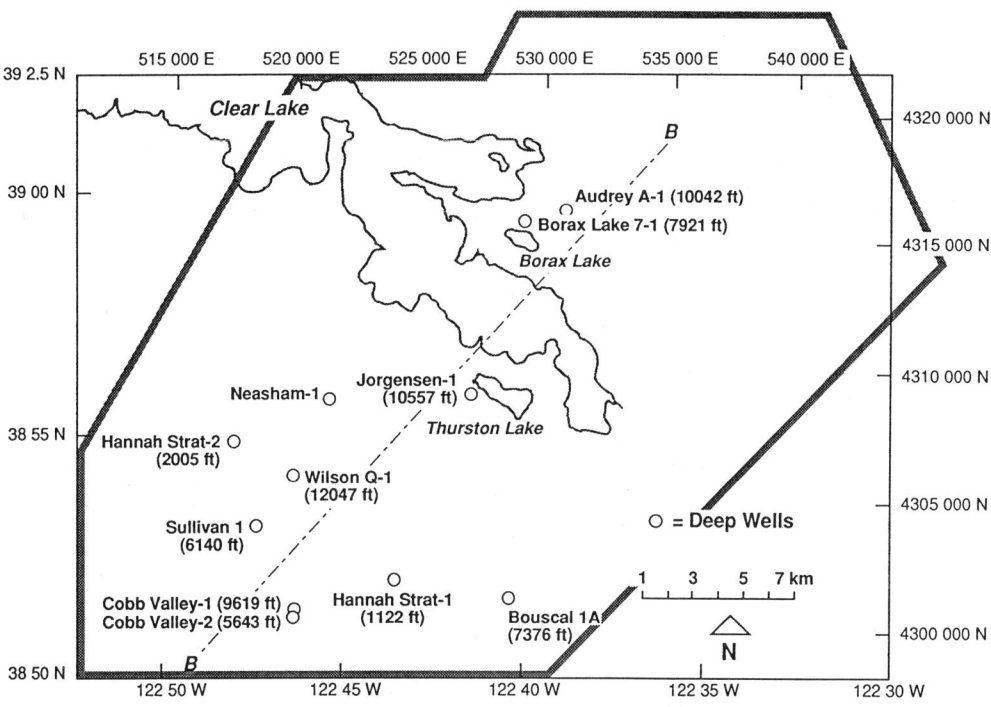

Fig. 2: Map of the central part of the geothermal anomaly, showing deep exploration wells (open dots) and the line of transect BB.

However the region is volcanically active, and the geology of the subsurface is changing, with most of the volcanic field being less than one million years old, and the youngest volcanic cones being less than 100,000 years old. The heat source is probably a magma chamber at a depth of 6km, and the heat flow from this would take about one million years to stabilize by conduction. So the present heat flow probably contains transients. If we estimate the subsurface geology along the section line BB, backwards in time for about one million years, we create a dynamic picture of the subsurface geology, of dimensionality L^2T^1 (multiple slices of Fig. 3). If the FEHM model can be adjusted to take these transient heat sources into account, we will have a succession of L^3T^0 models representing an L^3T^1 process.

In comparison, a crustal transect was developed for the Atlantic coastal margin by Unger et al. (1989). This contained only subsurface geological information, and did not address any active physical process. Another transect has been proposed, from Death Valley to Yucca Flat (Borns et al., 1990), which proposes to take into account coupled geologic, tectonic, hydrologic, and thermal processes. However the physical processes are to be determined by deduction from observations, and no synthesizing physical model is proposed. The Geysers - Clearlake transect is unique in that the field data from numerous sources will be unified by incorporation into an FEHM model capable of treating multiple, coupled physical processes.

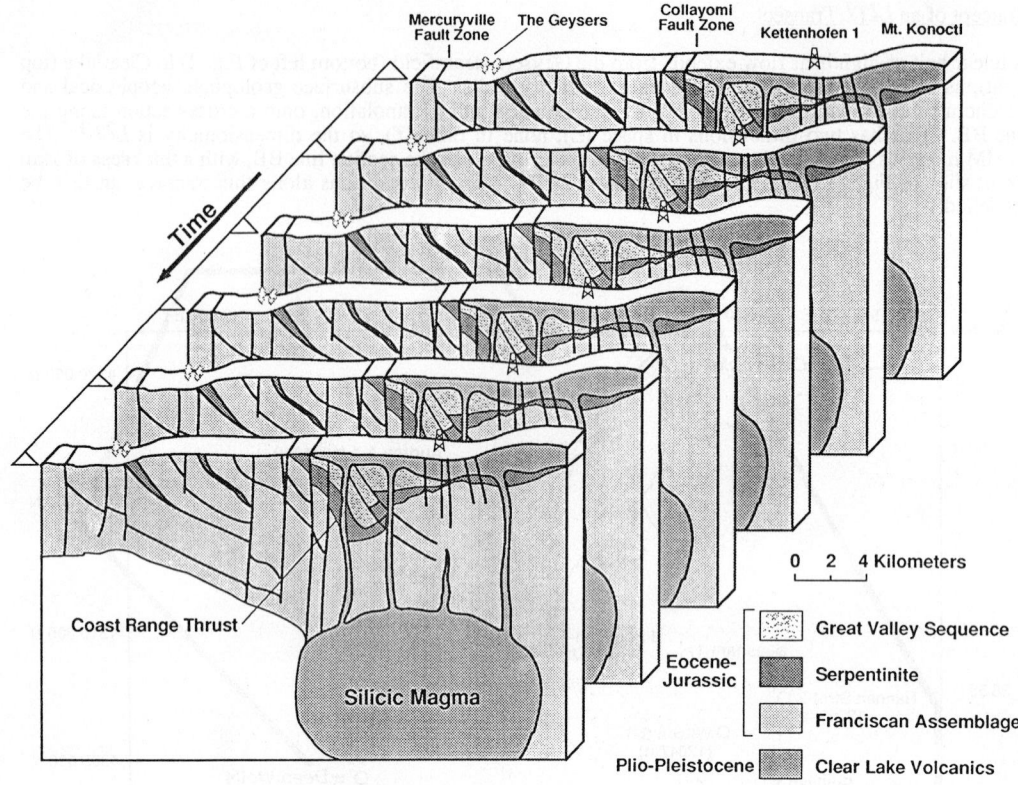

Fig. 3: Schematic diagram of the transect model. Each slice in this diagram is about 30km in length (along the section line BB of Fig. 2), about 17km in depth, and 1km in thickness. The front slice is the youngest, getting progressively older behind.

OVERVIEW OF THE EXPLORATION MODEL

Material Properties

The most important material properties are conductivity and permeability, and in order to predict their distribution in the subsurface, it is necessary to formulate a geometric model of the subsurface geologic structure. This model comprises three different geometric component systems, which are respectively, a lithotope, internal textural vector field, and several fault systems. These three components are illustrated for Franciscan melange in Figs. 4, 5 and 6. They are listed briefly below, and described more fully later on.

Lithotope. This is a boundary representation of the unique subdivision of the rock into lithogenetic polyhedra, as defined by Burns (1988). Each polyhedron is an instance of a lithogenetic rock unit, or "rock formation". Subsurface geophysical information, from gravity, magnetics, reflection seismology, earthquake tomography, and resistivity, is incorporated into the construct.

Textural Vector Fields. The Cretaceous Great Valley Sequence is a wellbedded, stratiform sequence. The lithogenetic polyhedra are tabular in shape, with quasi-parallel top and bottom formation contacts. Internal lithological contacts are simple tabular sheets interpolated between formation boundaries.

The underlying Jurassic Franciscan Complex is a melange in which the lithological contacts have been destroyed by deformation. The orientation of the lithological contacts and accompanying foliations may be represented by a vector field normal to the contacts.

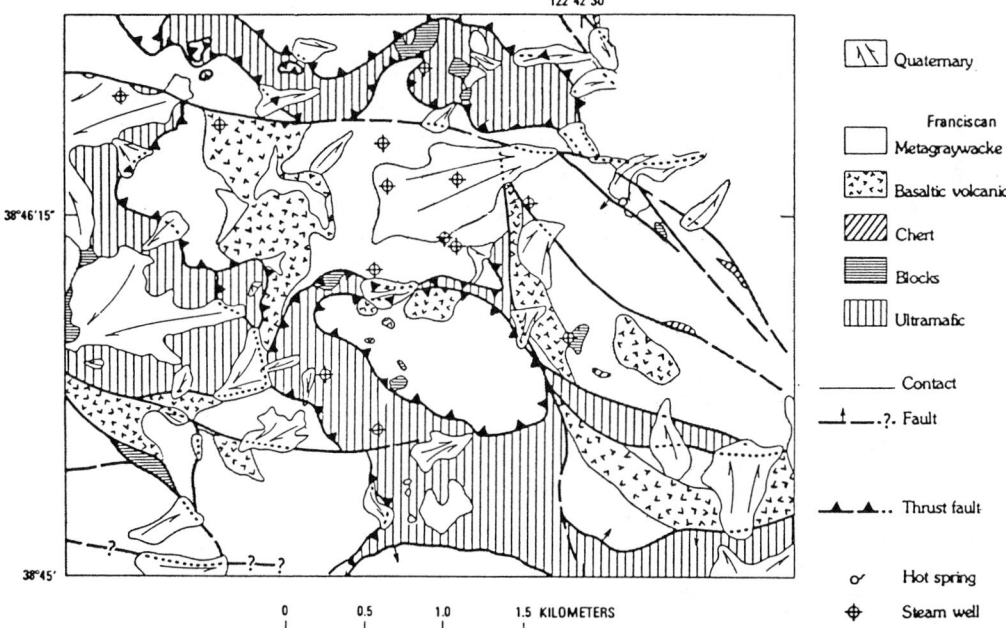

Fig. 4: Map of the Castle Rock Springs area, showing lithologic boundaries in the Franciscan assemblage. The three types of quasi-horizontal contacts are ordinary stratigraphic, melange, and thrust faults.

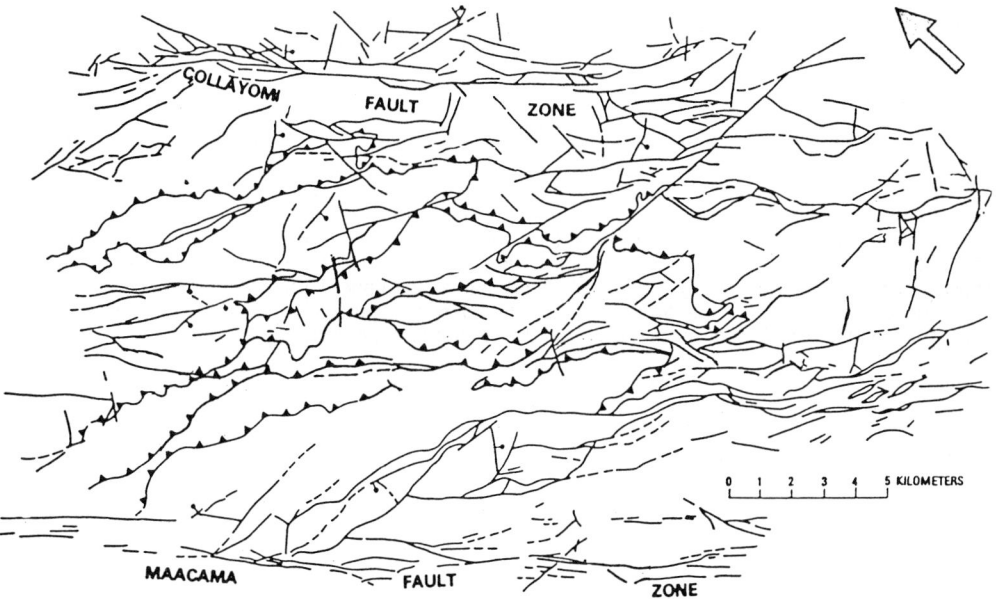

Fig. 5: Faults in the Geysers steamfield. There are two major systems, which are thrusts, and strike-slip faults of San Andrean affinity.

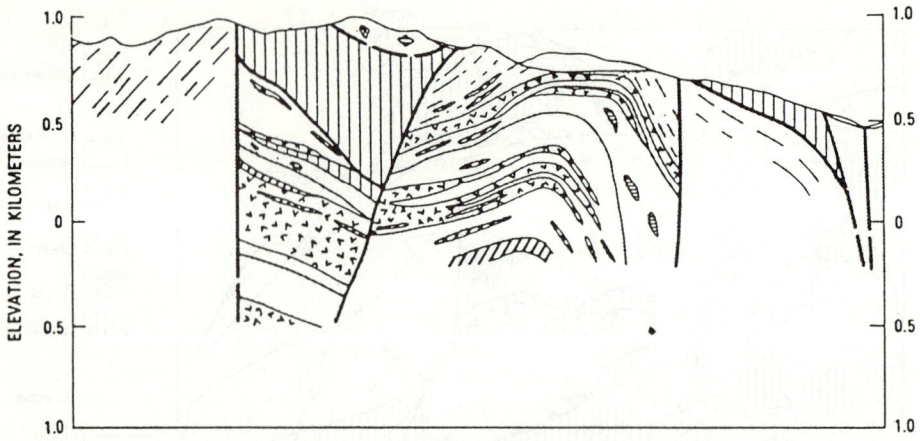

Fig. 6: SSW-NNE section of Fig. 4, showing the internal texture in
Franciscan melange indicated by lithological lenses in the greywacke.

Fault Systems. According to Wagner & Bortugno (1982), the thrust faults include disconnected fragments of the CBT (coastal belt thrust) and CRT (Coast Range thrust). There are also younger reverse faults, and yet younger strike-slip faults related to the San Andreas system. The latter are so young that they have been dated from their geomorphic expression (Bortugno, 1982).

In the post-Jurassic rocks, the mean values of conductivity and permeability are determined simply by the lithology inside each polyhedron. In the Franciscan melange, the lithology is not a deterministic, but a stochastic quantity. There is a defined probability of encountering lithologies such as metagreywacke, chert or greenstone, at any point. The mean values of material physical properties are replaced by a discrete set of probable values (Einstein & Baecher, 1981, 1983).

Material anisotropy is governed by the internal textures. The most important is that formed by small-scale lithological contacts. It is represented by a vector field normal to the contacts. The second-order properties of the conductance and permeability are axially-symmetric about the field vector.

Boundary Conditions

Temperature. Temperatures are obtained from two sets of wells. The first are deep exploration holes (Fig. 2), where the temperature was measured during drilling (mud outflows, or bottom-hole) and in subsequent logs. The logs vary from run to run, depending upon thermal effects caused by drilling, and determining the true temperatures in deep holes is a non-trivial problem in interpretation of thermal transients around wellbores. The second are patterns of shallow "gradient" holes, where the temperature is obtained from logs. The mean surface temperature of slopes with average solar aspect is about 12 deg.C. These temperature measurements specify a boundary condition on the upper surface of the FEHM model. The lower boundary condition is set at the top of the inferred magma chamber, at the inferred melting temperature; and at the edges of recent necks and dykes, at inferred cooling temperatures.

Geochemistry. Stable isotope and tritium data will be analysed to determine whether the thermal waters originated from large interconnected reservoirs or from small, isolated, fault-controlled systems of small volume. Stable isotope data will indicate recharge areas and reservoir processes, while tritium data will determine mean residence time and reservoir volume.

The thermal fluids, sampled in hot springs and wells, are usually (>90%) meteoric in this region, with some connate water, expelled from the unmetamorphosed (metamorphosing!) Cretaceous Great Valley Sequence. Magmatic water is difficult to find. Stable-isotope geochemistry can fingerprint the water source as meteoric, connate or magmatic; and indicate whether the fluid was transported as liquid or vapour. The isotopic composition of water can indicate source temperature. Tritium in water geochemistry can determine, with fairly large uncertainties, the size of any pools (reservoirs) along the flow path; and determine residence time in the reservoir.

Seismology. Natural earthquakes, where they occur along the strike-slip faults and in several clusters under the regional geothermal anomaly, appear to be confined to cold crust overlying magmatic rocks at depths shallower than 6 km. The location of the magmatic earthquakes is probably controlled by a combination of temperature (induced contraction stress) and fluid pressure, and probably indicates the proximity of sources of pressurized hot fluid.

Analysis of first motions of the earthquakes indicates stress conditions at the source. Inversion of the first motion data will provide estimates of in-situ stress (Bufe et al, 1981). This will help determine which faults are likely to be conductive under the current stress regime.

Heat-Mass Transport Processes (FEHM Model)

The synthesizing numerical physical model is described by Zyvoloski et al (1988). The flow of fluids through the crust is modelled using the equations of mass balance for the fluid phases, energy balance in the fluid-solid system, and Darcy's flux law for each fluid phase (liquid and vapour), in conjunction with the equations of state for the fluids. Solid deformation is modelled using static stress balance. The effect of fluid pressure on the rock is treated by poroelastic theory with an effective stress law. Thermal effects in the rock are modelled by first order thermoelasticity.

The equations of mass and energy balance are strongly coupled because porosity and permeability vary with displacement, rock moduli vary with pressure, and pore pressure affects displacement in the solid. The energy balance and solid deformation are strongly coupled through thermal deformation, temperature-dependent rock moduli and frictional heating. The resulting equations are nonlinear, coupled, partial differential equations.

Geothermal Regimes

The concept of geothermal regime was explained by Muffler et al (1979). A pre-computer interpretation of the geothermal regime at the Geysers is shown in Fig. 7. The material is assigned to one of three permeability classes, which are impermeable (serpentinite, greenstone, melange and meta-greywacke); fracture- dominated permeability (greywacke); and matrix- dominated permeability (young acid volcanics). The fluid phase is predominantly vapour at the Geysers, liquid elsewhere. The FEHM model aims to sharpen, refine, and extend this interpretation.

The geological and geophysical data will provide the material properties for the FEHM model. The temperature, geochemical, and seismic data will provide boundary conditions and interior constraints. The data is then transformed to SOE (spatial occupancy enumeration) form, for transfer to the FEHM program. The resolution (unit voxel) is 1 km per side.

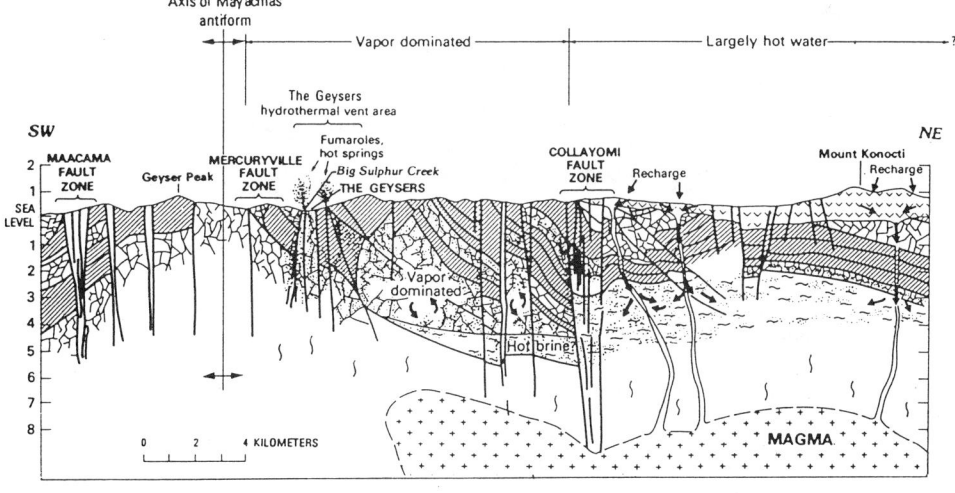

Fig. 7: Diagram of the geothermal regime at the Geysers.

The FEHM model will then be adjusted to fit the data. It is expected to delineate different regions as steamfields, hot springs, hot water aquifers and regions of hot dry rock, and yield quantitative information on each resource. The result will be guidance on which production method should be used for energy-extraction in each region, and the size and location of the resource accessible by different engineering methods.

We now examine some of the preceding topics in detail.

THREE DIMENSIONAL LITHOTOPE

Topology of the Geological Subsurface

The topology of the three-dimensional subsurface was shown by Burns (1988) to be readily generalized from the two-dimensional topology that has become familiar in digital mapping. The following description follows that treatment. A subsurface "cell" or unit occurrence of a rock formation is a lithogenetically-homogeneous volume in the subsurface. This is bounded by "faces" which are interfaces with adjoining cells. A face is a surface patch which is defined by continuity of curvature. Each face has a perimeter ring of "edges" which are space curves, and each edge is bounded by a pair of "vertices". The subsurface is then a construction based on geometrical entities of dimensionality 3, 2, 1 and 0, which are cells, faces, edges and vertices, as shown in Fig. 8(a). The symbols C3, F3, E3, and V3, will be used to denote these objects.

These objects are related to each other by adjacency relations. In a valid model, two adjacent edges are incident on a face, and two adjacent faces on an edge, and three edges on a vertex. After Coxeter (1973, p.68), use n(A, B) to denote the number of B-objects that are incident upon a single A-object. Then n(F3, C3) = 2, n(E3, F3) = 2, and n(V3, F3) = 3. These validity conditions are not the same for all three-dimensional constructs (Requicha, 1980; Weiler, 1985). These conditions are based upon geological considerations. Adjustment of a preliminary three-dimensional construct to meet these conditions is termed "reduction to normal form", defined below.

If the three-dimensional space is intersected with a surface, X2, which may be plane or curved, the result is a two-dimensional diagram, such as a cross-section, subcrop map or underground plan. Where x denotes geometric intersection, define a lithological "domain" by C2 = C3 x X2, "boundary" by F2 = F3 x X2, and "junction" by E2 = E3 x X2. The cross-section is then a two-dimensional construct of domains, boundaries, and junctions, as shown in Fig. 8(b). If X2 is the interface between the subsurface and atmosphere, the construct is an ordinary geological map, or outcrop map, otherwise, a subcrop map.

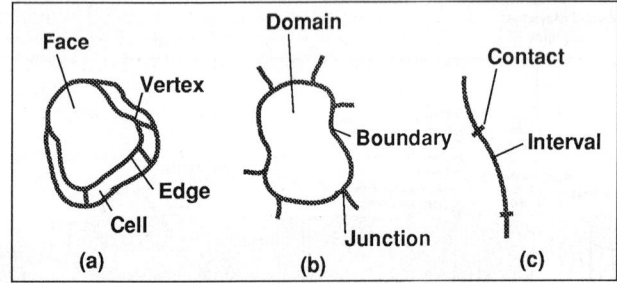

Fig. 8: Lithogenetic units in a polytope: (a) cell in three dimensions, (b) domain in two dimensions, and (c) interval in one dimension.

If the three-dimensional space is intersected with a line, Y1, which may be plane or curved, the result is a one-dimensional diagram, such as a welllog, traverse line or highway log. Define a lithological "interval" by C1 = C3 x Y1, and "contact" by F1 = F3 x Y1. The subsurface log is then a one-dimensional construct of intervals and contacts, denoted C1 and F1, as shown in Fig. 8(c).

The geometric objects in a 3D lithotope are summarized in Table 1. In a space of dimension d, the extensive lithological unit is Cd, which is a cell, domain, or interval. Two adjacent lithological units share an object Fd which is an interface. Two of these share an object Ed which is an edge or junction. And if d is three, three of these are incident on an object Vd which is a vertex.

Dimension of Space	Dimension of Object			
	3	2	1	0
3 subsurface	C3 cell	F3 face	E3 edge	V3 vertex
2 map, section		C2 domain	F2 boundary	E2 junction
1 log, traverse			C1 interval	F1 contact
0 index, key				C0 lithology, transition

Table 1: Tabulation of geometric objects in a lithotope.

Indexation

Bouille (1976) typed objects on a map by a subscript index set. A systematic indexation scheme helps to ensure a valid assembly. Bouille's scheme is readily extended to three dimensions. The index set $(i,j,...,l)$ refers to constituent rock formations. An object of any particular type may be repeated many times. An individual is distinguished by an additional index, which counts the occurrences.

For example, on a map, C2im is the mth outcrop of the ith lithology. The set $C2i = \{C2im: m=1,M2\}$ is a "rock formation" which occurs in M2 outcrops. It is convenient to put the formations in order. If formation C2i is older than formation C2j, assign the indices so that $i<j$. Then the notation F2ijp means that F2ijp = (C2im o C2jn) is the pth occurrence of an interface between adjacent outcrops of formation C2i and Crj, where the contact relationship is denoted "o".

Operations and Transformations

There are operators which, applied to a construct, alter its form. These include normalization and triangularization.

Normalization. A map is in Kempe-normal form (Appel & Haken, 1977) when no domain completely surrounds another and no more than three domains meet at a point. Intrusive plugs on geological maps sometimes have an isolated boundary. This can be linked to other boundaries on the map by introducing a "virtual" boundary (Bouille, 1976). A set of editing operations like this, which converts the field map to normal form, is termed "normalization", N. The inverse operation, N', is striking virtual boundaries.

Triangularization. This is the pair of operations that transform a Kempe-normal construct to another form, termed the dual. This is the reciprocation operation of Coxeter (1973, p.17).

Graph of the Cartographic Form

Bouille (1976) recognized three graphs in two dimensions. These were termed the "cartographic form", "spatial succession" and "process model" by Burns (1988). The graph of interest here is the "cartographic form", S, where S = <arc, node> = <F2, E2>.

The cartographic form is S = <arc, node> = <E3, V3>, <F2, E2>, <C1, F1> or <0, C0> for spaces of dimension 3, 2, 1, or 0, respectively, which run down the last two columns of Table 2. For example, the well-log form is a lineal thread S = <C1, F1> = <lithological intervals, contacts>. The degenerate form S = <0, C0> has no arcs, and is the set of nodes which are lithological transitions, where transition at a unit means the change from the formation below the unit to the formation above the unit. In two dimensions, this is the format of the Bureau of Census topologically structured "Tiger" files (Herring, 1987), while in three dimensions, it is a wireframe model.

Dimension	Dimension of Object			
of Space	3	2	1	0
3 subsurface	C3 cell	F3 face	<E3, V3> <edge, vertex>	
2 map, section		C2 domain	<F2, E2> <boundary, junction>	
1 log, traverse			<C1, F1> <interval, contact>	
0 index, key			<null, C0> <no arcs, transitions>	

Table 2: Components of the graph of cartographic form, eg. S = <arc, node> = <E3, V3>.

Construction of the Lithotope

In ordinary model construction, such as in CAD/CAM or digitization of geologic maps, the topology is completely known. The objects are known according to location and according to their topological character. So construction of a model is simply a matter of structuring the computer files. However the topology of a collection of geologic field data in 3D is not known, and has to be inferred from the data. This is a fundamental difference between construction of 2D and 3D models.

Each of the geometrical objects in the three-dimensional lithotope is constructed by inference from objects of lower dimensionality. This requires a search for objects of appropriate topology, which can be done by manual assignment if known, or by automatic search procedures. As an example of the latter, suppose the set of formations with indices {i, j, k, ...} were observed at scattered points in space. For each point, calculate a list of the nearest neighbours. For the ath point, this will be an ordered set Ka drawn from the preceding set. If the first three indices of Ka are all the same, the point is classified as in the interior of a cell. If the first two indices are different, the point is near a face; and if the first three are all different, the point is near a vertex. By searching through the data set in this manner, the topology can potentially be constructed.

The lithotope is, therefore, recovered from the data by topological inference from the lithology observed at the data points. The normal form is a consequence of a valid procedure.

FAULT SYSTEMS

Topology of Fault Systems

The proposed lithological search procedure will generate a faulted lithotope, that is, the surfaces will include faults as well as formation boundaries. The faults must be identified by field observations and distinguished as such in the construct.

Thrust Systems

The two thrust systems are widely exposed and recognized regionally. The surfaces are quasi-horizontal, offset by later faults. A method of reconstruction of discontinuous surfaces was described by Pouzet (1980), and similar techniques are available commercially.

Strike-slip Systems

The large offsets between the patches of thrust faults provides a measure of strike-slip displacements. The strike-slip systems are so young that the space curves can be found, traced by the fault surfaces on the topography (Bortugno, 1982). The appropriate description is edge-based (Weiler, 1985) or boundary loops (Bak & Mill, 1989). There are few subsurface intersections in wellbores. The surfaces are probably near-vertical.

TEXTURAL VECTOR FIELDS

Vector Field Normal to Layered Textures

The Cretaceous Great Valley Sequence is a wellbedded, stratiform sequence. Within each stratigraphic unit, a field of surfaces is formed by sedimentary laminae, bedding and minor lithologic contacts. The normal vectors to the surfaces constitute an orientation vector field, as described by Agterberg (1974, pp.488-508), and as illustrated in Fig. 9.

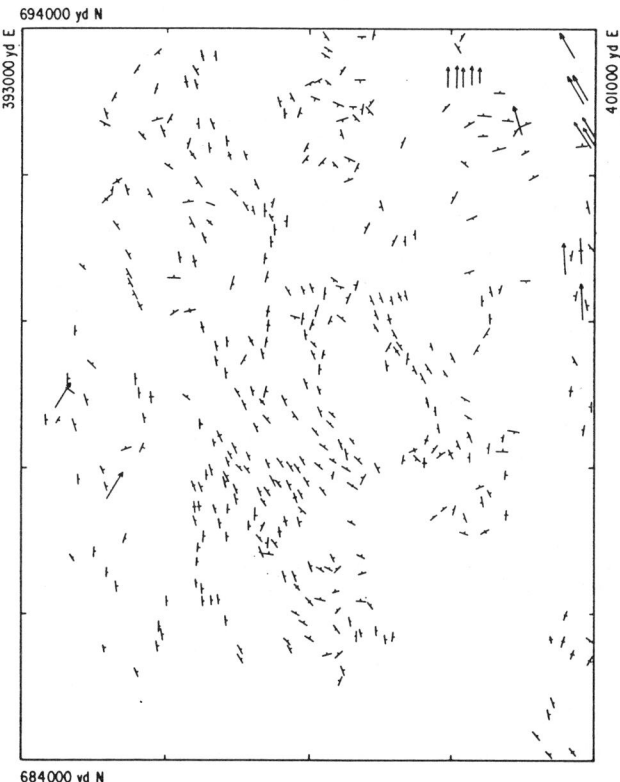

Fig. 9: Field of planar surfaces in a metamorphic terrain (Port Davey, Tasmania). The normals to these surfaces constitute a textural vector field.

The underlying Jurassic Franciscan Complex is a mixture of lithologies, such as greenstone, chert and greywacke, in the form of lenticular blocks. The complex is a melange in which the formation boundaries have been destroyed by deformation. Internal lithological contacts are of tectonic, not sedimentary or magmatic, origin. The lithologic contacts and accompanying foliations tend to form a layered texture, which may also be represented by a unit vector field normal to the surfaces.

Topology of the Vector Fields

The vector fields are reconstructed by extrapolation and interpolation. Interpolation methods were described by Agterberg (1974), Watson (1985) and Mendoza (1986). The fields are continuous within lithogenetic polyhedra. In fact, this is a criterion of "lithogenetic". The fields terminate on two opposite faces of the polyhedra. In sedimentary formations, these are the upper and lower faces. The tangent curves end in "attachment points" at the faces. The fields in adjacent polyhedra are different, that is, there is a discontinuity across faces.

The vector field may also have singularities, or critical points, inside the polyhedra, as illustrated in Fig. 10. These are axial surfaces of folds, traces of faults and mylonite zones, or zones of concentrated shear. The critical points are defined as points where tangent curves in the field start arbitrarily close to each other, but end up in substantially different regions.

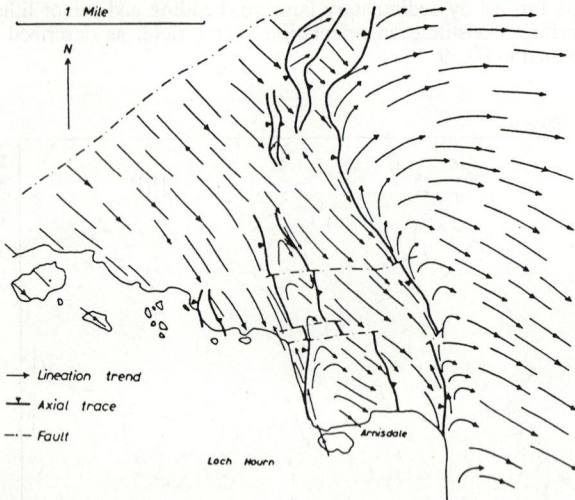

Fig. 10: Discontinuities in a textural vector field (Loch Hourn, Scotland).

Topology of Fluid Flow Fields

The representation of vector fields is at the frontier of scientific visualization. Helman and Hesselink (1989, 1990a) have pioneered the visualization of the topology of vector fields of dimensionality L^2T^1. The parameter in the flow field is the fluid velocity.

The reconstruction of the flow field starts with the classification of points on the boundary (attachment and detachment points). Critical points are then found, as those points where the magnitude of the flow vector vanishes. The critical points are classified according to the eigenvalues of the Jacobian matrix of the vector, with respect to position, at the critical point. The eigenvalues classify a critical point as an attracting node, a repelling node, an attracting focus, a repelling focus, a centre, or a saddle.

The field is reconstructed by following tangent lines in the field, from where it leaves one critical or boundary point, to where it reaches another. Each tangent line is represented by a linked list in Common Lisp. The tangent curves are integrated in Fortran routines called from Lisp. The field is reconstructed in two dimensions for an instant of time. Reconstructions for a series of successive instances led to the L^2T^1 visualization of Helman & Hesselink (1990b).

INTERFACING THE 3D-GIS TO THE FEHM MODEL

Topology of SOE Form

The subsurface geological model is transformed to SOE (spatial occupancy enumeration) form (Requicha, 1980), where each voxel is one cell of the FEHM model. Each cell is assigned a topology or character according to the predominant character in the geological model.

The topological character derived from the lithotope is whether the voxel is the interior of a polyhedron, or lies on a face, edge or vertex. The character derived from the fault systems is that the point lies on one of the fault systems or not. The character derived from the textural vector fields is whether the point lies on an attachment point, detachment surface, axial surface, or none of these.

Isotropic Conductances

The conductance parameters are tensors, and can be treated as the sum of two parts, a deviation, or isotropic part, or mean value, and a "difference tensor", or deviatoric part (Jaeger, 1962, p.90). The isotropic part of the permeability and conductivity inside polyhedra, at no special points, is specified by scalar properties of the lithology, such as sand/shale ratio and grain size. The value is estimated from calibration data.

In the Franciscan melange, the lithology inside an ordinary voxel is usually unpredictable. It can, however, be assessed on a probabilistic basis, using methods such as those of Einstein & Baecher (1981, 1983). The calibration data could be used to assign "most probable" conductances to voxels in the melange.

The permeability is, however, very heterogeneous, and is likely to be quite different at un-ordinary points. Low permeability zones are expected along faults of the strike-slip systems, possibly with an error-function distribution for several hundred metres either side of the fault surface.

This is a wrench-tectonic regime, and different generations of faults are likely to intersect in vertical lines. As a result, linear, low permeability zones are expected at fault intersections. The high areal frequency of hot springs (Fig. 1) may be attributed to this cause.

Deviatoric Conductances

To a good approximation, the deviatoric permeability tensor is axially symmetric, oblate, with the least principal axis normal to the layering, and the other two principal axes parallel to the surface. The ratio of the two extreme principal values is very high. The anisotropy is specified by the ratios of principal axes and the orientation of the normal to layering. So the textural vector field provides information on the orientation of the deviatoric conductance tensors at ordinary points in the vector field. The conductance tensors associated with singularities remain to be determined.

Resource Assessment vs Site Characterization

In engineering hydrogeology of shallow terrestrial fluvial systems, especially those associated with cleanup of toxic waste, the isotropic material properties are determined by a random spatial process, best treated by geostatistical methods (Turner, 1989). In this resource model of the Franciscan Complex, the variations are even greater. A volumetric average might not be valid, and it may be necessary to select a value from a discrete set of probable states.

The deviatoric components are a deterministic process in both cases. However the orientation is likely to be fairly constant in fluvial channels, while it is a complicated field containing singularities in the Franciscan. It is concluded that while there is fundamental similarity between this resource model at regional scale and site characterization at engineering scales, the resource model presents a higher level of complexity.

ACKNOWLEDGEMENTS

The author is grateful to Professors Herbert Einstein, Geoff Watson, Lambertus Hesselink, and James Helman, and to colleagues Drs. James Albright, Dave Duchane, Bob Potter, Fraser Goff, and George Zyvoloski, for helpful discussions. Ruth Bigio prepared some of the figures. Figures 4 to 7 are based on McLaughlin (1981, Fig. 9, p.18; Fig. 7, p.16; and Fig. 6, p.14). Figure 10 is based upon Agterberg (1974, Fig. 108, p.495), and is reproduced with permission of the author and Elsevier Scientific Publishing Company.

REFERENCES

Agterberg, F.P. (1974). *Geomathematics*. Elsevier, Amsterdam.
Appel, K., Haken, W. (1977). The Solution of the Four-Color-Map Problem. *Scientific American, 237*, 108-121.
Bak, P.R.G., Mill, A.J.B. (1989). In: *Three Dimensional Applications in Geographic Information Systems* (J. Raper, ed.), pp. 155-182. Taylor & Francis, London.
Borns, D.J., Sass, J.H., Schweickert, R.A. (1990). Proposed Study of the Basin and Range From Death Valley to Yucca Flat: *EOS Trans. AGU, 71*, 1012-1013.

Bortugno, E.J. (1982). Map Showing Recency of Faulting, Santa Rosa Quadrangle, California, 1:250,000. *Regional Geologic Map Series*, California Division of Mines and Geology, Sacramento.

Bouille, F. (1976). Graph Theory and the Digitization of Geological Maps. *J. Int. Assoc. Math. Geol.*, 8, 375-393.

Bufe, C.G., Marks, S.M., Lester, F.W., Ludwin, R.S., Stickney, M.C. (1981). In: *Research in the Geysers-Clear Lake Geothermal Area, Northern California* (R. J. McLaughlin, J. Donnelly-Nolan, eds.), Geol. Surv. Prof. Pap., *1141*, pp. 129-137, US Govt. Printing Office, Washington.

Burns, K.L. (1988). Lithologic Topology and Structural Vector Fields Applied to Subsurface Prediction in Geology. *Proc. GIS/LIS' 88*, *1*, 26-34. ASPRS, Falls Church.

Burns, K.L., Potter, R.M. (1990). Application of Hot Dry Rock Technology in the Clearlake area, California. In: *The National Energy Strategy -- The Role of Geothermal Technology Development*, Department of Energy, Washington DC, *in press*.

Coxeter, H.S.M. (1973). *Regular Polytopes*. 3rd ed., Dover Pubs., New York.

Einstein, H.H., Baecher, G.B. (1981). Probabilistic and Statistical Methods in Engineering Geology: In: *30th Geomechanics Colloqium*, Salzburg.

Einstein, H,.H., Baecher, G.B. (1983). Probabilistic and Statistical Methods in Engineering Geology, Specific Methods and Examples, Part I, Exploration. *Rock Mechanics and Rock Engineering*, *16*, 39-72.

Helman, J., Hesselink, L. (1989). Representation and Display of Vector Field Topology in Fluid Flow Data Sets. *IEEE Computer*, *22*, 27-36.

Helman, J., Hesselink, L. (1990a). Surface Representation of Two- and Three-Dimensional Fluid Flow Topology. *Proc. IEEE Conf. Scientific Visualization*, IEEE, Los Alamitos, *in prep*.

Helman, J., Hesselink, L. (1990b). In: *Visualization in Scientific Computing* (G. M. Nielson, ed.), videotape to accompany conference report, IEEE, Los Alamitos.

Herring, J.R. (1987). TIGRIS: Topologically Integrated Geographic Information System. *AutoCarto 8*, 282-291.

Jaeger, J.C., 1962, *Elasticity, Fracture and Flow*. 2nd ed., Methuen, London.

Mendoza, C.E. (1986). Smoothing Unit Vector Fields. *J. Int. Assoc. Math. Geol.*, *18*, 307-322.

McLaughlin, R.J. (1981). In: *Research in the Geysers-Clear Lake Geothermal Area, Northern California* (R. J. McLaughlin, J. Donnelly-Nolan, eds.), Geol. Surv. Prof. Pap., *1141*, pp. 129-137, US Govt. Printing Office, Washington.

Muffler, L.J.P., Costain, J.K., Foley, D., Sammel, E.A., Youngquist, W. (1979).Nature and Distribution of Geothermal Energy. In: *Direct Utilization of Geothermal Energy* (D. N. Anderson, J. W. Lund, eds.), GRC Spec. Rept., 7, pp.1-1 to 1-15, GRC, Davis.

Requicha, A.A.G. (1980). Representations for Rigid Solids: Theory, Methods and Systems. *Computing Surveys*, *12*, 437-464.

Turner, A.K. (1989). In: *Three Dimensional Applications in Geographic Information Systems* (J. Raper, ed.), Taylor & Francis, London, pp. 115-127.

Unger, J.D., Liberty, L.M., Phillips, J.D., Wright, B.E. (1989). In: *Three Dimensional Applications in Geographic Information Systems* (J. Raper, ed.), Taylor & Francis, London, pp. 137-148.

Wagner, D.L., Bortugno, E.J. (1982). Geologic Map of the Santa Rosa Quadrangle, California, 1:250,000. *Regional Geologic Map Series*, California Division of Mines and Geology, Sacramento.

Watson, G.S. (1985). In: *Multivariate Analysis* (P.R. Krishnaiah, ed.), Elsevier, Amsterdam, pp.613-625.

Weiler, K. (1985). Edge-Based Data Structures for Solid Modeling in Curved-Surface Environments. *IEEE CG&A*, January, 21-40.

Zyvoloski, G., Dash, Z., Kelkar, S. (1988). FEHM: Finite Element Heat and Mass Transfer Code. Rpt., LA-11224-MS, *Los Alamos National Laboratory*.

APPLICATION OF ROCK-CAD MODELLING SYSTEM IN CHARACTERIZATION OF CRYSTALLINE BEDROCK

PAULI SAKSA

Saanio & Riekkola Consulting Engineers Ltd.,
Laulukuja 4, 00420 Helsinki, FINLAND

ASBTRACT

The Finnish power company Teollisuuden Voima Oy studies crystalline bedrock in Finland for final disposal of high-level nuclear fuel waste. In evaluation of the varying lithological and structural conditions CAD-based ROCK-CAD system has been developed. ROCK-CAD is based on true solid modelling approach. One modelled volume consists of several mutually independent submodels. Mainly lithological, structural (fracturing) and hydraulical properties are modelled. ROCK-CAD is in operational use and experiences have been got from four sites modelled this far. The main uses of the software have been in general visualization, in planning of sopplementary investigations and in qualitative interpretation and model development done by the experts. Computerized models form also the basis for groundwater flow simulations and rock mechanical calculations. Two example drawings are presented and discussed.

KEYWORDS

Rock modelling; computer aided design; visualization; computers; nuclear waste disposal.

INTRODUCTION

The Finnish power company Teollisuuden Voima Oy (TVO) operates two nuclear power plant units at Olkiluoto, western Finland. Associated with energy production the waste producer is also responsible for the safe management of radioactive wastes. TVO studies crystalline bedrock at five sites for the final disposal of high-level nuclear fuel waste in the future. The preliminary site characterization program during 1987 - 92 consists of comprehensive geological, geophysical and geochemical investigations conducted both at the surface and in the boreholes. The studies as a whole are carried out in several phases within the years.

The size of a typical survey area is around 10 km^2 and drilled boreholes extend down to the depth of 1000 m. Precambrian age rocks are mainly granitic or dioritic in composition and cut by thin dykes and veins. The sparsely fractured intact rock is intersected by crushed and fractured zones forming a mosaic block structure. The purpose is to determine the lithological and structural geometry and character of the selected bedrock block. Its environment and the occurrence of the various rock types, geochemical circumstances, fractures and fracture zones preferable to groundwater flow are of main interest. The most important property is the movement of the groundwater through the rock.

Within the scope of the investigation project it was realized that three-dimensional (3-D) rock modelling system is needed for integrated analysis and visualization purposes. ROCK-CAD™ named new rock modelling system based on CAD-type approach was created.

Previously 3-D computer graphics and modelling has been used for example in mining geo-branch. For subsurface imaging one has principally two main courses to create a graphical illustration that represents a 3-D model. One method is to design an own database for rock structures and with a programmed application software calculate a graphical presentation. Typically this approach will utilize some 2-D/3-D general purpose graphics software package as one building stone. The benefits are the functions programmable for own specific needs and minimal dependency of outside software components. The evident drawback is the continuously required software maintenance work and possibly short lifetime of the investment. Second alternative is to apply commercially available present 3-D solid modelling CAD-softwares. The benefits are many software functions instantly suitable also for geoengineering field and higher reliability in software operation and support. Produced plots can be identical in either case but the methodology behind is differing.

If CAD-based alternative is chosen, true solid modelling software is the basic requirement for the task. However, general purpose CAD-systems as themselves can not handle many of modelling requirements met in geological work. Other requirement is open, convertible CAD-software and possibility to include own application modules into it. The special features of subsurface modelling are coordinate systems applied, efficient object generation, geological property handling and borehole/well inclusion.

TECHNICAL SYSTEM DESCRIPTION

The nucleus of ROCK-CAD is Prime Medusa 2-D and 3-D CAD-packages. The ROCK-CAD application software is a shell around Medusa that gives the tools to model the geological properties and the geometry. The ready-to-use CAD-modules take care of the assembly load and graphics presentation. Medusa 3-D has Romulus type modeller that operates in batch mode as its kernel component. Solid object is determined by its closed surface boundary representation (Mäntylä 1988).

The developed software has been kept simple in form and in minimum extent. Database contains only relatively small number of parameters for each object. Software linkages to associated general work report libraries or geographic databases (2-D GIS) has this far not been created. ROCK-CAD is now implemented in Prime 50-series environment. Application software is mainly FORTRAN coded. It is also convertible to UNIX-based workstations.

ROCK-CAD is so configured that each created volume entity to be modelled is set as an individual project. The model volume is set to be rectangular in shape. Modelling volume can consist of several different types of submodels. The basic system handles lithological (rock types), structural (fractured zones and fracturing) and hydraulic (hydraulic structures) submodels. This setting can be expanded or changed to specific needs. Submodels are either independent from each other or can be related and include common objects.

One submodel consists of a set of objects having both geometry definitions and related attributes. The maximum number of objects in one submodel is 1000, the limitation of which is not absolute because the set of objects can be grouped to form one object. Hierarchy of the objects is tree structured. Topographical variation as an individual object can be included as limiting upper surface boundary. In our case topographical undulation of upper model surface is usually so small that it practically can be neglected (planar).

Definition of Object Geometry and Attributes

There are graphical and numerical interfaces to get in geometrical boundary representations for each object. A high-level geometry description language (GDL) was developed to define object primitives numerically in a fast way. The software is installed in PC micros where one can generate the rock object files for Medusa. After the work is finished file transfer takes place to hosting CAD-system. Our choice of Macintosh micros has been based on other reporting, graphics presentation and desktop publishing needs but any PC micro or workstation is an obvious alternative. Graphically object volume is drawn via CAD-system graphical design interface. Object description is practically a "drawing". During graphical input the system prompt has been programmed to show world coordinates to allow cursor position tracing. Currently object definitions are in most cases generated in micro environment.

Each object or set of objects is linked with attributes. Attributes used are property of an object (fill pattern code or colour), its degree of certainty, geological age, descriptive name of an object etc. Fill pattern and symbol library is common for all projects. After definition phase is completed, model assembly takes place in CAD-system. The composition of one model is shown in Fig. 1.

Supplementary Functions and Output

From the user point of view the output forms are essential. In most cases graphics output is through vertical and horizontal cross-sections and perspective views. Defined object geometries extending outside the modelled volume are automatically clipped. Because the objects are real solids they are by presumed to be invisible. Hidden-line removal is sometimes by-passed to generate transparency in the rock.

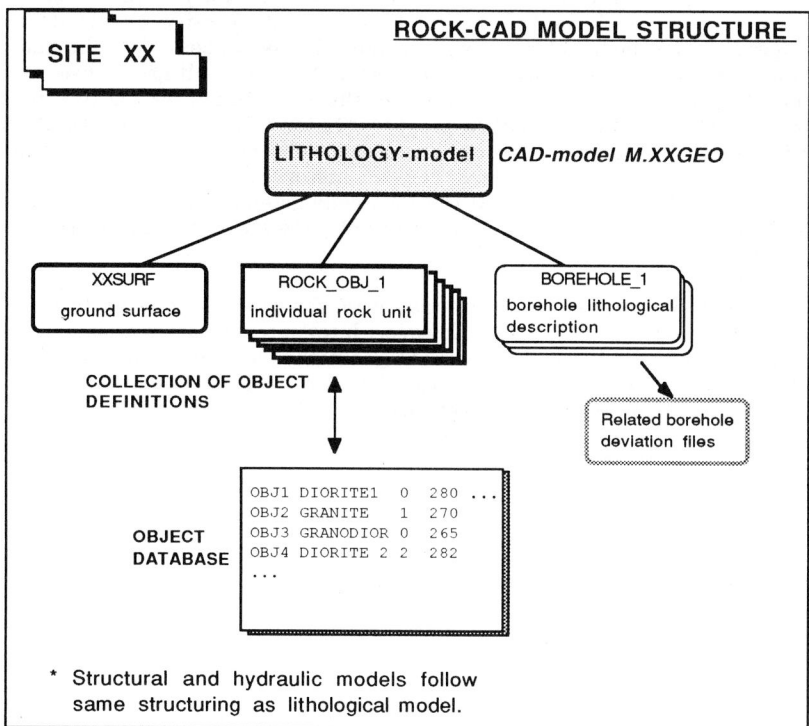

Fig. 1. Composition of the ROCK-CAD model.

Boreholes as important data sampling lines are essential in subterranean conditions and preserve special treatment. One or several boreholes can be projected onto taken cross-section plane with their property profile. Each borehole is described in a model plot as deviated 3-D pipe with varying property profile along it. The variation of the property along borehole is shown with fill pattern, colour and/or text labels. Strongly exaggerating diameter of a borehole pipe is a parameter possible to choose. In cross-sections rock objects can also be traced back to the database and identified by labelling them. This is important feature because often it is not straightforward in complex 3-D to know what the rock object is that appears in particular position on the plot.

Coordinate system in ROCK-CAD is national geographic coordinate system. Both input and output takes place in world coordinates. Coordinate grid is selectable and often generated into the final cross-section. Also local or user defined other coordinates are possible to use. Finally, drawing is labelled and appended to the document base. The pen plotter or laser printer are typical output devices. Often graphics and drawings are transferred to other computer systems in currently used CAD-data exchange formats (DXF, IGES).

DATA PROCESSING AND INTERPRETATION PHASE

The diverse data collection, processing and interpretation phases lead to modelling of geometry and properties of rock structures. Although these preceding phases before modelling are wide and generally out of the scope of this paper some factors involved are important to recognize.

The result of field investigations comprises large amounts direct and indirect geo-observations and indications. Their mutual accuracy, reliability and interpretability is strongly differing. Interpretation work is the phase to sort and classify observations and indications. It has a strong factor of entropy involved. It easily seems to split and result into innumerable details not necessary useful at all in modelling. Thus relevant observations must be identified and focussed on. On the other hand, during interpretation general genetic geological knowledge makes progress. Later that may give useful guiding in generation of geometric models when direct and exact observations are missing or just impossible to get. An example of knowing genetic history that is useful in modelling crystalline bedrock is presented in Fig. 2. If the rock body geometry is delineated in 3-D by ground surface observations alone it results in different geometries due to diverging status in the genetic succession.

In the beginning interpretation can be concentrated in, for example, one surface mapping method or logging results of one borehole. However, later interpretation process naturally expands and brings two- and three-dimensional geometry problems into consideration. The amount of the data that passes to integrated and conceptual modelling is strongly reduced while its information density is increasing.

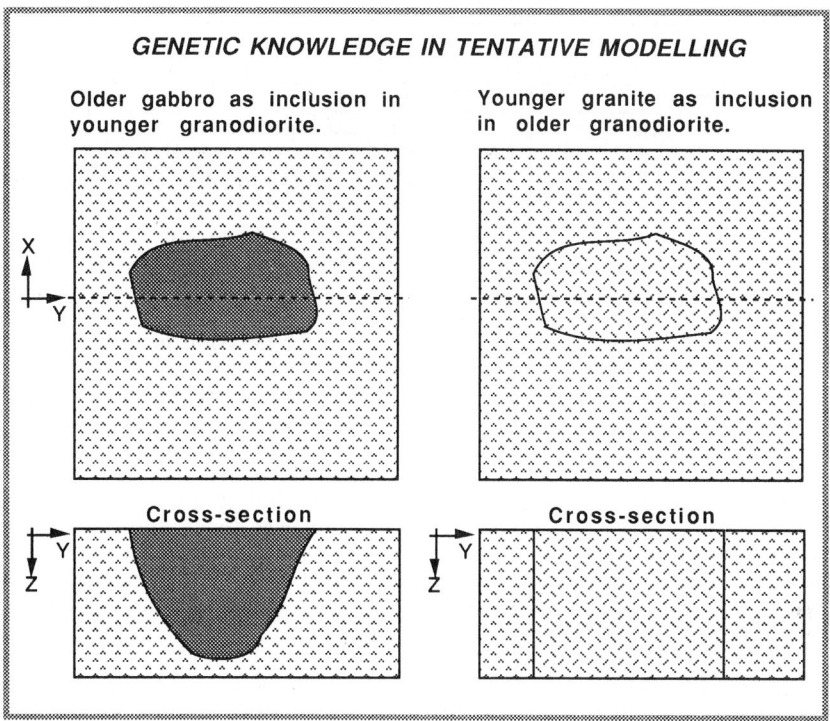

Fig. 2.   Two different models result from genetic and
lithological history.

MODELLING PROCESS

The modelling of the rock is integrating, synthesizing, simplifying and conceptualizing part of the data processing chain. Large reduction within the serviceable data must be made again. Modelling is also in a crystalline and igneous bedrock environment a sampling-limited problem. Structures are typically realized in our case through a few observations and they do have non-unique geometric or parametric solution. On the other hand, if much data is available from different sources, it may prove to be partly in internal disagreement. Modelling concepts that are used in petroleum exploration and production related work are discussed more by Jones (1988).

The development of 3-D modelling techniques requires necessarily classification and definition criteria to be determined and documented. Classification principles can be difficult to set: for example cut-off type criteria for the detection of a fracture zone. Conceptualization of the rock structures takes place through such a definition-limited data sorting.

Established and unchanging nomenclature of the rock structures within the modelling work team is desirable to be achieved. Definitions for the fracture zones and structures in general or rock types and names serve as examples.

The use of ROCK-CAD system during different stages of investigations is depicted in Table 1.

Table 1. ROCK-CAD functions during investigation program.

Phase of the survey project	ROCK-CAD use
Reconnaissance and standard investigations	• Tentative conceptualization • Creation of tentative volume model • Planning of supplementary investigations • General visualization purposes
Supplementary investigations	• Conceptualization (fine tuning) • Planning of detailed studies/verification • 3-D model updating, adjustment and analysis
Finishing and verifying investigations	• Visualization for decision purposes • Design of underground rock caverns • Documentation of the state-of-the-art rock volume knowledge

After the decision to model any site is made, the volume that is illustrated has to be set. Typically, model volume is rectangular in shape. In our case topographical undulation of upper model surface is usually so small that it is neglected. However, topographical variation is included as upper surface boundary if elevation differences are significant when compared to model yx-dimensions. Rock structures that extend to or are situated outside modelling volume are automatically clipped.

During starting phase of the field survey project tentative lithological, structural and hydraulical submodels are created. Conceptualization of structure and rock type classes to be used takes place first time. A tentative model is used to plan borehole locations and orientations to be drilled. In planning of supplementary investigations geometrical structure alternatives recognized and to be solved is considered. The purpose of the models during starting stage is to serve as quiding template for cross-hole investigations. Because model creation must be relatively quickly accomplished structure geometries are then generated within ROCK-CAD system numerically by PC numerical toolbox. On the other hand, the geometrical outlook is strongly simplifying and angular reflecting correspondingly the degree of geological knowledge.

During supplementary investigations the models are updated and more detailed structure classification procedure can be conducted. Geoengineering oriented assessment and modelling work employs primarily vertical and horizontal cross sections. Perspective and axonometric views are frequently used in presentation purposes. Special cuts of the model with arbitrarily orientated plane are sometimes produced in complex geological environment. Coordinate grid is appended to calculated cross-section graphics and objects are identified. Quite often smoothing is applied to smear out angular appearance.

Finishing and verifying phase covers documentation and fine tuning of the models. Often local small subvolume cubes are detached from site model to focus on inspection of some details. Visualization for general information needs and decision purposes is done. Analysis of the uncertainties related to the spatial information reliability and the amount of the investigated rock volume has to be done by the experts. Concerning the reliability or the degree of knowledge all modelled structures are classified as certain, probable or possible. In finalizing modelling unknown subvolumes not screened by any survey method and "blind sectors" method by method are necessary to evaluate. This work has not been conducted this far within site investigation procedure and modelling.

APPLICATION EXAMPLES AND EXPERIENCES

ROCK-CAD has been in operational use since 1989. Currently submodels have been created concerning four investigation sites. Usually lithological submodel is compiled first because lithological maps are directly available. Tentative structural submodel is formed later. Hydraulical submodels has not been made into CAD-system as own individual submodels. The reason is that hydraulical submodels have been concluded from structural ones by parametric substitution. The main difference between structural and hydraulical submodels is depth dependence of hydraulic conductivity that can not be directly deduced from observed fracture intensity or other fracturing properties used in structural submodels.

Two Examples

Example plot from Veitsivaara lithological submodel is in Fig. 3. The viewpoint is from the southeast. The model is cut "open" with inclined plane. The bedrock in Veitsivaara, aged up to more than 2800 Ma years, consists mainly migmatitic banded gneisses with amphibolitic inclusions. It has gone through a polyphasic Archaean deformation. Dominating rock types are gneiss and coarse/medium grained granite as a younger plutonic body. The host rock is intersected mainly by metadiabase dykes but also felsic granitic and pegmatitic veins are present. The thickness of the metadiabase dykes is in the range 5 to 30 metres. Due to the genetic history of the metadiabases their depth extension is large. The drawing shows also the main orientations NW-SE and E-W of the dykes. During brittle deformation phase metadiabases have been faulted which is also visible in Fig. 3. Metadiabases have mostly been approximated with plate like bodies. This picture is illustrative example of ROCK-CAD use in general purpose presentations. However, it is difficult to utilize it in any planning work or in numerical manipulation.

The second example is from Romuvaara site The Archaean age geology is similar to Veitsivaara site discussed earlier. Due to polyphasic deformation history, geometrical shapes of the lithological units are complex and varying. The bedrock structures are interpreted on the basis of their intersection and folding relations as representing six plastic deformation phases. After that sharp faults developed during at least four further movement phases which can be regarded as representing brittle deformation causing fracturing. Identified fracture zones are not controlled by lithological units or boundaries expect the metadiabase dykes. Some major water conductive zones have been identified. They coincide mainly with the major fracture zones.

Local scale vertical cross section within the site and taken from the Romuvaara structural submodel is presented in Fig. 4. The section view focuses on the borehole 2 with fracture structures in its neighbourhood. The intact rock matrix is described by fracture mesh that indicates orientation of the different fracture sets. Line types in fracture mesh describes the intensity in each fracturing set. Three classes of fracture zones are marked. Five planar minor and major zones are visible. They are modelled according to gained knowledge from the borehole 2 investigations but also results from gephysical surface investigations and from other boreholes are extrapolated into this cross section.

The section plot is used when long distance hydraulical test pumpings are considered and planned. It is also helpful in determining the packer and measurement interval positions when borehole packer system is installed for long-term head monitoring in the rock. In addition, the section drawing forms useful material when large distance 3-D test pumpings are interpreted qualitatively. The major fracture zone that has been investigated with test pumpings is labelled as number 1 in Fig. 4. The illustration describes quite well the complexity of the structures that are met in crystalline bedrock although the section shown here has been selected as being relatively simple one. It can be noted that topographical variation is rather small even in this locally zoomed in view.

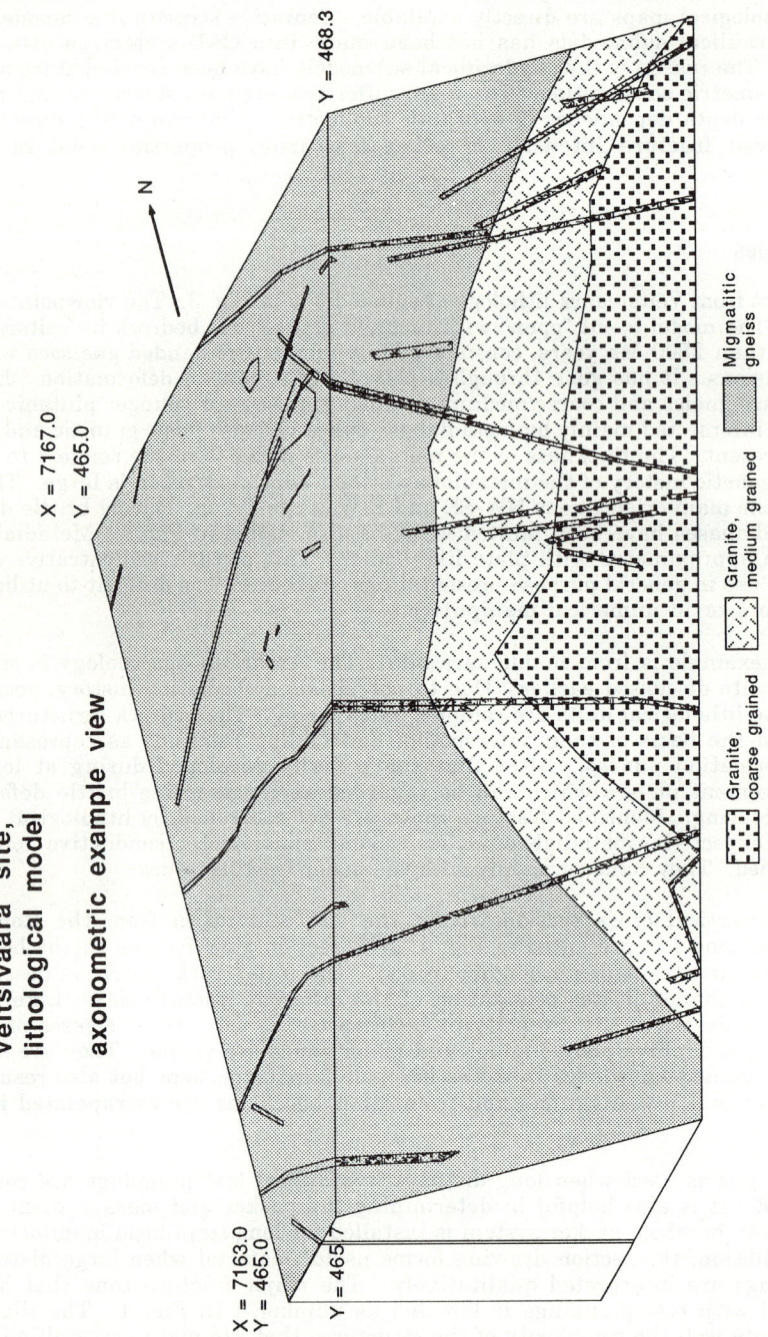

Fig. 3.   Axonometric view into Veitsivaara site lithological submodel.

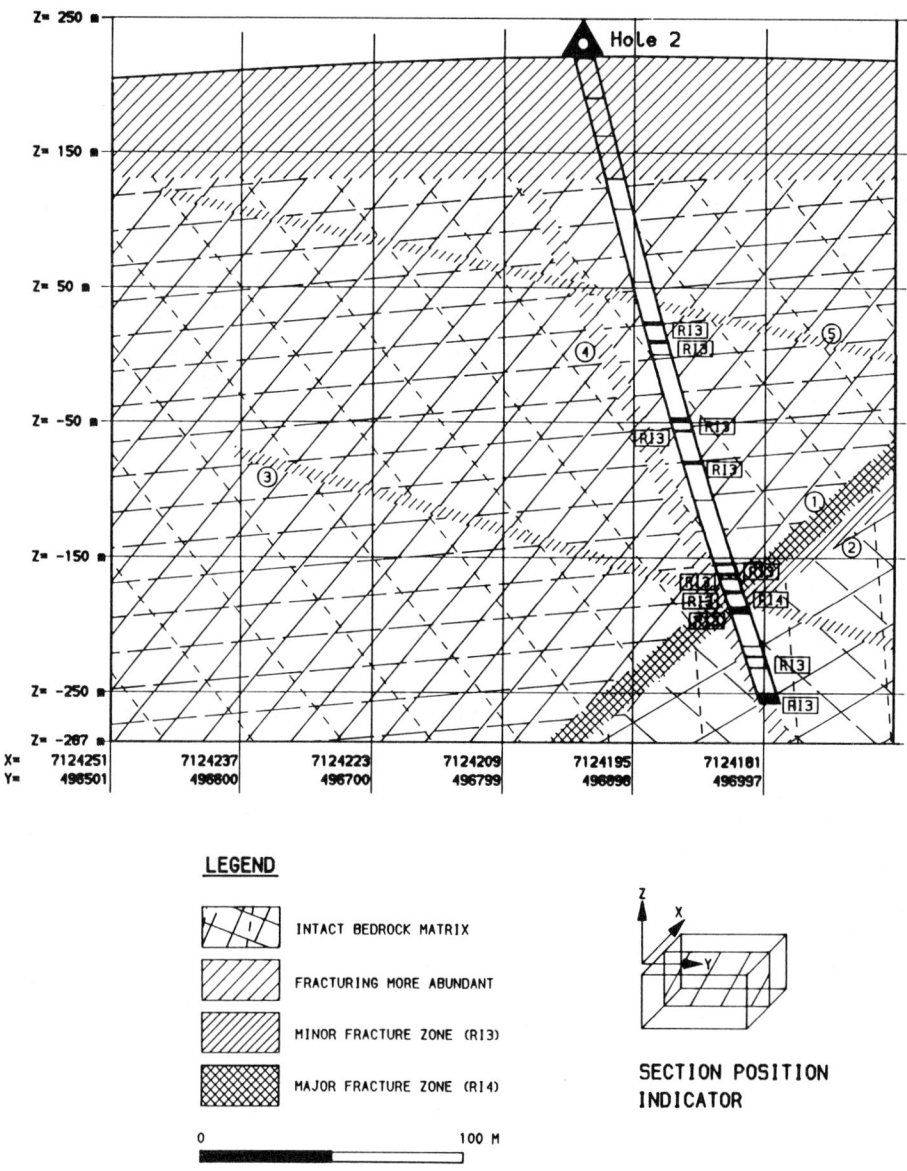

Fig. 4. Local cross section view of the structural submodel
at Romuvaara site. Five fracture structures are
numbered.

General Experiences

Experiences gained from four sites indicate that ROCK-CAD system is necessary if complex geological conditions has to be studied and illustrated in detail. Time consuming pseudo-3-D and error-prone manual compilation of drawings has been replaced. The modelling process is iterating by nature - after two-three cycles relatively stable conceptual model will be achieved. Rock modelling is strongly simplifying and sampling-limited in engineering field. Thus a well-balanced composition between simplification, artistic and visually acceptable graphical outlook but still honouring real degree of knowledge should be reached. No special "painting and polishing" efforts have been use to object geometries or drawings to make them look more detailed than what is really known by the observations. Fill patterns and symbols applied to have been selected to conform established practise.

The main use of the ROCK-CAD system is to assist the conceptual modelling and graphical visualization of rock conditions. Spatial relationships are easily recognized and can be discussed by experts. The visual output feeds creative thought work in iterating modelling. The site model can also be used to familiarize known rock conditions for authorities, non-geoexpert decision makers and for ordinary people.

Planned rock caverns, tunnels, shafts etc. are analyzed later within ROCK-CAD model. Location and shape of underground rooms may be changed due to realized rock conditions. Three-dimensional rock model provides a best-estimate prediction for construction work or other kind of evaluation.

Models are further transferred to 3-D calculations. The most important one is hydraulic flow model simulation. ROCK-CAD model forms also the basis for small volume 3-D rock mechanical calculations. Because the creation of three-dimensional geometry is a complicated task, standards to endorse 3-D CAD model transfer between different computer systems are still required. On the other hand, post-processing is often used for graphical drawings. Graphics is transported to desktop publishing system for reporting. Large variety of the existing 2-D graphics drawing and painting programs allow easy editing of the pictures for publication and presentation needs.

CONCLUSIONS

CAD-based modelling approach is applicable to geologic modelling and has been used to visualize 3-D conditions met in the crystalline bedrock. Many examples indicate that geologically oriented 3-D modelling software is necessary tool in efficient and objective modelling efforts. Models aid in planning of investigations and point out also effectively where data-limited interpretation problems are present. In the near future 3-D model transfer between different computer systems should reach more standardized level. Computerized realistic outlook rock model is beneficial basis also for the later stage design of the rock caverns.

REFERENCES

Jones, T.H. (1988). Modeling geology in three dimensions. Geobyte, 3, 1, 14-20.

Mäntylä, M. (1988). An introduction to solid modeling. Computer Science Press, Rockville.

SUBJECT INDEX

Lecture Notes in Earth Sciences